HOW TO
BE YOUR OWN ARCHITECT

2ND EDITION

HOW TO BE YOUR OWN ARCHITECT

2ND EDITION

MURRY C. GODDARD AND MIKE AND RUTH WOLVERTON

SECOND EDITION
FIFTH PRINTING

Printed in the United States of America

Library of Congress Cataloging in Publication Data

Goddard, Murray C.
How to be your own architect.

Includes index.
1. Architecture, Domestic—Amateurs' manuals.
2. Architecture, Domestic—Designs and plans.
I. Wolverton, Ruth. II. Wolverton, Mike. III. Title.
NA7115.G54 1985 728'.068 85-22214
ISBN 0-8306-0790-0
ISBN 0-8306-1790-6 (pbk.)

TAB BOOKS Inc. offers software for sale. For information and a catalog, please contact TAB Software Department, Blue Ridge Summit, PA 17294-0850.

Questions regarding the content of this book should be addressed to:

Reader Inquiry Branch
TAB BOOKS Inc.
Blue Ridge Summit, PA 17294-0214

Contents

Preface

Architects tend to keep a very low profile. They are continually and quite profitably at work in the community, but what it is exactly brings in all those profits is beyond the experience of most of us.

If someone asked you point blank, "What does an architect do?" you'd probably answer immediately, "An architect designs buildings." You might even volunteer the information that without an idea in the architect's mind, none of the great buildings would exist.

If, however, someone questioned you for details you likely would relate that an architect makes plans and drawings that are used in the construction of buildings.

Traditionally, architects are concerned with the enclosure of space. Besides enclosing space, most architects also work with the space to make it appropriate for its function as well as aesthetically pleasing.

The services of an architect can be divided into six sections: the building program, preliminary design, cost estimate, working drawings and specifications, bids for construction, and construction.

During phase 1, the building program, the architect collects all the facts pertinent to set up the requirements for the proposed building. Client and architect consult at length, and the results are written down and called the *program*.

In phase 2, the architect will use the program as a springboard for ideas. He will sketch out possible designs until he finds one that satisfies him. The architect will present this to the client for approval and, if necessary, modify the work to please the client.

Phase 3 is the cost estimate. Although it is too early to predict exact construction costs, the architect will at this time present the dollar version of his design to the client, normally within 10 percent of the final figures.

Phase 4 is the busiest phase for the architect. Working drawings are prepared. These large-scale, detailed drawings will be used by the contractor during the entire construction of the proposed building. The working drawings will be accompanied by written specifications of materials to be used, from the nails in the studs to the pewter soap dish in the master bath. The client, of course, approves or rejects all plans, drawings, and specifications before they are set out in final form.

Phase 5 is the time when construction bids are taken. Copies of the approved drawings are distributed to contractors, who then submit their bids to the architect. Usually the lowest bid is taken but there are exceptions. A contract is written at that time. The contractor agrees to build the house according to the drawings and specifications, and the client agrees to pay the contractor the amount stated in the bid. Usually payments are made on a monthly basis, and are based on the amount of work completed during that time.

Phase 6 is the house construction time. The architect will visit the construction site regularly in order to check with the contractor and make sure that the drawings and specifications are followed precisely. The architect will check the contractor's monthly statements for accuracy before he turns them over to the client for payment. When the house is completed, the architect's job is done.

The usual fee an architect will receive for his services is a percentage of the construction costs. The percentage will vary in accordance with the complexity of the building. Private homes are considered the most complex of jobs in architectural circles and will demand the highest percentages: 12 percent or higher.

In phase 1, the building program, you can get a head start on any architect you might employ in this initial stage. The architect has to find out the requirements of the proposed home. You and your family already have a pretty good idea of what those requirements are.

You know, for instance, that your two young sons love to share a room and won't have fits if separate bedrooms for them are not forthcoming. You also know that a large playroom, preferably far away from adult quarters, is a necessity for your kids during winter. You know that, for those same two youngsters who have a great affinity with mud, a mudroom is the only solution if you and your spouse want to keep the rest of the living quarters in a halfway presentable condition during the spring and autumn months.

Phase 2, while it may seem quite overwhelming at first, is another instance where you already have a lot of knowledge, even if you are not consciously aware of it right now. You can tell immediately whether you like modern, contemporary architecture or prefer something that looks mildly Colonial. You also have intuitive feelings on such questions as ranch versus multilevel homes, open and separated living areas, and such. When you give the matter serious thought, you'll find that you have a pretty clear idea of what you want this proposed house to look like—similar to something you saw in a magazine, or visited during an open house, a friend's home, or a house you saw on television.

Phase 3, the cost estimate, involves inquiring in the right places for costs of labor and materials and determining what you can afford.

Phase 4 is the most technically demanding. Here you will have several alternatives—learning how to draw your own plans which is the most creative of all and most enjoyable if you are inclined that way, adapting other people's plans to suit your own design aspirations, or simply using ready-made plans that you can buy. Once you know how to read a plan or elevation drawing, the rest is simple.

Phase 5, taking the bids, is again a matter of assembling the right information, using the telephone, and coming up with three or four contractors who will submit the bids. Next you sit down with the contractors, one by one, and go over their figures and what they mean by their specifications. Then you award the bid to the contractor who offers you the best deal.

Phase 6, the construction supervision of your house, will be time-consuming, but not as much as you might think. Every person who has a house being built spends a lot of time kibitzing on the site, investigating what the workers accomplished since the last night. When you are your own architect, you might make a few more notes than most and, instead of calling your architect and bringing points to his attention, you can call up your contractor and complain to him directly. It is simply one less stage of command. Most people are quite observant when their dream home takes on reality and they can spot a mistake in a specification instantly. Who but the owner cares that much? Phase 6 is a natural for you.

You might think that we have oversimplified things. In a sense that is true. You will have to learn some new skills or brush up on some you already have. The satisfaction you will obtain from seeing your own ideas become reality will more than compensate for the effort and time involved. Additionally, there is the very real saving in money, which everyone can appreciate.

Mike and Ruth Wolverton

Preface to First Edition

Drawing your own house plans is a matter of creativity and practicality. You do it because you want a special house that speaks of you and your needs, because professional planners are expensive and because you want to see some marks on paper become a real live honest-to-goodness house that you designed.

Fortunately, professional planners aren't the only ones who can turn out workable plans, and that's what this book proves. If you know something about how houses are put together, and you understand a few basic drafting techniques, you can draw your own professional-quality house plans that will meet the requirements of any building authority in the country.

This book teaches you more than just drafting skills. You get every scrap of information you need to make intelligent decisions about house construction and design. You learn about building materials, architectural styles, structural requirements, building codes, plumbing, millwork, framing, and even construction costs. And it's all fully illustrated, including plenty of sample house plans.

To make sure that you understand the whole planning process—from the preliminary floor plan right up to the final millwork drawings—I show you step by step how I drew all the plans for two very handsome houses.

You learn how to draw every kind of plan for any kind of house—even a dream house.

Murray Goddard

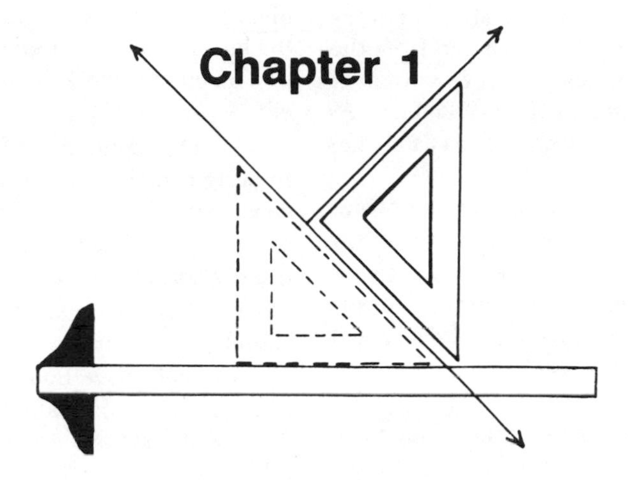

House Plans

House plans are scaled and dimensioned drawings of a house or any of its parts. *Specifications* (hereafter called *specs*) are written descriptions of materials and procedures that cannot be shown in picture form. All the house plans and the specs constitute the *working drawings* that tell the builder where and how to build a particular house.

You may hear house plans called *blueprints*. Blueprints are duplicate copies of the original, fragile drawings made in pen or pencil or semi-transparent tracing paper. *Tracings* are similar to a photographic negative from which any number of copies (blueprints) can be made. Plans used on the job are blueprints because they are cheap to replace if worn or torn.

Before the final house plans are worked up, there has to be a lot of thinking about what the house is to be. There will have to be discussions about style, materials, costs, and family requirements. Sometimes, making preliminary decisions means looking through house plans clipped from magazines. Rough freehand sketches will have to be made. Eventually a fairly accurate floor plan and drawing of the proposed house have to be produced. These become the *preliminary sketch*. Only after the preliminary sketch is carefully examined can it be developed into the working drawings.

These steps must take place whether you employ a professional to do the planning or draw them yourself.

Basically there are three types of house plans:

☐ *Permit plans* contain the minimum information required to get a building permit. They are the least costly and are used by speculative builders who, with minor changes, use the same plans over and over for many houses. These plans can be used by an owner-builder doing all or part of the work himself. Such plans cannot be used in many cities.

☐ *VA-HUD* (Veterans Administration/U.S. Department of Housing and Urban Development) plans are intended for the builder who is using VA-HUD approved financing. If you are a veteran or otherwise eligible, you may be interested. The VA-HUD minimum construction requirements are given in a looseleaf book, *Minimum Property Standards*, which contains useful information for every planner and builder. It can be purchased from the Superintendent of Documents, U.S. Government Printing Office, Washington, D.C. 20402. The VA-HUD book is not as well illustrated or as easily understood as the old FHA No. 200.

☐ *Complete Plans* are usually for the owner who expects to take competitive bids, thus every detail is included. Whether you are planning your own home or acting as your own contractor, drawing complete plans will teach you a great deal about the construction of your house.

Whatever type of plans you draw, they must contain enough information so that the location, size, and construction of your house can be understood by the builder and the permit-granting authorities.
Usually, the following house plans are required:

Plot plan.
Foundation plan.
Floor plans.
Elevations.
Construction sections.
Cabinets and millwork.
Electrical-mechanical.

When a house becomes larger and more complicated, additional plans and longer specs may be needed to explain details. These may include a *roof framing plan* or large-scale plans of stairs, fireplaces, and millwork (millwork is finished wood, such as built-in furniture, fabricated in a millwork shop). The more detailed and accurate your plans, the more likely they are to be carried out as you want.

It is true that some able and competent workmen are afraid of properly detailed plans. They dislike the necessity of frequent referral to dimensions and prefer to get the general outline and "fly free." This attitude may arise because they don't understand plans. Thus it is important to make plans as decipherable as possible.

The above plans can be defined in this way:

☐ The plot plan is a map of the building site giving its *legal description*, that is, where it is, how the house will be located on it, and other information on outside work such as drives, walks, walls, etc.

☐ The foundation plan tells the builder how to bring construction to first-floor level (if of slab construction) or to the top of foundation walls (if of wood construction). The foundation plan is usually traced over the first floor plan.

☐ The floor plan(s) shows the house as it would appear if it were sawn through horizontally about 6 feet above floor level with the roof lifted off. You see solid walls with open spaces where doors and windows occur. You look down on the kitchen cabinets, bath fixtures, and other built-ins in their places. The floor plan helps you think about the location of your furniture. It helps you ensure that there is sufficient wall space for the larger pieces, that door swings are not hampered, that there is passage room.

☐ Elevations are drawings of the sides of the exterior of the house. They show how the house will appear with doors, windows, roof, and other features as you intend them to be. Elevations are worked out in conjunction with the floor plan and the 1/4-inch scale construction section. Elevations are the easiest of the drawings to make because dimensions, locations of openings, and details have already been worked out. The scale is the same as the floor plan, 1/4 inch = 1 foot. On permit plans only the front and one end elevation need be drawn unless unusual work occurs on the back and opposite end.

☐ Construction sections are cross sectional views of the house drawn to a larger scale (1/2 inch = 1 foot or larger). If there is a variety of construction, a section must be drawn for each different type. The drawings will show footings, foundation walls, exterior and interior walls, ceilings, roof, and stairs. The section drawings should name and specify the size, grade, and other details of the materials to be used and show how they are placed.

The 1/2 inch scale section cannot adequately show how materials are to be used at cornices, fireplaces, moldings and the like. Larger scale sec-

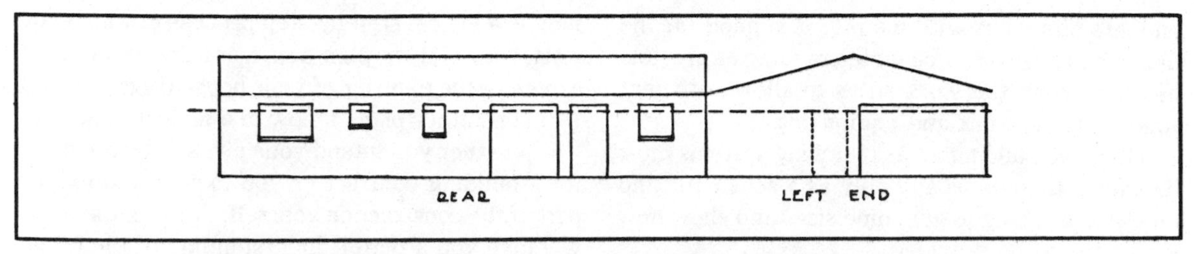

Fig. 1-1. After you complete the floor plan, the other drawings can follow in a logical order.

tions must be drawn. Use 1 1/2 inches = 1 foot, 3 inches = 1 foot or even full scale. Such details can be very helpful to an owner-builder doing his own work.

☐ Cabinet and millwork drawings for kitchens and bathrooms give the location and dimensions of counters, wall cabinets, vanities, and adjacent appliances and fixtures. Use the 1/2 inch = 1 foot scale or larger.

☐ The electrical plan is usually included on the floor plan for small houses and permit plans. Otherwise it is drawn on an undimensioned floor plan traced from the complete floor plan and shows the location and capacity of the electrical outlets, switching arrangements, and built-in electrical equipment. For most house plans the arrangement of the circuits is left to the electrical contractor, but if an owner is doing the work himself he should have the necessary knowledge to figure his own circuits in accordance with the National Electric Code or local code.

Some permit authorities may require complete details of all heating, ventilating and air conditioning work. Unless the planner is experienced in such matters, the subcontractor who submits bids for that work usually does the necessary engineering

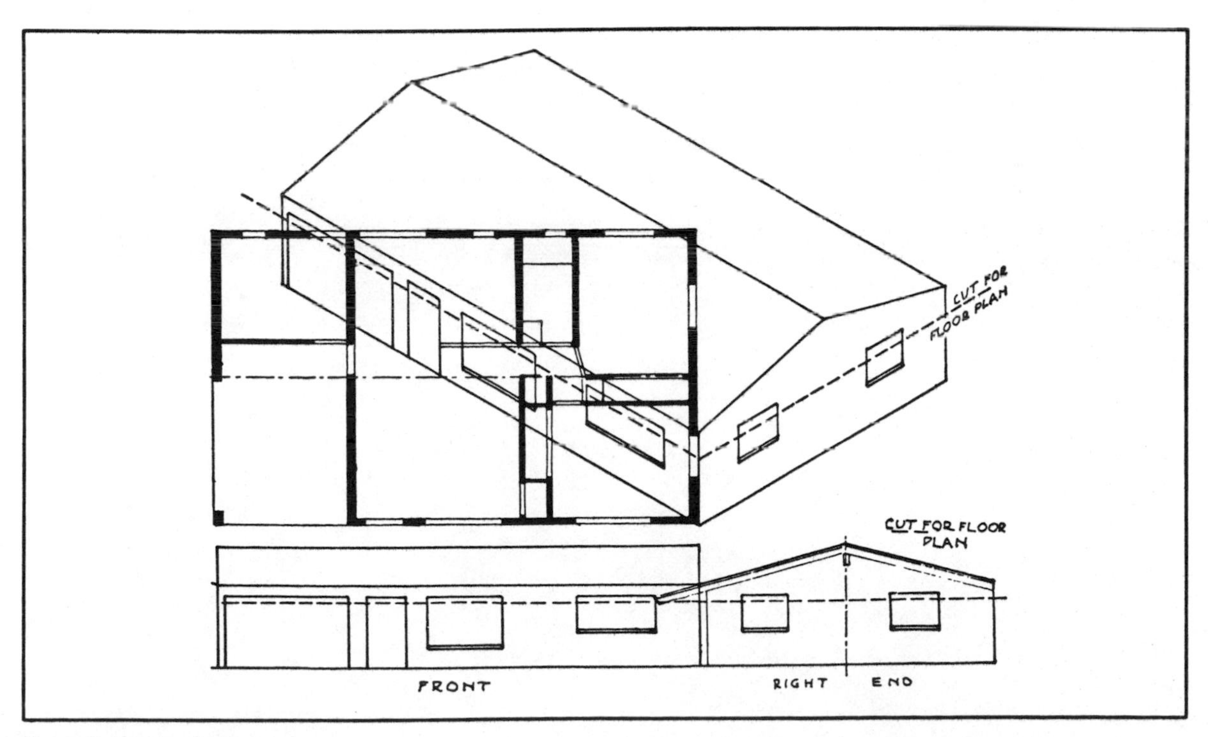

Fig. 1-2. House plans.

and submits plans with his bid. It is good for the planner to have as much understanding as possible of this part of the work so as to allow sufficient space for ductwork and equipment.

In some building areas plumbing systems must be shown though usually only as a schematic and unscaled drawing to give pipe sizes and show how venting is to be provided.

If roof construction is complicated by intersecting ridges and valleys, it may be desirable to include a roof framing plan with any necessary details to make its construction understandable.

Does all this sound complicated? Actually, planning is a logical step-by-step process in which you start with a floor plan and from this work up a sketch of the exterior of your house (Figs. 1-1 and 1-2). The other plans follow in a logical sequence.

Whether you intend your plans to be the basis for a building contract or you expect to do all or part of the construction yourself, drawing the plans will give you a better understanding of the building process.

Whether you intend your plans to be the basis for a contract or you expect to do all or part of the construction, drawing the plans will give you a better understanding of the building process.

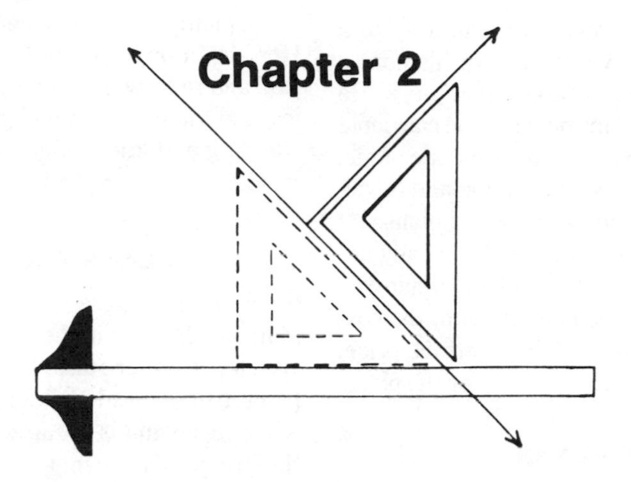

Where House Plans Come From

House plans are generated in a lot of different ways. Many people draw their own; others seek professional help.

ARCHITECTS

If you prefer to go to an architect who does house planning, tell him in detail what you want and what you can spend. He can prepare plans, take bids from contractors, supervise the construction, and deliver a completed home ready for occupancy.

An architect is a professional building planner with specialized education and is usually legally licensed or certified.

For planning and building a house, most architects charge a fee of 10% or more of the building costs. Of course, you pay all building costs.

HOME DRAFTSMEN

The next possibility will be home draftsmen or designers. You can find them in your local telephone directory. They are usually less educated in design techniques but can produce plans to fit your needs and may be able to save you some money.

Some home draftsmen, hereafter called planners, may quote you a flat price based on the time it takes to produce the drawings for the type of house you want. Others charge according to the size of the proposed house. The hourly rate is fairest to both planner and owner, for the time required depends on how well you know what you want and how many changes you make after planning is underway.

In dealing with planners and architects, do not hesitate to ask about charges. Get his fee schedule, and if you employ him get a written statement of his charges which requires your signed acceptance. Expect to make a down payment of 15% to 20% of the probable cost of the plans. The down payment covers the cost of preliminary studies. It is not unusual for an owner to decide not to complete working plans after the planner has spent hours of preliminary studies, especially if the owner finds that he cannot afford the dream house that he asked for.

BUILDERS

Many prospective home owners go directly to a builder. They may have visited one of his homes for sale or have heard of his satisfactory work. He usually has a number of plans on which he can quote his price. He can build on your lot or on one of his, for he often owns land or works with a land developer. His price will include the cost of plans.

If you want something different, he usually has a professional planner prepare drawings to suit you. And if you have him do the construction, the cost of plans may be included in the contract price. Check with the builder as to payment for the plans.

STOCK PLANS

If, in your search through home magazines and plan books, you find a plan that seems to fit your needs, it is usually possible to purchase sets of blueprints. These are generally less costly than plans drawn especially for you, but because they are not designed specifically for your region and local building codes, it is often necessary to have them modified. In addition you will need a plot plan showing the placing of the house on your particular lot. At first it is better to buy only a single copy of a stock plan; you must be certain that it can be used. Here is an opportunity to use your own time and skill and redraw the plan yourself. The careful tracing and modification will give you a better understanding and knowledge of the house you hope to build.

PLANS AND THE LAW

Some states require the examination and licensing of home planners in the same way that they license architects. Other states leave the approval of house plans to the local building authorities. But even where licensing is required, house plans drawn by the prospective owner are generally accepted for a building permit if they are competently drawn. Nevertheless, it will be best to check with your local authorities as to who may draw house plans.

Even though your local authority may insist on a seal, you can usually find a licensed professional who, for a small fee, will allow you to come into his office and use one of his drawing boards to retrace your work under his inspection. He will then legally put his stamp on your plans.

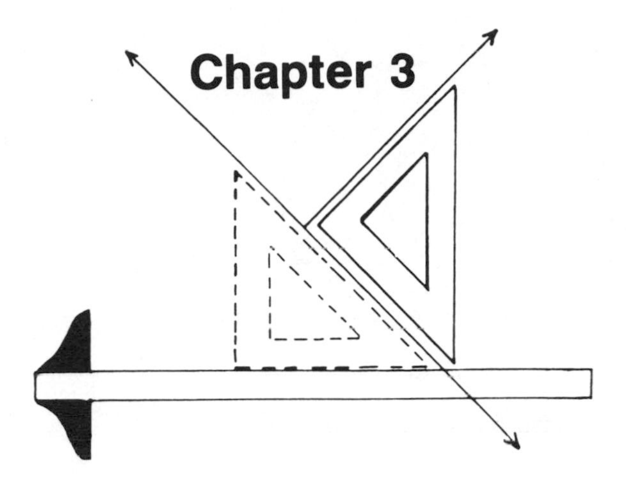

Drafting Tools

The professional house planner or draftsman has a variety of tools, books, data tables, and building material catalogs to help him. Building material yards and home improvement centers are glad to supply you with information and catalogs. You may write for others advertised in building magazines and plan books. Mail-order catalogs such as those from Sears give much helpful information as well as prices of the many building products they sell. These catalogs are particularly useful when you need to know the dimensions of a piece of equipment or furniture.

EQUIPMENT AND MATERIALS

The following is a list of the essential tools and materials you will need to do competent house planning. How they are used is described in subsequent chapters.

☐ A drawing board (Fig. 3-1). Most small house plans can be drawn on 18- × -24-inch paper which requires a board at least 20 × 26 inches. If you are likely to plan very large houses, you should use 24- × -36-inch paper and a board at least 31 × -42 inches. The larger board may be handier to work on even though you are using 18 × 24-inch paper. There is more room for tools, calculation pads, etc.

You can make a satisfactory, though heavier, board from 5/8-inch or 3/4-inch smooth-surface plywood. The raised grain of fir plywood makes too rough a surface, but it can be covered with a plastic-surface paper made to cover drawing boards. This paper can be found in drafting supply stores. It is tinted for eye comfort. Secure it to the surface with double-surfaced transparent tape.

If you want to make a plywood drawing board, have it cut straight and true (30 or 32 × 48 inches) at the lumberyard. Use a fine-tooth saw for smooth edges. A small board is easier to move about, but a large board is better to work on.

☐ A T square, 30 inches for small drawings, 42 inches for the larger ones. A maple T square, school grade, is quite satisfactory but those with transparent plastic edges are better.

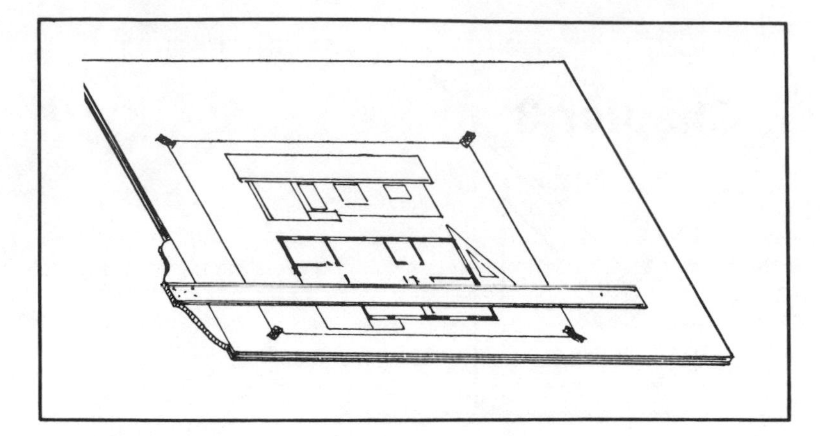

Fig. 3-1. A typical drawing board.

☐ A 12 inch 30°-60° transparent plastic right triangle (Fig. 3-2).

☐ A 10 inch 45° transparent plastic triangle (Fig. 3-3).

☐ A 6 inch 30°-60° transparent lettering triangle (Fig. 3-4).

☐ A 10 inch transparent plastic adjustable triangle (Fig. 3-5). Could be omitted but is a time-saver.

☐ Two irregular curves (Fig. 3-6). A #20 and #18.

☐ A transparent plastic plan template (Fig. 3-7). Timely #T-35.

☐ An architect's 12 inch triangular white plastic scale. (Fig. 3-8).

☐ An engineer's 12 inch triangular white plastic scale.

☐ A small pencil compass (Fig. 3-9). Better grade 3 1/2 inch and 6 1/2 inch compasses are easier to work with. As an alternate you may but a small case of drawing instruments though you may never use some of them.

☐ A metal erasing shield.

☐ A dusting brush (Fig. 3-10). Draftsman's type has a wide, soft bristle. Dime store paint brush will serve.

☐ Long-bladed shears for trimming paper.

In addition to the tools listed above, you will need the following materials:

☐ Four drawing pencils (Fig. 3-11). One each 2B, HB, 2H, and 4H. They may be the common

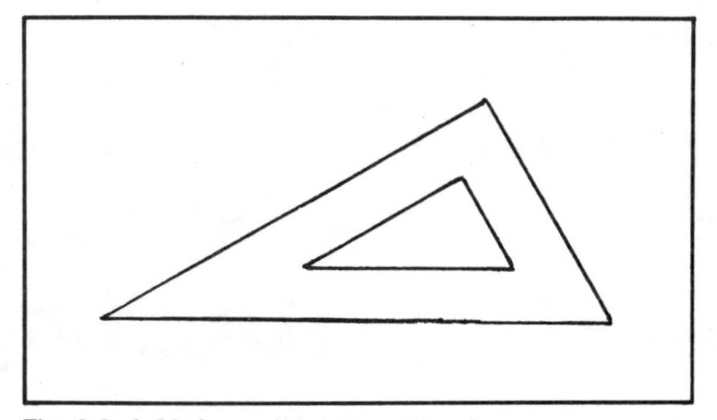

Fig. 3-2. A 30-degree/60-degree triangle.

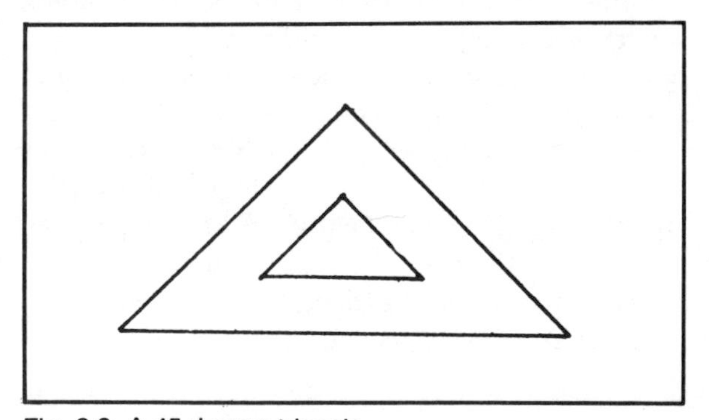

Fig. 3-3. A 45-degree triangle.

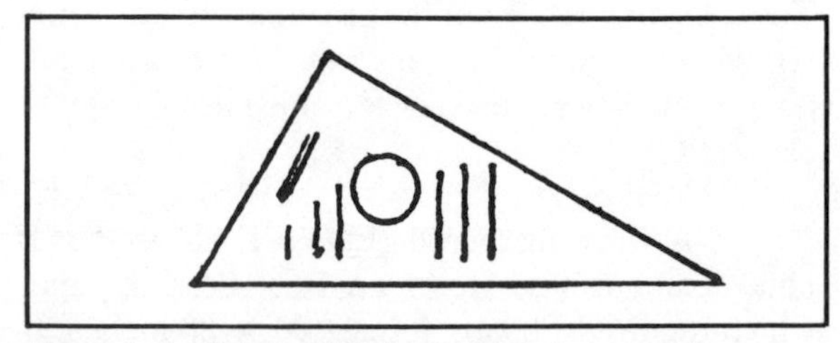

Fig. 3-4. A 30-degree/60-degree lettering triangle.

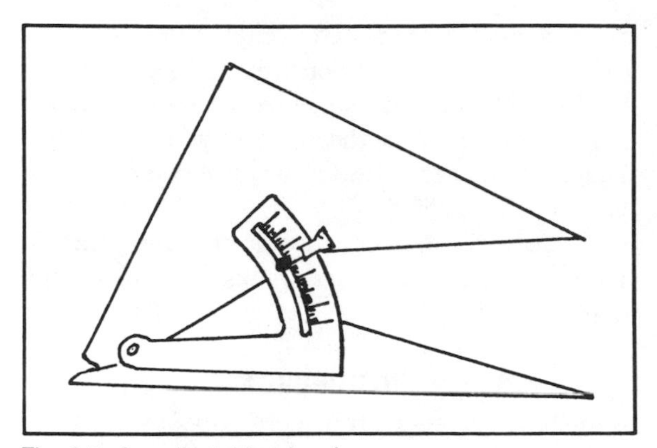

Fig. 3-5. An adjustable triangle.

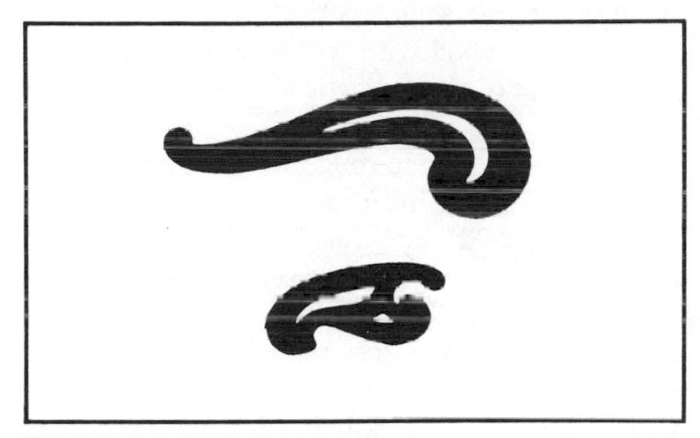

Fig. 3-6. Two irregular curves.

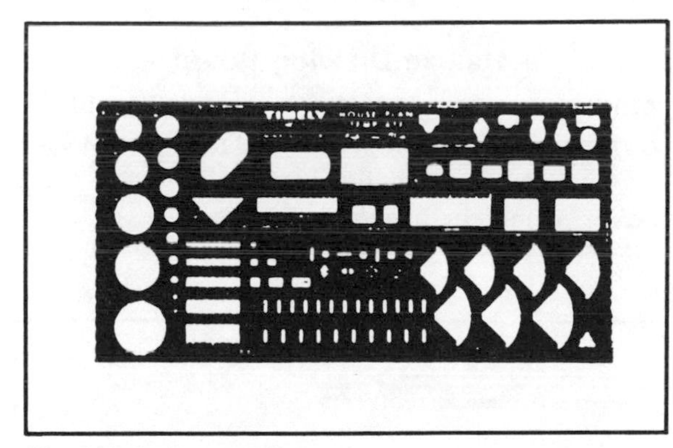

Fig. 3-7. A plan template.

wood-cased types or loose leads held in mechanical holders.

☐ Two pencil erasers, a wedge-shaped soft green one and a paper-cased (Blaisdell) soft green one (Fig. 3-12).

☐ Emery boards for pointing lead (those used for filing finger nails). Professionals usually have power-driven erasers with a lead pointing attachment.

☐ A black ink ballpoint pen such as Bic's.

☐ A black ink felt tip pen.

☐ A roll of 3/4 inch painter's masking tape.

☐ A package (24 sheets) of 18- × -36-inch drawing paper. This is throwaway paper. Any tough, erasable, low-cost light paper will do.

☐ Twelve sheets 24 × 36 inches of high grade tracing paper, sometimes called vellum. Cheap school grade tracing paper is not satisfactory. Finished drawings are made on this paper and are like photographic negatives from which duplicate copies (blueprints) are made.

Tracing paper also comes in 20 yard or 50 yard rolls, 24 or 36 inches wide from which sheets of any size may be cut. This paper is costly so don't buy more than you will use.

☐ A pad of cross-ruled (1/4 inch squares) yellow paper.

These are all that will be needed to produce professional-grade plans. But the quality of your drawings depends on you. Don't expect your first endeavors to be flawless, so redraw as professionals do when they botch a job. You will learn by doing.

EASY-TO-MAKE DRAWING BOARDS

For those who do not have desks, we have designed a portable drawing board/storage setup that will see you through royally in any and all house planning

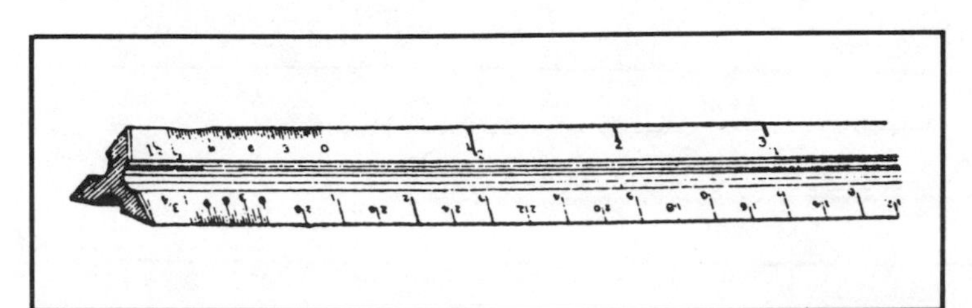

Fig. 3-8. An architect's triangular scale.

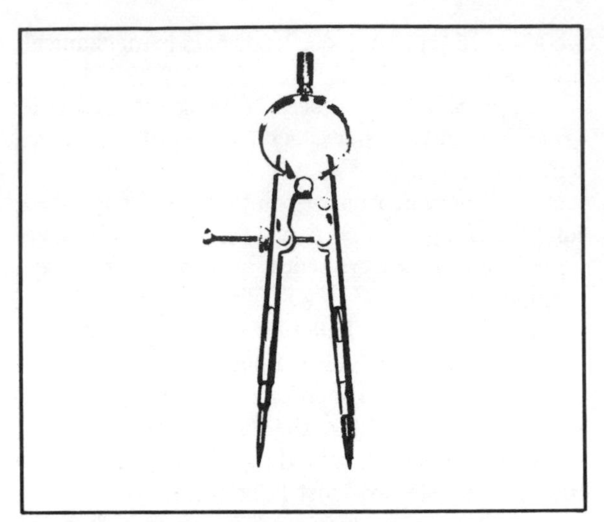

Fig. 3-9. A pencil compass.

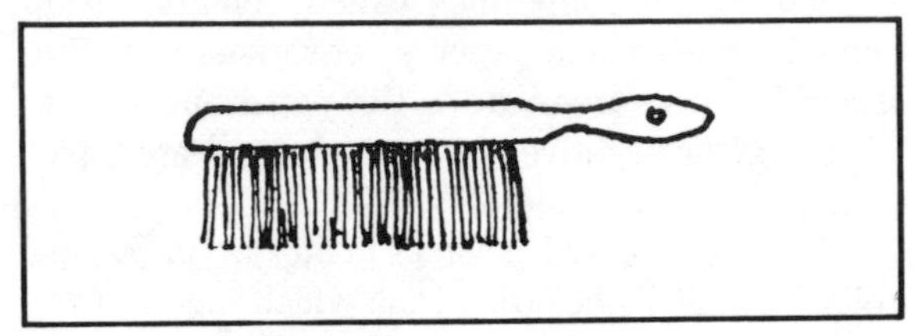

Fig. 3-10. A dusting brush.

activities. And it can be stored in the back of a closet. We have come up with two models of the drawing board so you can take your pick.

Economy Drawing Board

This board is quick and easy to make. You'll find it a joy to have and you will find lots of uses for it even after the house planning is completed.

Materials. One board, roughly 18 × 24 inches (but you can get by with one that is 14 × 20 inches) cut from 1/2-inch marine plywood. Two lengths of 3/4-inch elastic twice the width of your board plus 1 inch. Wood sealer and clear plastic spray or adhesive-backed vinyl.

Tools. Sandpaper, a needle and heavy thread, a staple gun and staples, or tacks and a tack hammer.

Instructions

If the board is relatively smooth, sand it thoroughly on both sides, first with coarse, then medium, and finally with fine sandpaper.

Seal the wood with wood sealer. When dry, give it a coat of clear, plastic varnish as protection and for easy cleaning.

If the board is too rough for sanding, and you don't have a plane, cover the board with solid-color, adhesive-backed vinyl. See Fig. 3-13.

For either board, sew or staple the ends of the elastic securely together.

Attach one elastic loop to each end of the board. Fasten it down on top and bottom as shown in Fig. 3-14. There is your drawing board that can be wiped clean in an instant!

Deluxe Drawing Board

Materials. A 1 × 10 (4 feet long), 3/4 outside corner trim (3 feet long), a 2 × 2 (9 feet long), a 1 × 3 (37 inches long), 2d finishing nails, wood glue, sealer, small corrugated fasteners.

Tools. Saw, C-clamps, carpenters' square,

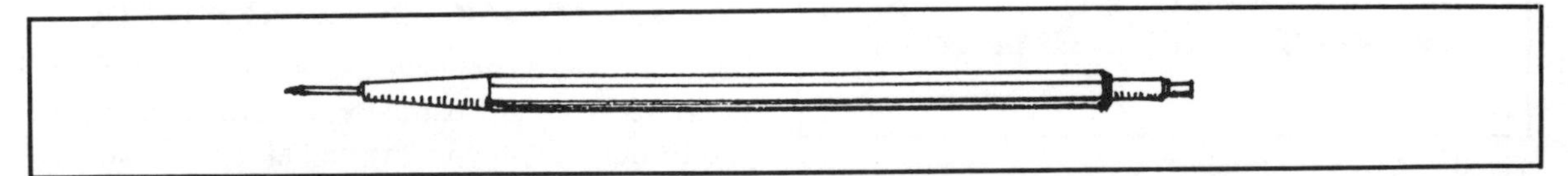

Fig. 3-11. A drawing pencil.

Fig. 3-12. A paper-cased pencil eraser.

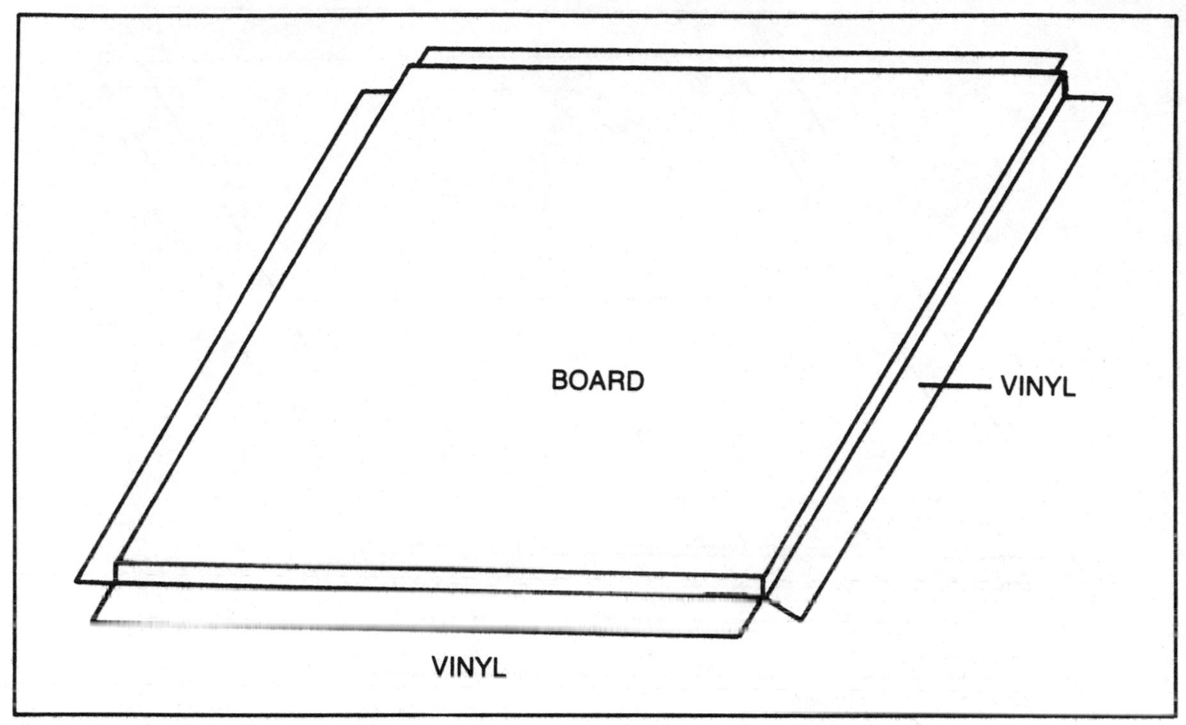

Fig. 3-13. Covering the drawing board with vinyl.

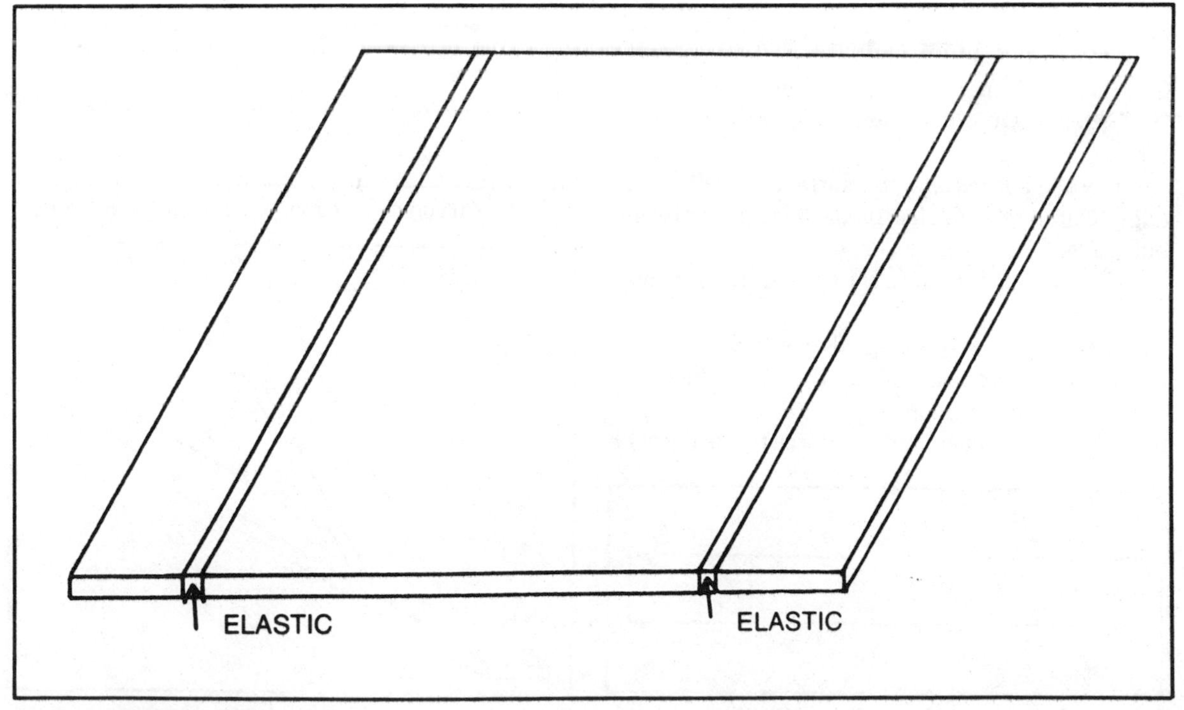

Fig. 3-14. Attaching elastic bands.

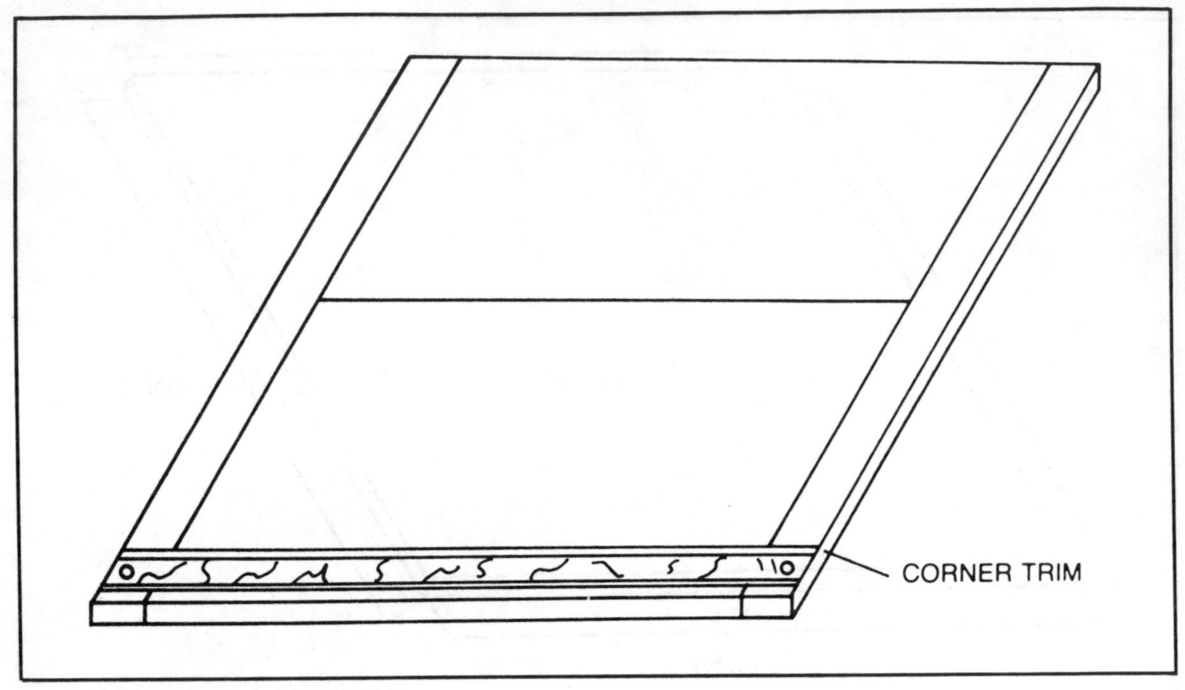

Fig. 3-15. Constructing a deluxe board.

ruler, hammer, plane protractor, sandpaper.

Instructions

Cut the 1×10 into two 2-foot-long pieces.

Spread glue on one long edge of each of the boards.

Lay the boards on a flat surface, edge to glued edge, clamp with C-clamps (at least four clamps, but six is better) and let dry.

Cut two pieces of 1×3 (18 1/2 inches long, each).

Spread glue on one long edge of each 1×3 piece and clamp to the board along the short sides. See Fig. 3-15. Let dry.

Reverse the board and reinforce the seams with

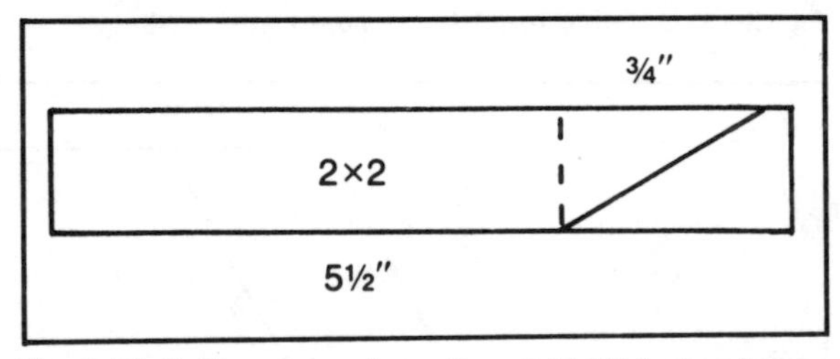

Fig. 3-17. Cutting an angle on 2 × 2 (5 1/2 inches long).

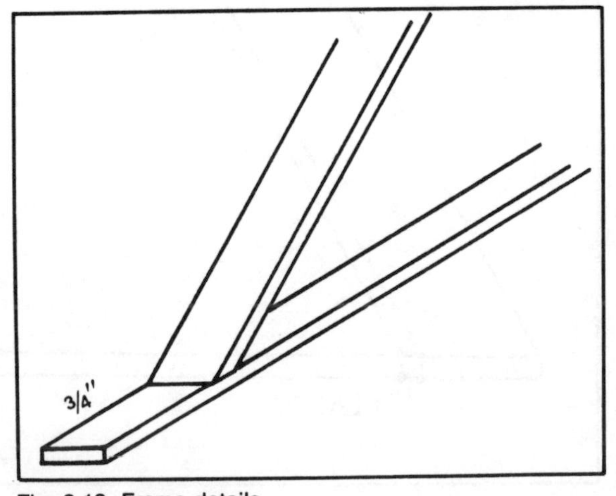

Fig. 3-18. Frame details.

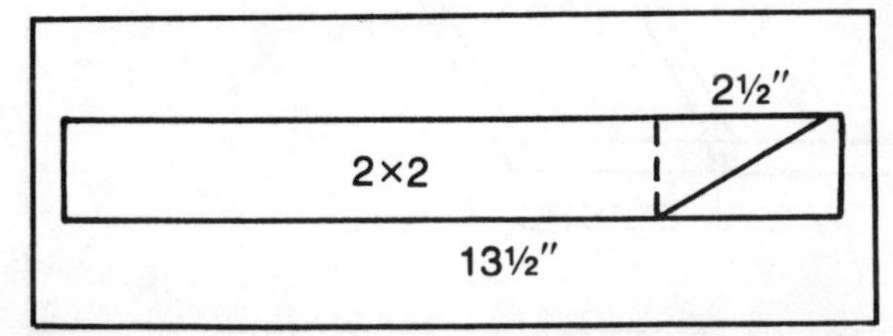

Fig. 3-16. Cutting an angle on a 2 × 2 (13 1/2 inches long).

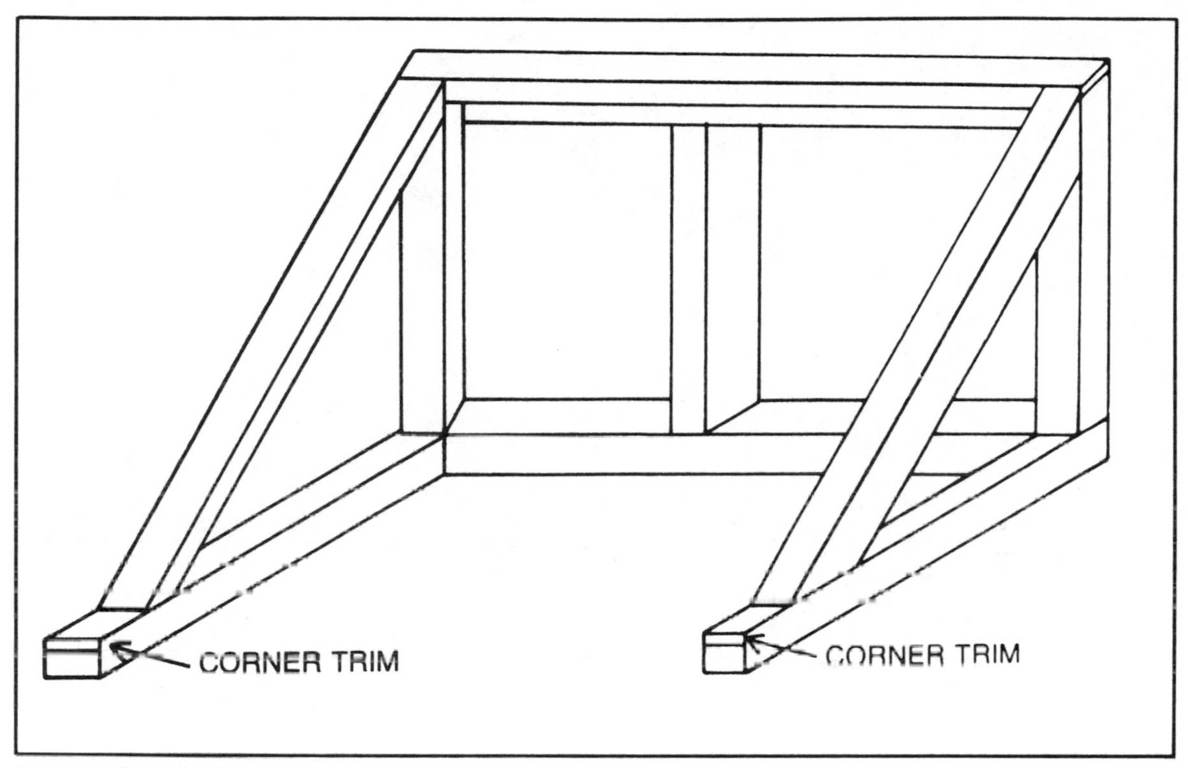

Fig. 3-19. Constructing a frame.

small corrugated fasteners.

Check corners and sides with your carpenters' square. Corners should be right angles (90 degrees) and sides should be straight. Plane and sand if necessary.

Sand the entire board smooth, and round corners slightly.

Cut the outside corner trim to fit one long edge of the board (about 29 inches).

Glue and nail to the board forming a shallow well at bottom of the board.

Seal the board with a wood sealer or clear varnish.

For the stand, cut two pieces of 2×2 22 inches, two pieces 13 1/2 inches, two pieces 14 1/2 inches, two pieces 5 1/2 inches and one piece 4 1/4 inches.

Mark off the 13 1/2-inch pieces 2 1/2 inches from one end.

Draw out a rectangle and mark the diagonal as shown in Fig. 3-16. Cut along that line.

Mark the two 5 1/2 inch pieces 3/4 of an inch from the top. Again draw a rectangle and diagonal. Cut on the diagonal. See Fig. 3-17.

Nail and glue the 14 1/2-inch pieces to one of the 22-inch pieces forming a U shape.

Nail and glue the 5 1/2-inch pieces to the corners of the U.

Mark the middle of the 22-inch piece and nail and glue the 4 1/4-inch piece in place.

Nail and glue the other 22-inch piece across (as shown in Fig. 3-18).

Nail and glue the 13 1/2-inch pieces to the frame, forming a triangle on each side. The cut end goes to the front 3/4 inches in from edge of the base. See Fig. 3-19.

Cut two pieces out of the scrap of the outside corner trim (each 1 1/2 inches long).

Nail and glue to front of frame as shown in Fig. 3-19.

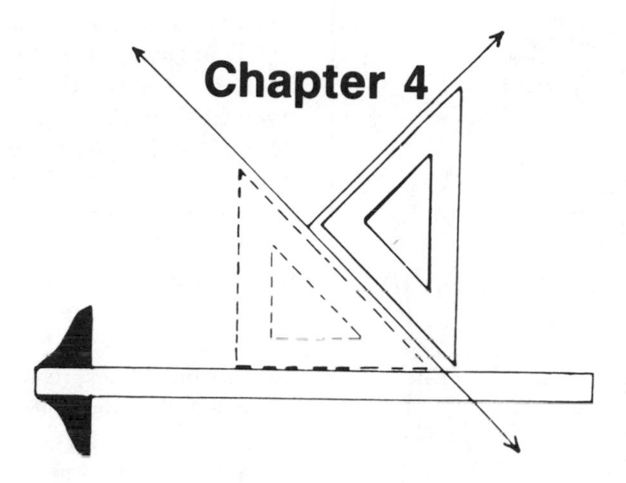

Using Drafting Tools

Learning to use drawing tools takes practice. But fortunately most of the skills involved can be mastered in a relatively short time.

THE DRAWING BOARD

When drawing at home, place your drawing board on a convenient table or desk with is about 30 inches above the floor. You should sit directly in front of the board, and the board should be tilted to give you a good view. If you are using a 24 × 36 inch board or larger, two 3-inch long metal door stops with rubber bumpers at each top corner affords a satisfactory slope (Fig. 4-1). You can also use a couple of 3 inch high blocks.

Place the board where the light is good and comes from the upper left so that the working edges of the T square and triangles will not be shaded. Whether daylight or artificial, the light should be shielded so that you do not look directly into it. Many draftsmen wear eye shades.

Be comfortable when drawing. The height of your chair should be such that your elbows can rest on the board. If you must use a common kitchen chair which is 18 inches or 19 inches high, you should raise the seat about 5 inches or 6 inches. Find a cushioned way to sit at the right height.

The surface of the drawing board is important. If the board is of soft pine or basswood, the drawing paper can be tacked directly to the board. Otherwise you should use a plastic drawing board cover. It can be held in place by staples or double-surface sticky tape at the top and bottom edges. Such a cover may expand slightly under use and may need readjusting occasionally to eliminate bulges.

When soiled, drawing board cover can be washed using a bit of regular scouring powder.

THE T SQUARE

The fixed-head T square is the commonly used instrument for drawing horizontal lines. Hold it firmly against the left edge of the board with the left hand. When it is in position to draw the line, slide your left hand over until the heel of your palm rests on the board with your fingers holding the T square

Fig. 4-1. An improvised drawing board.

firmly in place as you draw the line with the other hand.

TRIANGLES

Though the thinner school grade triangles will do, the thicker professional grade triangles are somewhat easier to use. The thin ones tend to slip under the edge of the T square and are more difficult to pick up. Triangles also come in tints which afford some contrast against the drawing paper.

Triangles are used to draw vertical and angular lines. First move the T square into position as directed above. Then place the triangle against the edge of the T square so you can move it along the T square into exact position with your fingers while the heel of your hand keeps the T square firmly in place (Fig. 4-2). Always draw against the side of the triangle nearest the head of the T square. Draw upward in most cases; this allows you to keep your eye on the line as it is being drawn. As when drawing horizontal lines, rest the pencil lead against the triangle and incline the pencil slightly to keep the point and line in sight.

You will find other uses for triangles, such as for drawing diagonal lines at 30°, 45°, and 60°. Sometimes you will use the triangle (and T square) for a series of evenly spaced parallel lines called crosshatching (Fig. 4-3).

The adjustable triangle is used to draw diagonal lines other than the 30°, 45°, and 60° lines available on the fixed triangles. The circular scale of this triangle covers only 45° but reads 0° to 45° and 45° to 90° to cover a full 90° quarter circle.

The adjustable triangle is especially useful when drawing the lot survey for a plot plan.

PLAN TEMPLATES

The house plan template, or guide, is a thin plastic

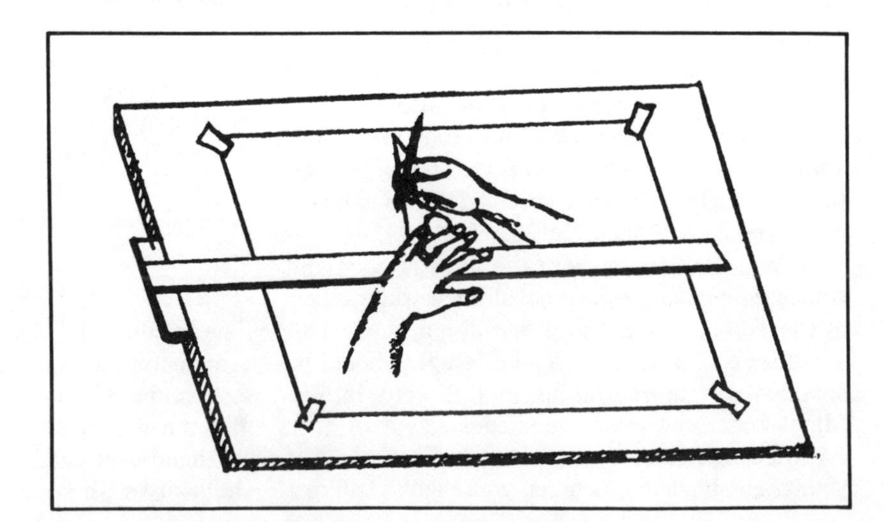

Fig. 4-2. Drawing lines with a triangle and T square.

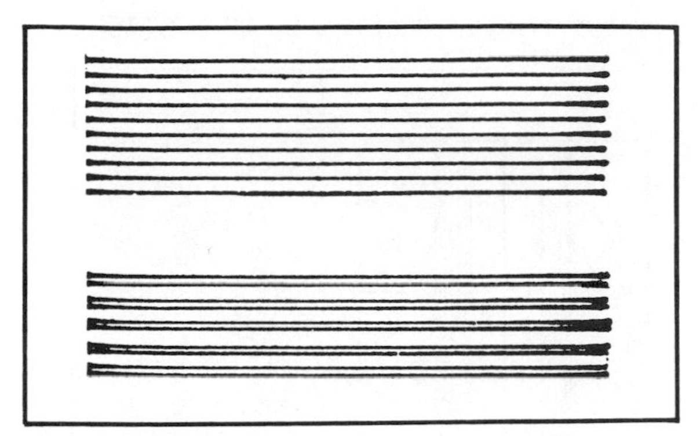

Fig. 4-3. Crosshatching.

sheet, die cut (to scale) to the outlines of the common plumbing fixtures, appliances, door swings, and electrical symbols. Simply position the desired cutout on the floor plan and run your pencil around the inside edge.

There are other templates available. You can buy one showing 1/2 inch scale plumbing fixtures (Timely #T-37); it's useful in detailing bathroom elevations. Timely also makes a handy 2 × 5 inch template for drawing small circles, hexagons, triangles, and squares. This is sometimes an advertising giveaway at drafting supply and blueprint stores. Ask for one.

SCALES

House plans are scaled drawings of the actual house as measured in feet, inches, and parts of an inch (call fractions). A scale, such as the architect's 12-inch triangular scale, allows you to easily draw a sketch of a house to the appropriate scale.

Professional planners generally use flat, bevel-edged scales which are more convenient. These are made in 6 inch, 12 inch and 18 inch lengths. Whatever the shape, scales are used the same way.

The architect's 12 inch triangular scale has 11 different scales on its 6 edges, ranging from full scale to the smallest scale (3/32 inch = 1 foot). Most of the scales are "open divided"; that is, only one foot division is subdivided into inches and the other foot divisions are just marked and numbered (Fig. 4-4). The numbering reads from zero at the divided end to as many feet as can be indicated in the 12 inch length.

The 10 open scales start at opposite ends so that each edge, or face, contains two scales.

To use a scale, place the appropriate one flat on the drawing so you can read along the edge. Keep it parallel and close to the line to be measured; you should just barely be able to see the line. If, for example, you want to mark off 26 feet 8 inches using the 1/4 inch scale, bring the 26 foot mark accurately to the starting point. From there move your pencil point along the scale past the zero foot mark and make a mark on the paper at 8 inches *beyond* zero. On 1/4 inch and smaller scales you can only measure to the nearest inch. On the larger scales the inch marks will be further divided for greater accuracy.

Correct use of the scale is very important, for inaccurate drawings can cause costly mistakes in actual construction. It is a rule that a builder must *never* take a dimension by attempting to scale from a drawing; therefore, every needed dimension must be written on the drawing. Further, the sum of the intermediate dimensions between two points must equal the total.

Laying out *fractions* of an inch when using a small scale requires visual estimation. This practice can lead to error. It is always best to mark off overall dimension rather than make several smaller measurements within the overall dimensions.

A civil engineer's scale is often useful too. These scales, both triangular and flat, are fully divided. The triangular scale includes scales from 1 inch = 10 feet to 1 inch = 60 feet. Land survey maps are drawn with the engineer's scale. It is also used for determining the approximate dimensions of an unscaled drawing.

Figure 4-5 illustrates ways of using a scale to divide a straight line into equal parts.

COMPASSES

An elaborate set of drawing instruments is not necessary for house planning, and a case set will contain many you may never need. Most used will be the small bow pencil compass for drawing circles up to 2 inches or 2 1/2 inches in diameter. A dime store compass will do that too, but a better instrument is worth its cost. The larger 6 1/2 inch com-

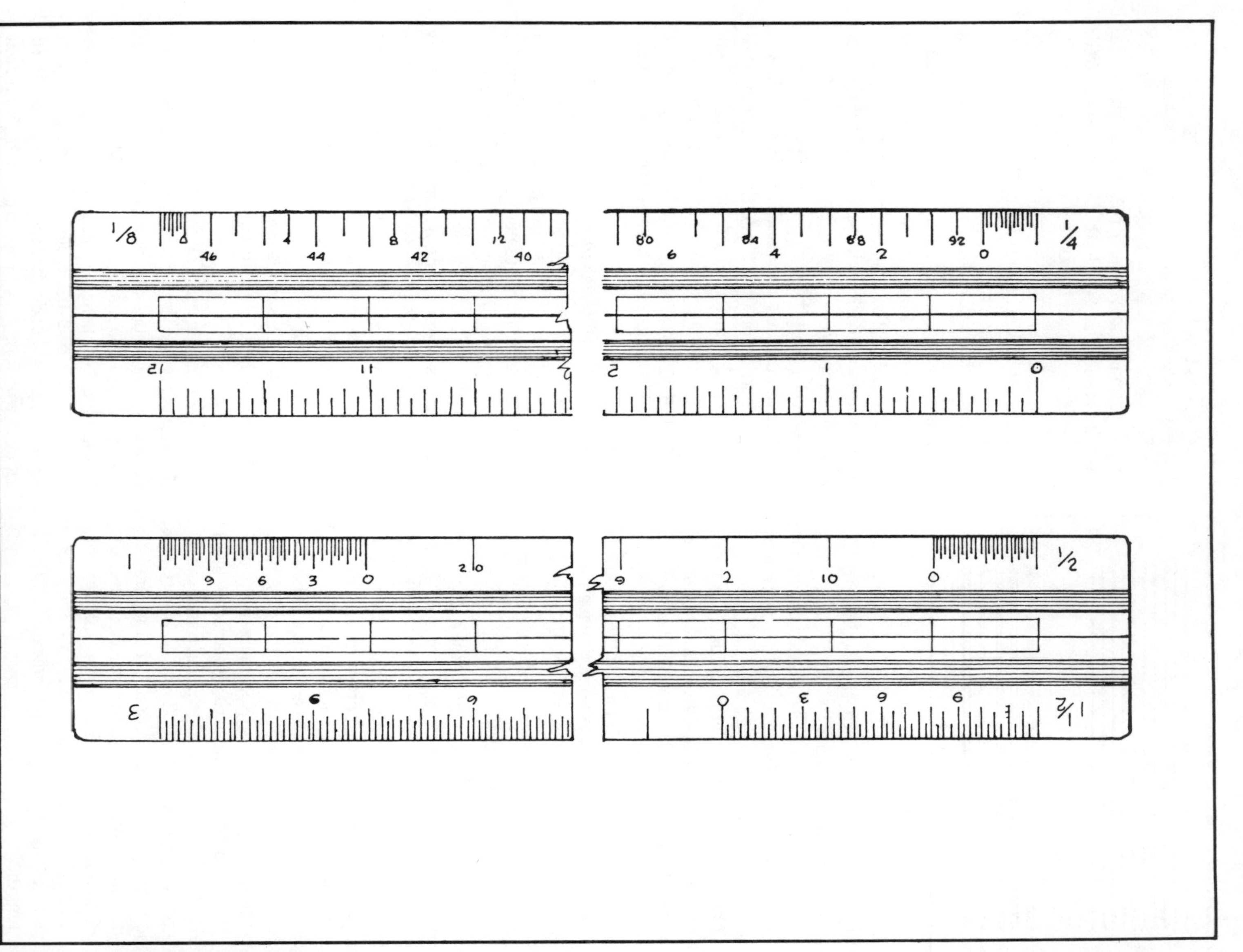

Fig. 4-4. The scales of an architect's scale.

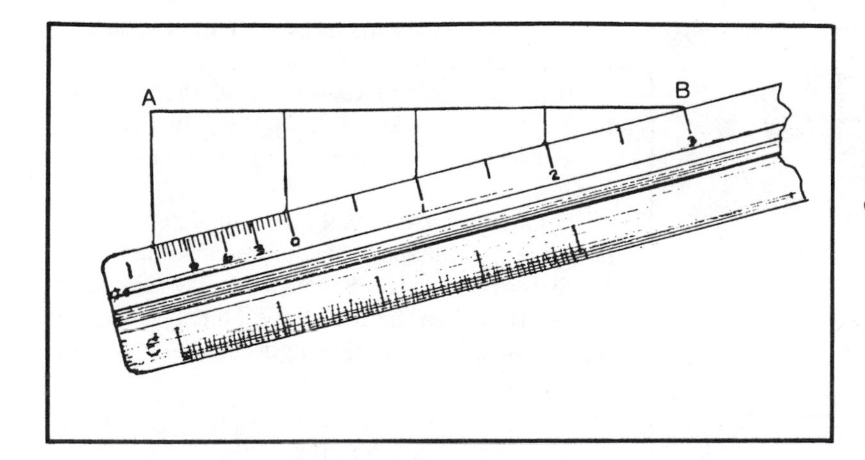

Fig. 4-5. Dividing a line into four equal parts.

pass gets occasional use.

A compass is held with the fingers and is rotated about a fixed center (the needle point). Always lean the pencil point slightly in the direction of the line. A short lead should be clamped in the compass and must be sharpened on a sandpaper file so that the point is wedge shaped. Figures 4-6 and 4-7 illustrate some of the uses of the compass.

ERASING SHIELDS

The erasing shield is a thin piece of metal with holes of various shapes. When you must erase, place the shield on the drawing and erase through the holes; this ensures that you erase only what should be erased.

BRUSHES

It is important to keep drawings clean. The movement of T square and triangles over the paper tend to smudge the drawing. Be especially careful with the final tracings, for dirt can show up on the blueprints. Use the brush after *each* erasure.

DRAWING PENCILS

The drawing pencils are of first importance.

Pencil leads may be encased in wood and exposed for use by cutting the wood away with a sharp knife or a mechanical sharpener; or loose leads may be clamped in a mechanical pencil. Whether the pencils are wooden or mechanical, the lead must be pointed by either rubbing them on a

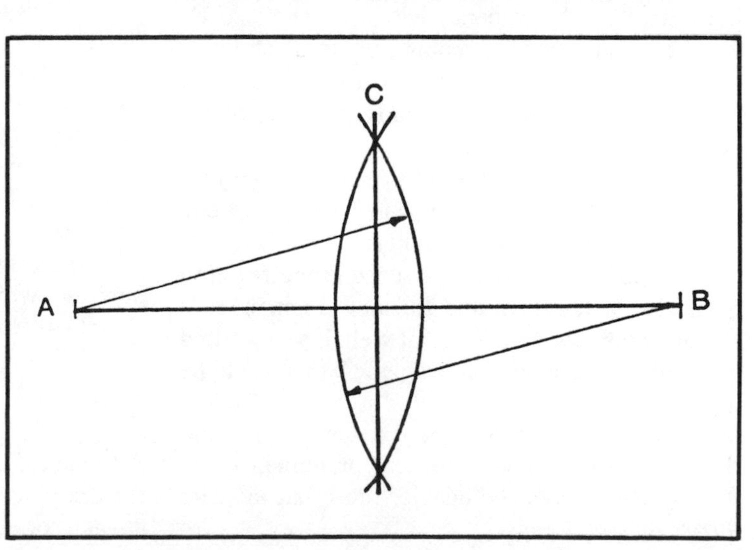

Fig. 4-6. To divide line AB into two equal parts, first set the compass point at A and draw an arc beyond the midpoint of AB. Then set the compass point at B and draw another arc, also beyond the midpoint. Draw a straight line between the two intersections of the arcs; the line divides line AB into halves.

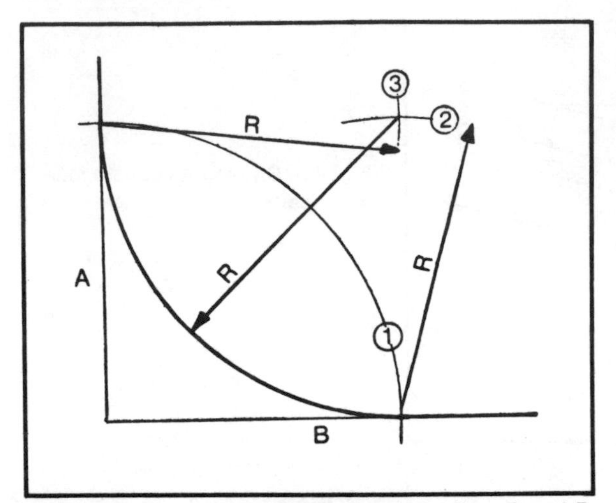

Fig. 4-7. To draw an arc that is tangent to lines A and B, first place the compass point at the intersection of A and B and draw an arc (1) that intersects both A and B. Make the arc of the desired radius R. From the intersection of 1 and B, draw an arc (2) of radius R. From the intersection of 1 and A, draw another arc (3) of radius R. From the intersection of 2 and 3, draw an arc to meet A and B.

sandpaper file or inserting them in a mechanical pointer.

There is yet another kind of pencil. It requires no sharpening because it uses very thin leads. This type is great when it works, but it seems more prone to trouble than the more sturdy type.

Drawing pencil leads come in 17 different degrees of hardness, from the very soft 2B to the very hard 6H. The choice depends on the pencil's use.

The 2B is used for making broad black lines, such as drawing freehand floor plans on cross-ruled pads, for rendering or sketching pictures, or for heavy-line title lettering.

The HB, hard and black, is used for minor title lettering or for bringing certain lines into prominence by making them blacker.

The 2H will be the most used grade for both preliminary drawings and finished tracings. It is strong enough to hold its point well. If you wanted only one lead in a mechanical pencil, this would be it.

The 4H is used for fine dimension lines and their extensions. These are less prominent lines, distinguished from the heavier lines that indicate a part of a building.

Drawing pencils are six sided and thus are less likely to roll of the tilted board.

It is important to keep the lead sharp and to draw with even pressure. Lean the pencil slightly in the direction of the line so its point can be kept in sight.

Humidity can affect the strength of pencil leads as well as the dimensional stability of some paper. A line drawn as 26 feet 8 inches one day may scale 26 feet 10 inches the next day. Thus it is important to include dimension figures, which do not change.

PAPER

Most *finished* plans are drawn in pencil on transparent tracing paper, sometimes called vellum, from which blueprints can be made. For the amateur planner it is best to draw dimensioned floor plans, front and end elevations, construction sections, and other details on cheaper drawing paper and then trace these preliminary drawings onto tracing paper. Mistakes and corrections may leave the first drawings smudged. Redrawing (tracing the old onto fresh tracing paper) does away with the poor and dirty work.

The 18 × 24 inch sheets of school grade buff or white paper are usually satisfactory for preliminary drawings. Any white paper that is fairly stable (not subject to stretching when humidity in the room is high) can be used. It should not tear too easily and be able to withstand erasing.

Tracing paper should be of good quality. Thin, tissuelike tracing paper is not satisfactory. When purchasing tracing paper at a drafting supply store, ask for the paper that is most used by professional architects. Tracing paper can be had in cut sheets or rolls.

Professionals often use a low cost tissue-thin buff or white paper for freehand soft-pencil sketches. It is not generally used by amateurs.

DRAFTING TAPE

All types of drawing paper can be held in place on the drawing board with short pieces of sticky tape at each corner. Supply stores sell drafting tape, but ordinary masking tape is usually cheap and satisfactory.

PENS

A fine-line ballpoint pen may be used to draw permanent lines on the preliminary drawing. These will remain when pencil lines are erased. Black, red, or blue is okay. A medium wide-line black-ink felt-tip pen can be used to draw border lines.

CALCULATORS

A real time-saver when figuring is a small electronic calculator. You can buy one for about $10. Without a calculator, calculating the size of girders or estimating materials might take an hour. The job can be done in less than a minute with a calculator.

A glance at the calculations in the chapter on solving structural problems will indicate how helpful an electronic calculator can be. However, if need be, you can plan a house without one. Remember that most of the great buildings and bridges now standing were planned before electronic calculators were invented.

HOW TO MAKE SCALE DRAWINGS

You'll need a drawing board, but there's no need to trot to the nearest art supply store and plunk down a bundle for their prefab deluxe model. Just consult the Appendix of this book and you'll find three versions of drawing boards.

Alternatives would be to use a large cutting board or a wide shelf of the proper dimensions. Make sure that the sides of the board are straight so that your T square will ride properly. Also check the angles at the corners; at least two of them must be 90 degrees (Fig. 4-8).

It's nice to have a drawing surface that slants. You can use the stands described in the Appendix. If you use an improvised board, you can prop up one end with something of the suitable height. We have found that telephone books or catalogs will work in a pinch. There is no law that you have to work on a slanted board. If you like, you can have your board nice and flat on the table top and we won't say a word.

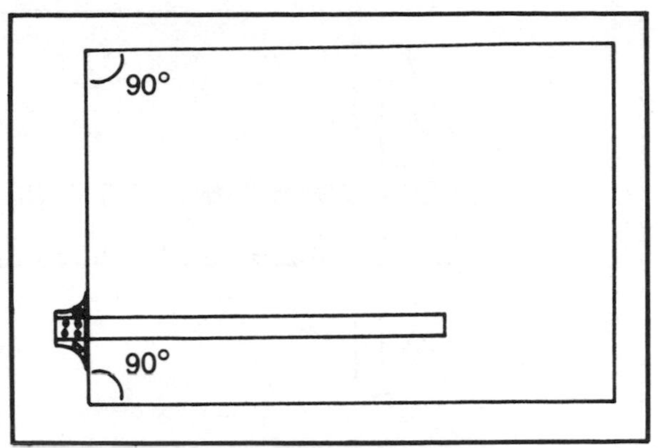

Fig. 4-8. Drawing board and T square.

You can use any kind of paper that's heavy enough to take a nice line and that will allow you to erase. For your sketches and exercises you might like to try regular white duplicating paper. Such paper is usually the least expensive available other than newsprint and light manila, which isn't suitable. For work drawings, you can get regular drawing paper in sheets or rolls. Select paper that has enough grain to take a pencil line readily, a hard surface so it won't be easily grooved by your pencil, and lets you erase your mistakes without anguish or hard labor.

You'll need something to attach your paper to the drawing board. You can use thumb tacks, drawing pins or drafting tape. Thumb tacks are readily available and easy to use. If you're careful and use the more expensive ones, with thin heads and steel points, you won't have any problem. Map pins have a tendency to lose their heads at the slightest tug. So use them only in emergencies.

Drafting tape has been designed specifically for the purpose of attaching paper to drawing boards. While drafting tape looks like a twin to regular masking tape, don't be fooled by appearances. Drafting tape comes off the drawing paper easily and cleanly. Masking tape might stick or leave a residue.

The T square (Fig. 4-9) is the staple of drawing house plans. You can get T squares in various length, weights, and prices. While a professional draftsman will do well by investing in a super-duper-extra-deluxe version of this basic tool, you

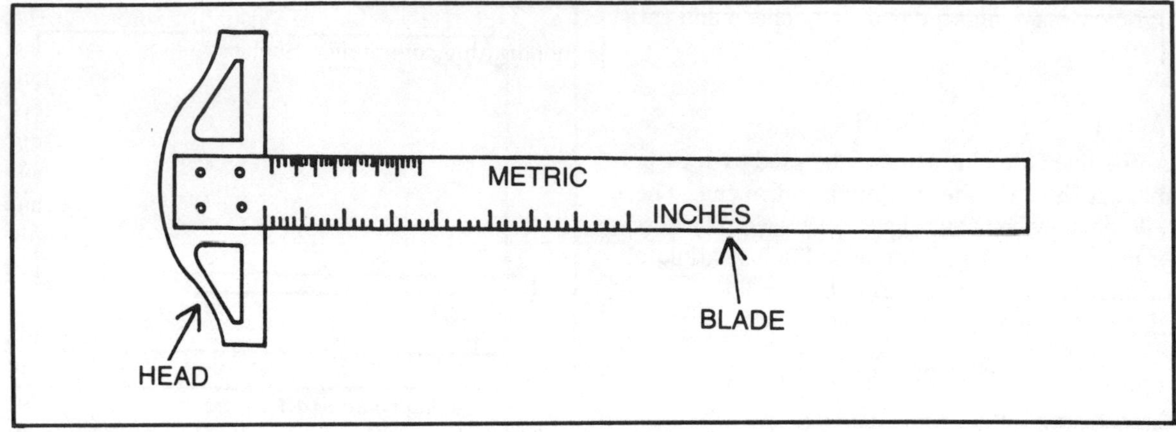

Fig. 4-9. The T square.

can do well with a quite moderately priced one. Expensive T squares are made out of hardwood and they are often steel edged. The costly models have an adjustable head square that the ordinary person never needs. Some more expensive models are equipped with transparent edges.

On par in importance with the T square, and used in conjunction with it continually, are the triangles. Triangles are usually made out of clear plastic and they should be stored flat to prevent warping.

We like the kind of triangles that are calibrated; they save a lot of time because you can draw and measure at the same time. We also love the kind of triangles that are marked in inches along one side and in metric on the other. Even if you have never used metric measurements and are only hazily aware of the conversions, you'll find that you can be much more accurate if you use the metric scale for measuring. The metric increments are much smaller and, because measuring is based on the 10 scale, you never have to experience that sinking feeling that we often used to get when we tried to measure a fifth or a tenth of an inch.

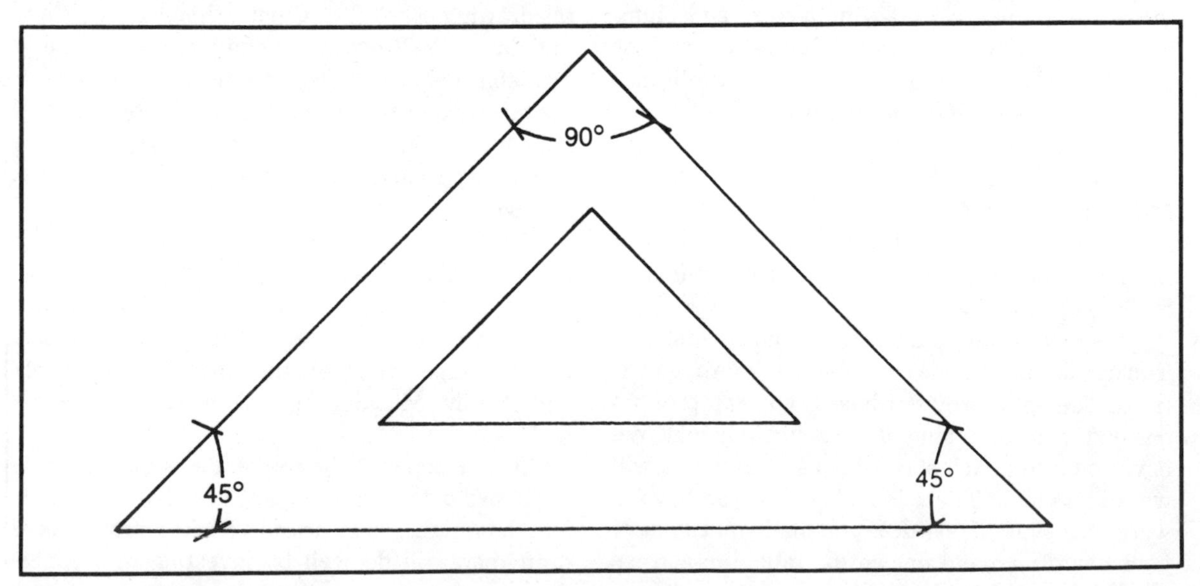

Fig. 4-10. A 45-degree triangle.

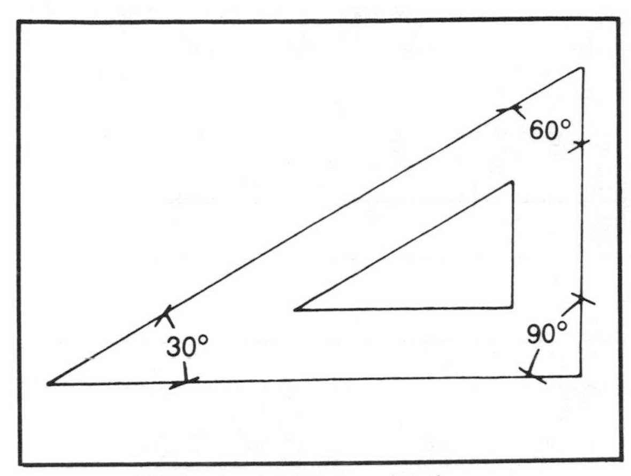

Fig. 4-11. A 60-degree/30-degree triangle.

You'll need at least two triangles: one that has 45-degree angles and one that has 30-, 60-degree angles. The triangles can be bought in 6-inch, 8-inch and 10-inch sizes. We find the 8-inch size quite satisfactory. See Figs. 4-10 and 4-11.

You'll need what most people term rulers, but what the purists of the drawing world refer to as *scales*. For simple measuring you can use any regular ruler you pick up at a variety store. We do recommend that you select one with steel edges. The plastic kind tends to nick and curve before you have it out of the sack.

Scales refer to the calibration rather than the instrument. In the U.S., we are in the process of converting to metric like most of the rest of the world. In the meantime, you will still find many rulers marked off in inches along one side and centimeters on the other side. As with your triangles, it is really easier to draw accurately if you use the metric divisions for measuring. If you use your ruler with the dual scale, you'll have an instant equivalent measurement either way (Figs. 4-12 and 4-13).

Architect's scales are usually esoteric to the nondrafting community. Such scales are three sided and each surface has its own use. As to calibration, architects' scales are divided into proportional feet and inches. These scales have divisions indicating 1/8, 1/4, 3/8, 1/2, 3/4, 1 1/2, and 3 inches to the foot. They are usually "open divided" which means that the units are shown along the entire length, but only the end units are subdivided into inches and fractions (Fig. 4-14).

Civil engineers' scales are divided into decimals with 10, 20, 30, 40, 50, 60, and 80 divisions to the inch. They are used mainly for plotting maps, but they can be used whenever you need to divide an inch by tenth. We feel that a simple metric/inch ruler will do just as well. Other tools that are nice to have are curved rulers that are usually referred to simply as curves or "French curves." You need them if you want to draw a curved line that is not a circle arc. The pattern of these curves are parts of ellipses and spirals as well as other mathematical curves in various combinations. To use them, you take the peck and hunt approach. You keep trying various portions of your French curve in the space to be drawn until you hit upon the one you need. It's much easier done than said and quite helpful. Also, it gives your drawings that last finished look. The cost for a French curve is only a dollar or so.

To draw a decent circle you need a decent compass. As you very well know, you can pick up a compass at a local drugstore for about 50 cents. These contraptions have a small pencil attached at one side with a small screwed on band. Whatever you do, don't buy one of these. They are an abomination to anybody who strives for even minimal accuracy.

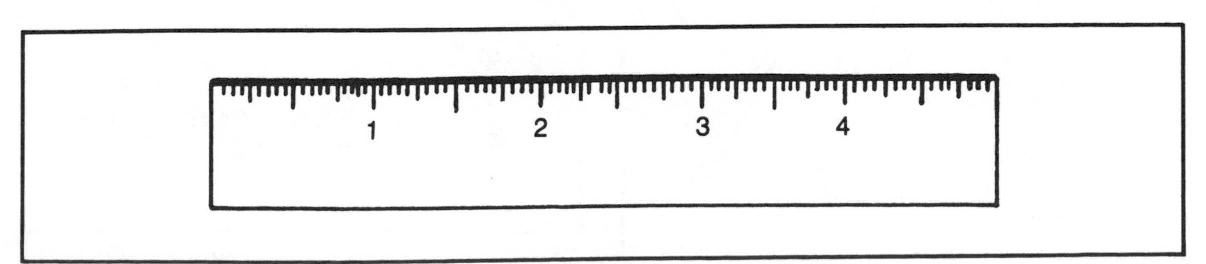

Fig. 4-12. A ruler with inch calibration.

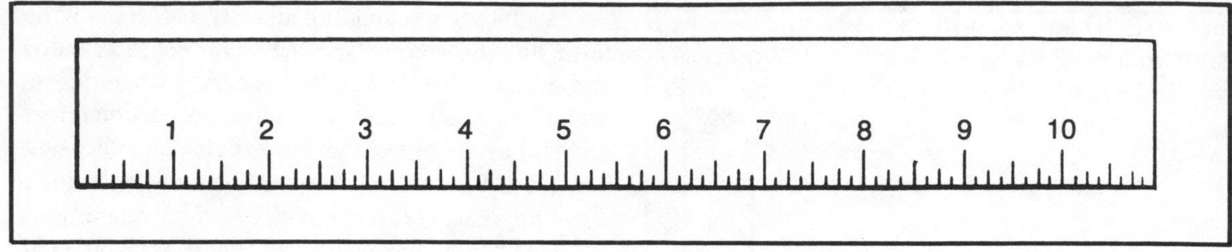

Fig. 4-13. A ruler with metric calibration.

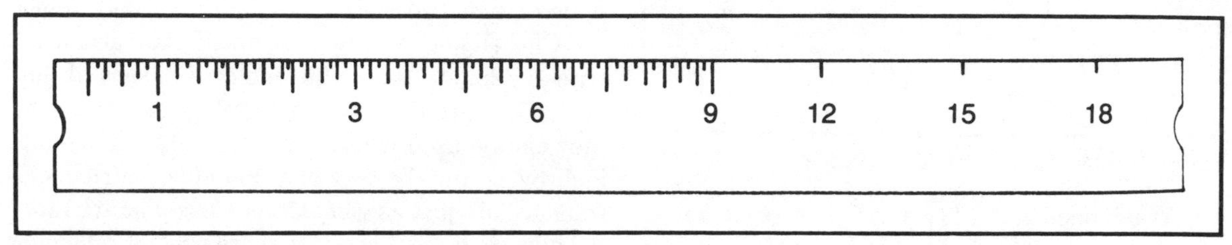

Fig. 4-14. Architects rule with an "open divided scale."

A decent *pencil compass* need not be expensive. Usually you'll find a good one for two or three dollars. A pencil compass, in this connotation *does not* refer to an actual pencil attached to your compass. It means the pencil point that is inserted in one leg of the compass. Often these pencil compasses have an extra steel point that you can put in instead of the pencil point. This transforms your compass into a divider (Figs. 4-15 and 4-16). A 6-inch model is great.

LEAD

POINT

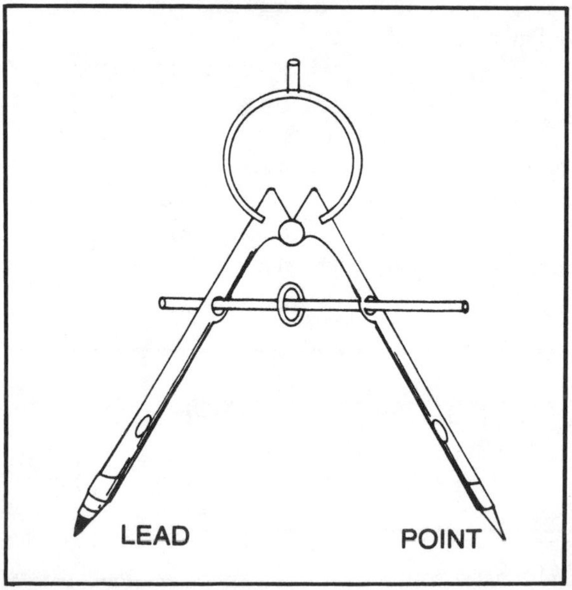

Fig. 4-15. A compass/divider.

You use dividers to transfer accurate measurements, to divide into segments (particularly on a curve), and to indicate very accurate positions. Some compasses and dividers have extension pieces, cross-set screws called *bows*, and pen attachments. These compasses are usually beautifully made and a joy to use. They are also quite expensive. Because you won't be using the pen attachment or the extension bars, there is little reason for the added costs.

Pencils are graded with a number and a letter. The grading is according to hardness. The H series,

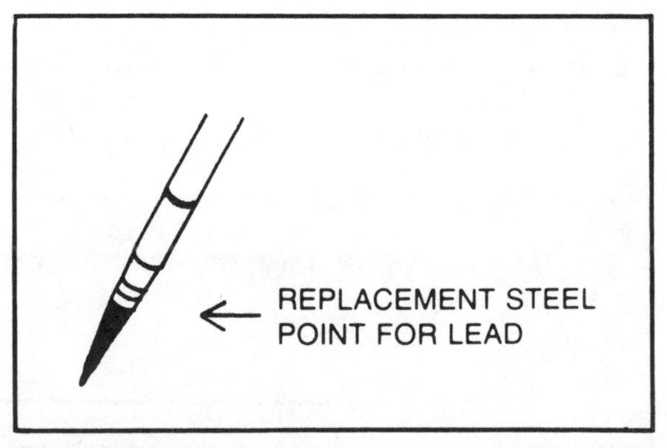

Fig. 4-16. Transforming a compass into a divider by replacing lead with a second steel point.

that is H, 2H and so forth up to 9H, proceeds logically from medium hard to hardest. The process is inverted when it comes to the B series. This series goes from 6B (very soft) through B, HB, and finally F (which is the middle). Then you pick up good old H and sanity again.

The soft B grades are primarily used for sketches and rendered drawings. The H family is preferred for instrument drawing. The more involved and exact the drawing the higher the H. You will be well equipped if you have one or two specimens from the B line plus a couple of Hs (but not harder than 4H). Even 4H is fairly fine and it leaves a deep mark on the paper when you try to erase it. It does, however, make a nice accurate line.

In addition to the usual wood-encased lead pencils, semiautomatic drawing pencils are available. The lead tips are graded exactly like the ordinary pencils and are interchangeable within the same pencil. Some of these automatic pencils have a built-in point sharpener. This is a blessing and so is an attached eraser at the other end.

There are some small accessories that are quite inexpensive and they give you a lot of comfort and convenience. The first and most important of those is your trusty eraser. You'll need several erasers for separate functions. A ruby pencil eraser, the large size with bevelled edges, is the standard model. It not only removes pencil lines like a breeze, but is actually more efficient than so-called ink erasers because it will, with a little elbow grease, remove the ink from paper or cloth without chewing up the paper or cloth in the process. The other kind of eraser you'll need is an Artgum eraser to clean up your paper when you're done and remove any finger marks or smears. Another kind of eraser is a secretary's pencil eraser. This type looks like a pencil and you unwind it slowly as needed. The tip can be sharpened. The kind we use has a white eraser base that easily takes off thin and heavy lines. We find it particularly handy for small alterations. You can erase precisely the spot you want without disturbing the rest. It is very useful when you do your lettering.

There is another kind of eraser that comes in a plastic case shaped like a pencil. This one is red gum like the ruby eraser. While the shape is handy, it doesn't do quite as nice a job as our white favorite.

A neat gadget that complements your eraser collection is an erasing shield. It consists of clear plastic that has cutouts on it: curves, straight lines, circles, wedges, what have you. You select whatever shape it is that your mistake resembles the closest, put your shield over the rest, and erase away. Not only does it protect your other lines from becoming extinct by association, but it also prevents extra finger and hand prints on the paper. You hold the eraser shield down with one hand and touch it instead.

You'll need some kind of a pencil sharpener. We prefer the small hand-held pencil sharpener that has two different-sized openings. While it is a bit more awkward than the wall model, we find we can get just the kind of point we want on our pencils without breaking them off. The trick is to start the sharpening process in the larger opening and then finish off in the smaller one.

If you're really a nut about getting a proper point on a pencil—and those highly trained draftsmen and mechanical drawing specialists will tell you that you need different points for different jobs such as a flat or wedge point for straight line work or a conical point for curves—you'll need two more gadgets. One such tool is a small sandpaper pad with a wooden handle. The other is a file; a nail file will do.

To get a wedge point, sharpen a pencil as usual and then use a pen knife or an X-Acto knife, and make two long cuts, on opposite sides of the pencil, flattening the end of the lead with sandpaper or a file. Then trim in the edges of the tip to make the wedge point narrower than the diameter of the lead.

For a conical point, proceed with the mechanical-sharpening and then refine the point with sandpaper or a file.

Keep your pencils sharp. It makes quite a difference in the width of the line. Sometimes you have to worry about such things in a finished plan. Make a habit of it from the start. Get rid of graphite and wood shavings immediately. If you leave them on a desk or drawing table, they can smear your drawings, hands, clothes, and whatever else is in range.

To put paper on a drawing board, use a T square to ensure that the paper will be parallel to board. This is essential because otherwise you won't have a line parallel to the edge of your paper. If the paper is much smaller than your board move it within 3 or 4 inches of the left edge of the board and several inches from the bottom edge. A T square will be most stable and rigid as close to the head as possible while it flips around a bit at the other end. Get close to the source. As to the distance from the bottom of the board, you need room to slide your T square down below the edge of your paper if you want to draw a line near the bottom of the paper.

Put your T square on top of your paper with the head riding on the left edge of your board. Carefully adjust the paper so that the top edge is even with your T square. Gently pull the T square down a few inches and secure the top corners with thumb tacks or drafting tape. Repeat the maneuver for the bottom edge of the paper, but this time pull your T square up instead of down (that is onto the paper not off it) before you secure the bottom corners. Now it should be all nice and square (Figs. 4-17, 4-18, and 4-19).

While you can learn to perform this little drill in a few moments, one of us—who is notorious about trying to find short cuts and absolutely abhors wasting time and energy—has come up with this idea. You proceed as stated, but instead of putting on paper and tape you pull down the T square and draw a good, dark line along it with a thin-point marker. You do the same at the bottom. Next you measure in 4 inches from the left edge of the board, top and bottom line, and connect the marks with a straight line. Put a piece of drawing paper on the board between the lines and mark off the right-hand edge top and bottom line again. Connect these marks. Presto! You have a permanent frame in which to position your drawing paper without going through the entire routine each time. It works like a charm if you're extra careful when you put your lines on the board and make the lines as thin as possible.

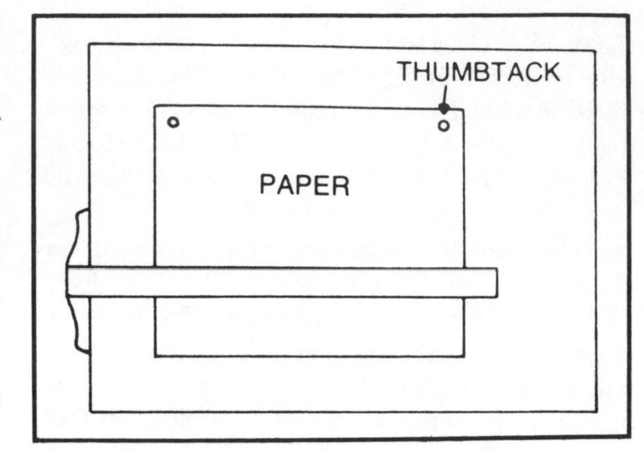

Fig. 4-18. Fastening corners with thumbtacks.

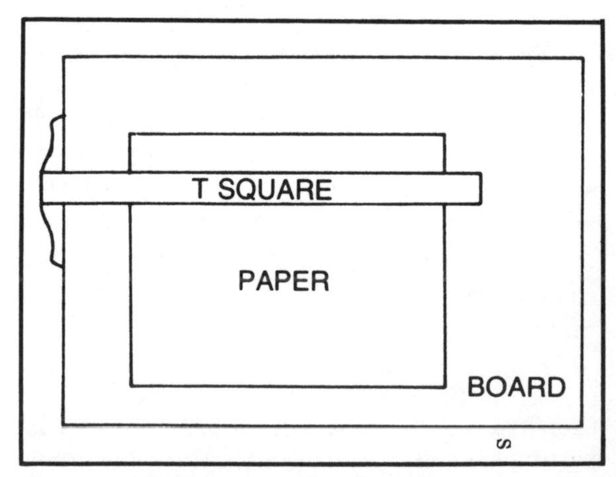

Fig. 4-17. Aligning paper with a T square.

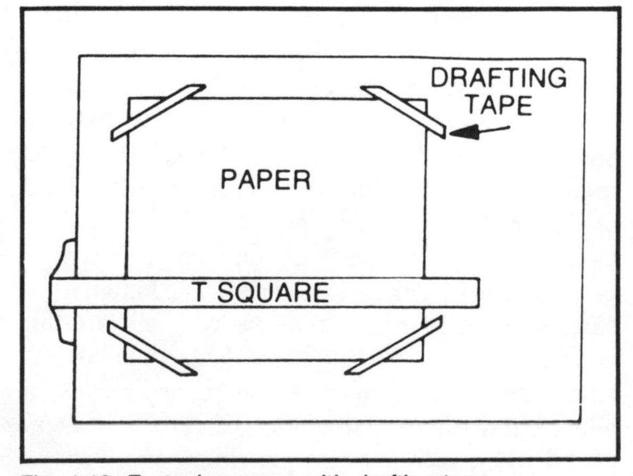

Fig. 4-19. Fastening paper with drafting tape.

A variation on the theme is to proceed as above, but use colored plastic tape for the framing. Make sure it adheres well. When you position your paper, be consistent about aligning the top edge with top edge of the tape or the bottom edge. It doesn't matter which, but don't switch around. This is even more crucial on the sides where a quarter of an inch can certainly cause your paper to be askew. See Figs. 4-20, 4-21, and 4-22.

Your T square does more than line up your paper on your drawing board. Actually, your T square is the foundation on which you start your drawing. All your horizontal lines are drawn along the T square. To do this, hold down the head of the T square with your left hand, slide the T square in position, and change your hand to hold down the extension while the right hand holds the pencil and draws the line along the edge of the T square.

When you draw a line along the T square, slant the pencil in the direction you are drawing the line (to the right) and keep the point of the pencil as close to the T square as possible. This is important. Hold your pencil lightly, but close to the edge of the T square and do not vary the angle at which you hold it till you come to the end of the line. It's always amazing to us how far off you can get from the straight and true by not observing the above procedures. That can mean a lot of erasing and doing things over.

If you're left handed just reverse the procedure. Put the head of your T square along the right-hand edge of your board, draw your lines toward the left—again slanting your pencil in the direction in which you're drawing—(to the left), while keeping the point of the pencil as close to the T square as possible.

Fig. 4-21. Taping a drawing board for paper position.

T squares do more than help you draw horizontal lines. They provide a base for drawing verticals as well. To do this, use your triangles. Let them ride on top of your T square with the perpendicular edge nearest the head of the T square and toward the light. (Don't try to draw in the shadow of your hand). Draw your perpendicular or vertical lines from the bottom (of the T square) up to the top. See Fig. 4-23.

When you draw these verticals along the straight edge of your triangle, you have to anchor

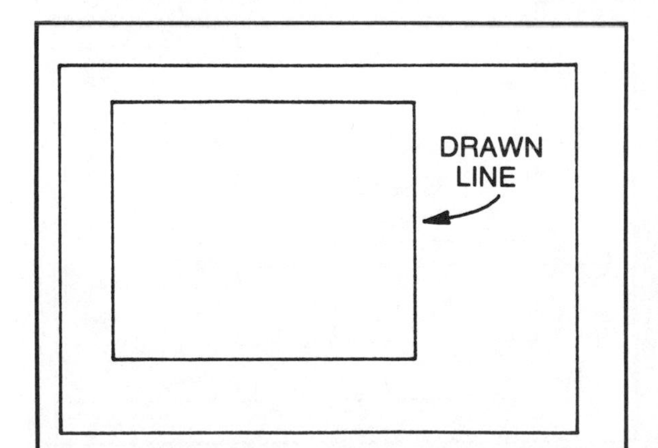

Fig. 4-20. Putting permanent position lines on a drawing board.

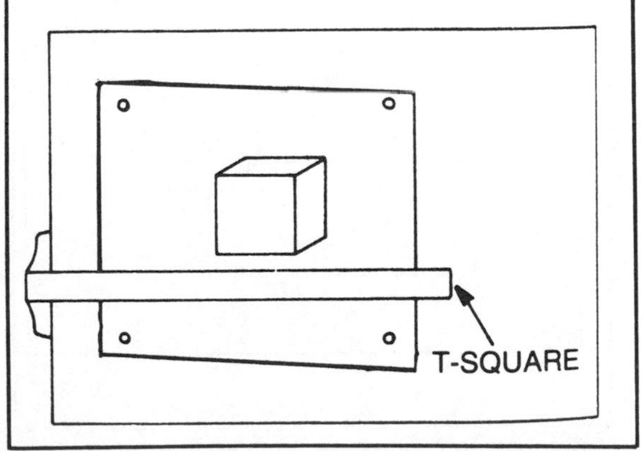

Fig. 4-22. Oops! That's what happens if you don't follow directions.

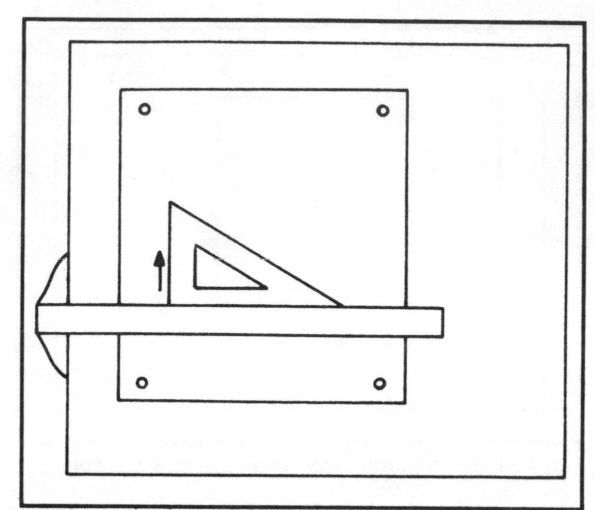

Fig. 4-23. Using a T-square and a triangle to draw perpendicular lines.

your T square against the board with your left thumb and little finger while the rest of the fingers of your left hand adjust and hold the triangle in position. Your right hand, of course is occupied with holding and guiding the pencil. To make sure that you have the proper contact with the board before you start your line, listen for the double clicking sound that means that the board and T square head have connected. It will take only a slight pressure of the thumb and little finger toward the right to hold the T square in place. When you draw that vertical line, the pressure of all your fingers will not only hold the T square, but also the triangle in place. Actually it's all quite simple. In this case a try is worth a thousand words. So go do it.

Be sure that, as you draw your vertical line, you keep your pencil point as close to your triangle as possible. Just as with the T square, the triangle has to be securely aligned with its guiding edge, which is the T square. Please don't let it flop around and expect exact work. Another trick to ensure accuracy is to keep the T square below the lower end of the line that you are drawing. That avoids working at the extreme corner of the triangle.

There's quite a lot more that your good old T square and triangle combo can do. If you want to draw lines at a 45-degree angle to the T square, all you have to do is put your 45-degree triangle on the T square and draw along the angled end. For your 30-degree and 60-degree angle, proceed likewise with your 30- and 60-degree triangle. See Figs. 4-24 and 4-25.

If you want to be really fancy and draw 15-degree, 75-degree, or even 105-degree angles,

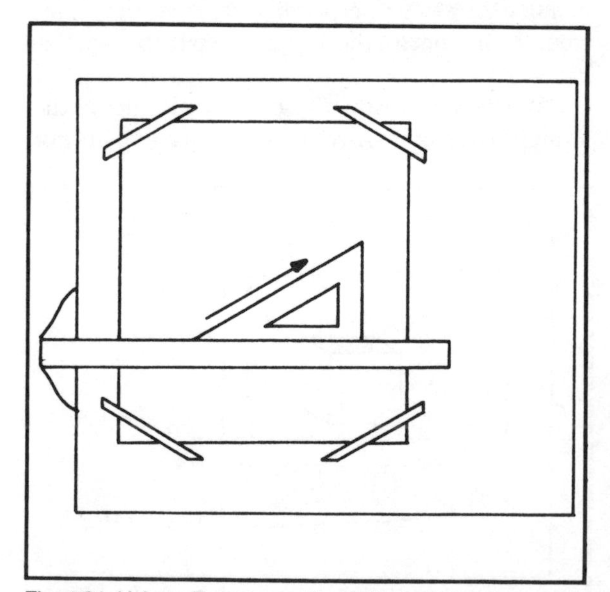

Fig. 4-24. Using a T square and a 45-degree triangle to draw slanted lines.

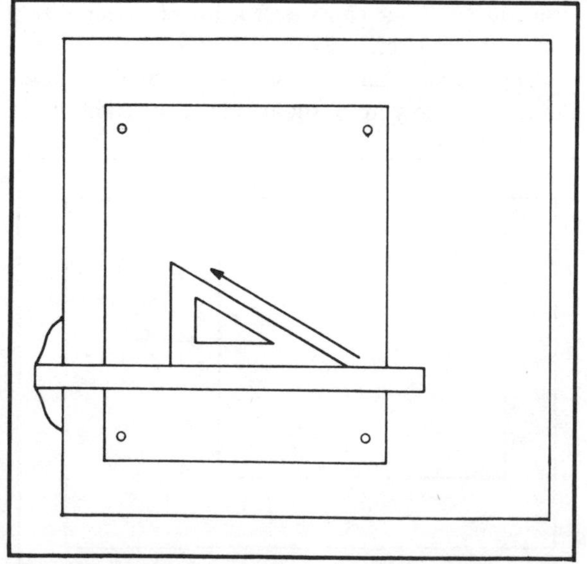

Fig. 4-25. Using a T square and a 60-degree/30-degree triangle to draw slanted lines.

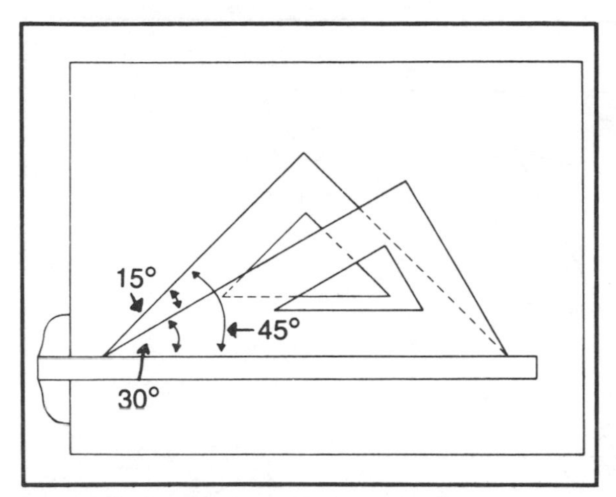

Fig. 4-26A. Constructing a 15-degree angle.

you can do that too. You can draw any angle that is a multiple of 15 degrees by using your two tri angles in combination with your T square. See Figs. 4-26A, 4-26B, and 4-27.

You can draw any number of parallel lines by using your T square and your triangle. Hold your triangle against the line you want to make a parallel line to, hold your straightedge as a guideline, and slide your triangle along the guideline until it is in position for drawing the new line. See Figs. 4-28 and 4-29.

To draw a perpendicular line, place the triangle with one edge against the T square. In a tight place it is sometimes easier to place the triangle against another triangle. Keeping the base steady, rotate the triangle and move it into the perpendicular position (Fig. 4-30). A shortcut would be to set the triangle with the hypotenuse against the guiding edge, fit one side to the line, slide the triangle to the required point, and draw the perpendicular as in Fig. 4-31.

When working with triangles be sure that your light source comes from the right or you'll be working in your own shadow. If you're left handed reverse all the instructions about holding down the T square and the triangle so that your right hand is doing the holding against the right edge of the board and your left hand is free to do the drawing from right to left with your pencil properly slanted. In this case, you should make sure that your light source is to the left of the board.

Measuring and marking accurately isn't a bit more tedious than doing it in a slipshod manner and the results are vastly different. The trick to doing accurate measuring is to have a sharp pencil to do the marking and to mark with a fine line—not a dot. Take your measurements directly from your ruler or calibrated T square. It is best to use calibrated triangles so you won't have anything extra to fool with.

When it comes to using a compass, the thing to remember is that the pencil lead should be adjusted to be of the exact same length as the pin or needle point so that your compass is vertically centered. For measuring, set your compass directly on the scale and adjust there.

To draw a circle, set your compass on the scale and measure off the distance you want for the radius. Next, place the needle point where you want the center of the circle to be. Holding the compass by the handle, draw the circle in one sweep. Hold the handle between thumb and forefinger and incline the compass slightly in the direction of the line. If the line is too dim, you can make it darker by repeating the procedure. See Fig. 4-32.

Using the French curve is a bit different in that you might have to turn and adjust the French curve every little bit to come out with what will look like a continuous and smooth line. If this sounds contradictory, we can't help it. Again a try is worth a thousand words. A good trick to remember is to draw in a proposed curve lightly by hand after you've established the points that are your guides.

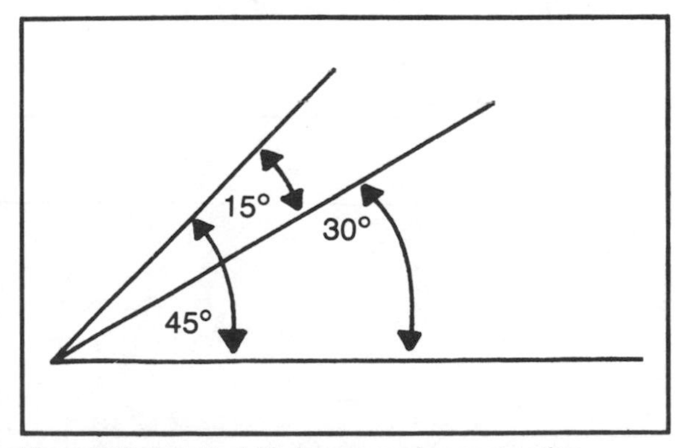

Fig. 4-26B. Detail of 15-degree angle construction.

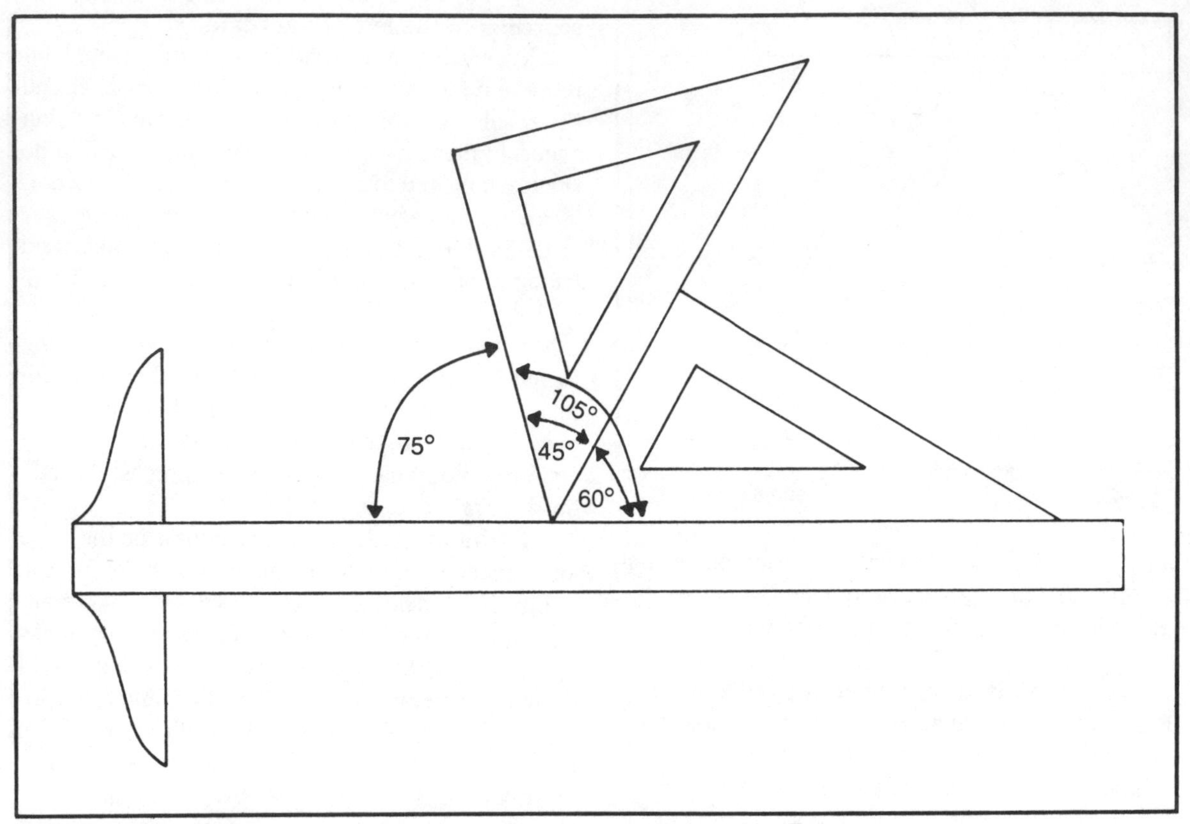

Fig. 4-27. Drawing 75-degree and 105-degree angles with two triangles and a T square.

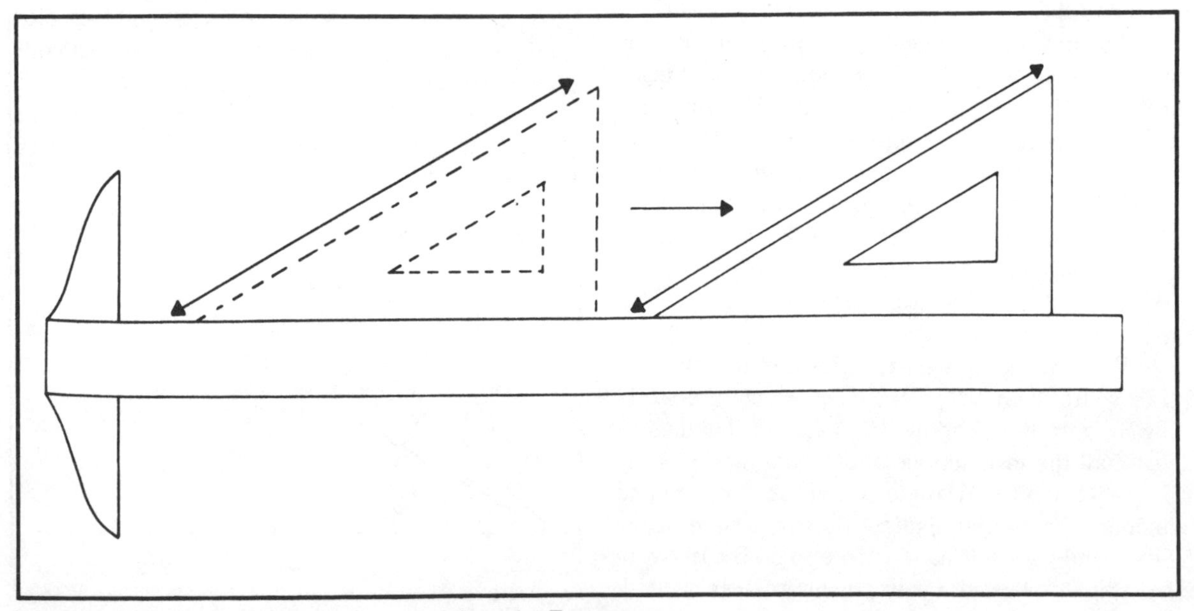

Fig. 4-28. Drawing parallel lines with a triangle and a T square.

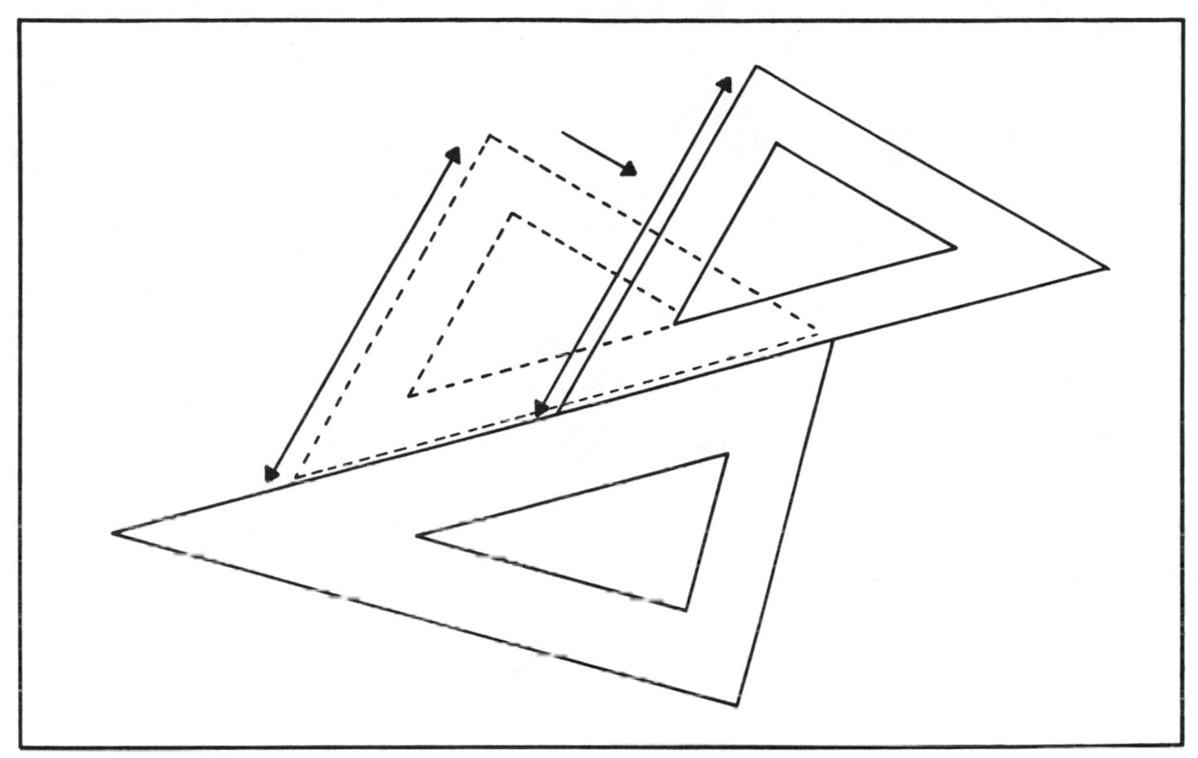
Fig. 4-29. Drawing two parallel lines with two triangles.

Then apply the curve to it after selecting a part that will fit the proportion of your proposed line most closely. Be careful to place the curve in the right direction. If your curve increases, use the end of the curve on your French curve that increases (not the other end that will decrease). Stop just a tad short of the point where the curve and the line will meet and shift to another little space on the curve that will make the transition smoothly. See Figs. 4-33A and 4-33B.

When you shift a curve, be sure that you always adjust the new portion of the curve you plan to use so that it will coincide with the end of the curve already drawn. Fooling around with curves can be a lot of fun and some people get to be quite expert at finding the right curve at a glance. Don't worry if you don't get the hang of the French curve. You can do an awful lot of great plans and other drawings without ever having to cope with the curve.

Plastic templates are a great help when it comes to making house plans. This is particularly

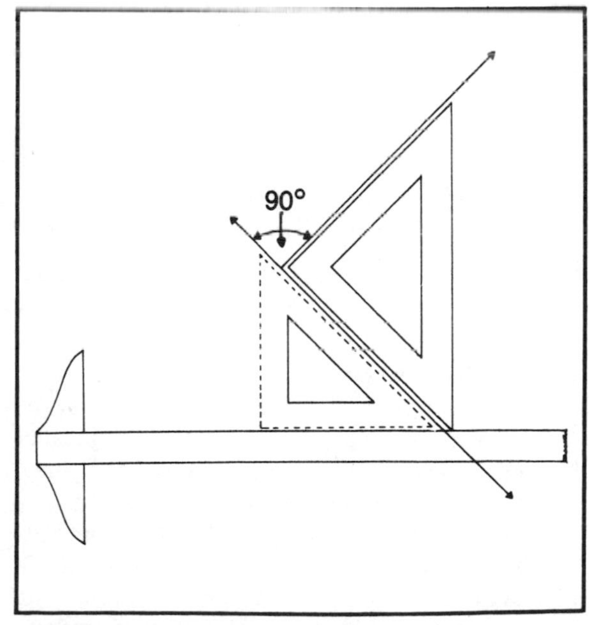

Fig. 4-30. Drawing a perpendicular line (method number one).

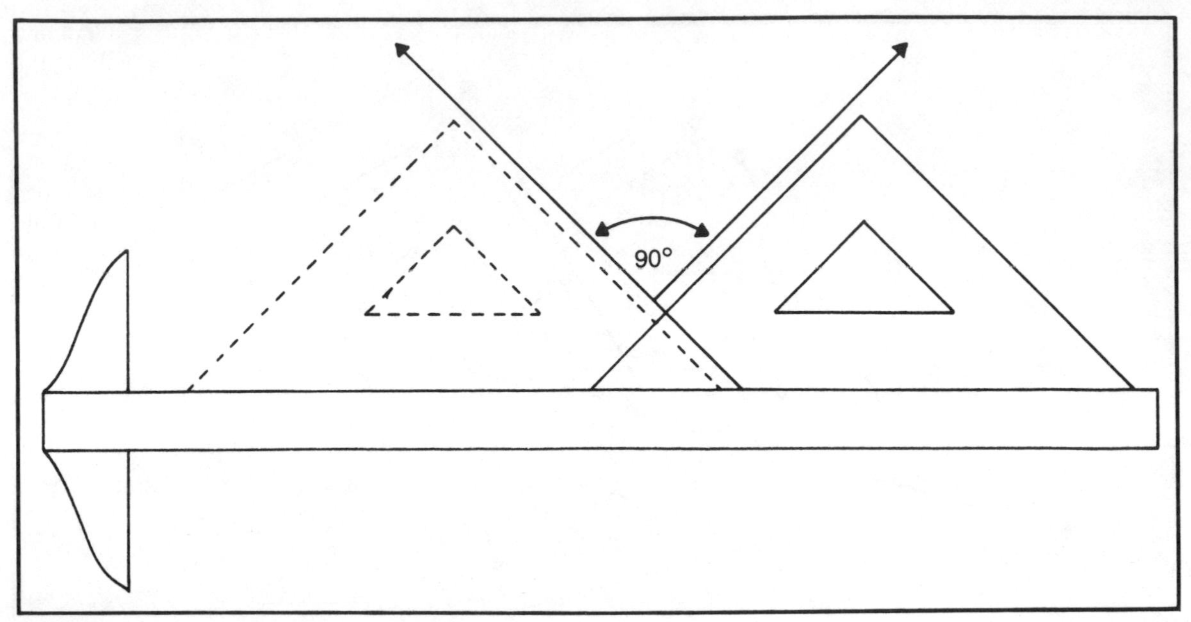

Fig. 4-31. Drawing a perpendicular line (method number two).

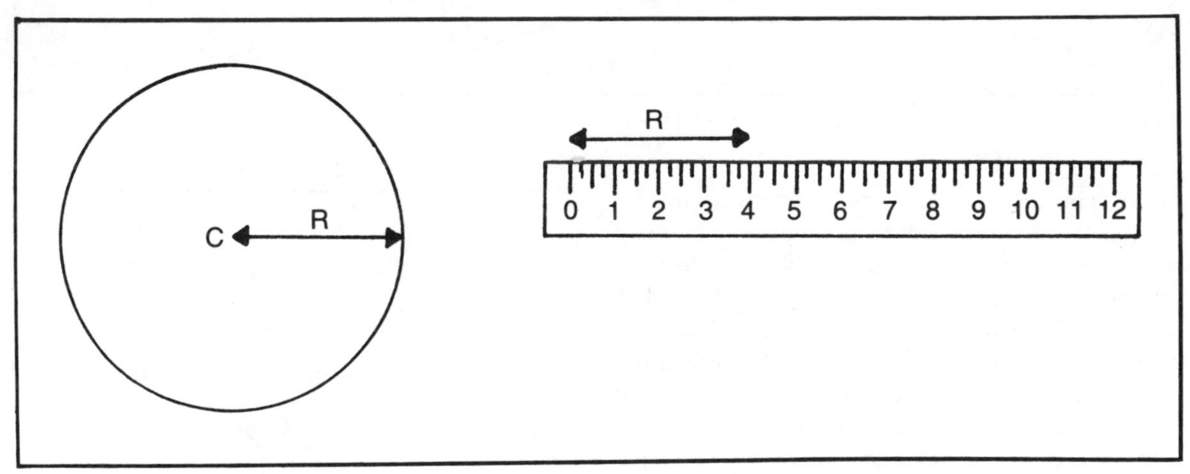

Fig. 4-32. Drawing a circle with a compass.

true for electrical plans and plumbing plans. What you do is simply to use the cutout places on the template as stencils and run your pencil point around the inside rim. Be sure to keep your pencil point in contact with the edge of the stencil or you won't get the benefit of regularity in size and line.

There are a number of these templates available and you can pick and choose from among them. Some people get a lot of fun out of working with templates; it's a bore and a bother to others.

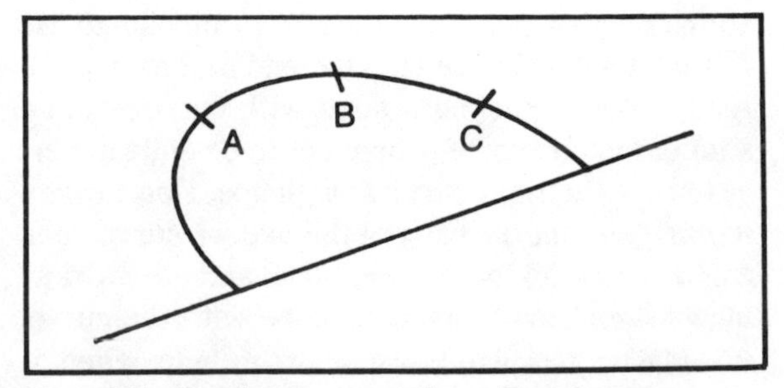

Fig. 4-33A. Drawing a curve using a French curve.

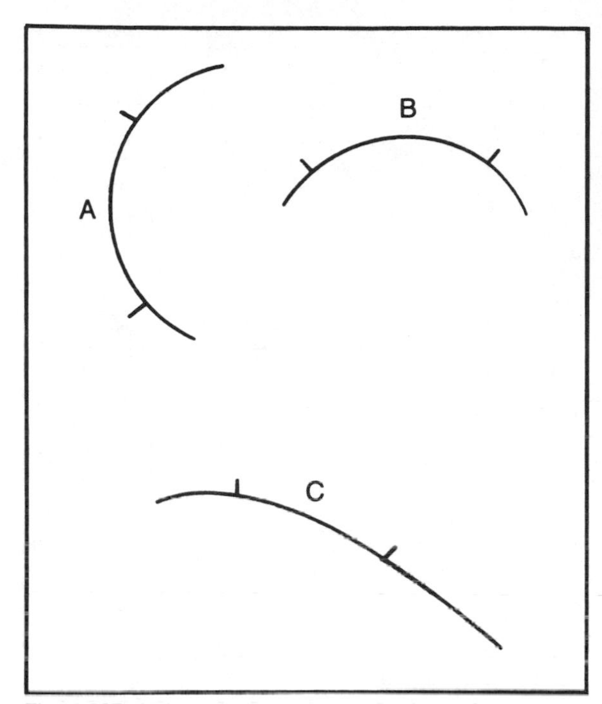

Fig. 4-33B. Using a French curve.

Suit yourself as far as templates are concerned. They are definitely not essential and you can do a fine job without them.

ERASING TECHNIQUES

A soft pencil line you want to eliminate or change is the simplest to work with. Just rub a soft pencil eraser lightly over the offending line. For heavier lines, use your ruby pencil eraser. If the paper is grooved by the line, it helps to run your thumbnail over the line to smooth out the paper. If your thumbnail isn't used to this kind of work, use a tongue depressor, a round ballpoint pen (with the tip down, naturally) or any other like instrument that will flatten the groove. If you'll slip your triangle under the paper, you'll have an extra-fine backing surface.

If your offending line is close to some that you intend to keep, use your eraser shield. First of all make sure that the eraser shield is clean on both sides. Horrible things can happen if graphite or eraser shavings are still sticking to it when you use it.

Select the opening that fits your problem and rub through the shield with your eraser, holding the shield down firmly. If you need to erase an ink line, which you won't have to do if you follow our format, use your ruby pencil eraser. Work the eraser along the line and across it patiently until the ink is off. Use your triangle as backing. Use correction fluid sparingly if at all. It really does show up as a blob on the paper and then by the texture of the line drawn on top of it. Try as you might, the line won't match exactly in thickness or tone.

Whether you erase pencil or ink, be sure to get the eraser crumbs off your drawing. Blowing is the old-fashioned way, and not highly recommended. A soft, clean cloth or a small paint brush is best.

ITS TIME TO DRAW

Put your paper on the drawing board with thumb tacks or drafting tape (Figs. 4-20 through 4-22). Now follow our step-by-step procedure for drawing a square, a rectangle with lines in it, various kinds of lines you'll be using for different purposes, circles, and a cube.

Drawing a Square

Your square will be 5 inches by 5 inches.

Put your T square on the board and draw a horizontal line where you want your square bottom to be.

Measure off 5 inches and mark a line on both ends with a small vertical line, as in Fig. 4-34.

Put your triangle on top of your T square that is now a bit below the drawn line with the perpendicular side lined up with the mark on your line (Fig. 4-35).

Draw a line; then move the triangle down to the other mark and draw the second line.

Measure both lines and mark the 5-inch length.

Move your T square up to the marks and draw a line between them. Presto! Your first square (Fig. 4-36). Erase any extra length and you're done.

Drawing a Rectangle With Lines In It

Your rectangle is going to measure 7 inches for the horizontals and 4 inches for the verticals.

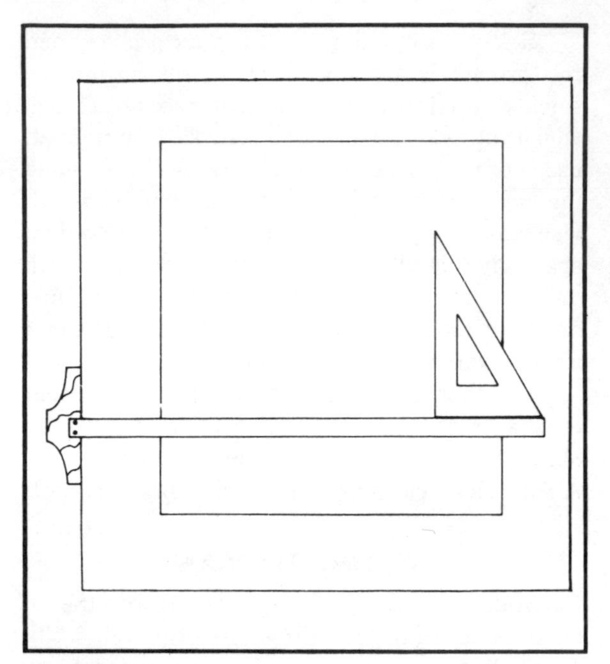

Fig. 4-34. Drawing a square.

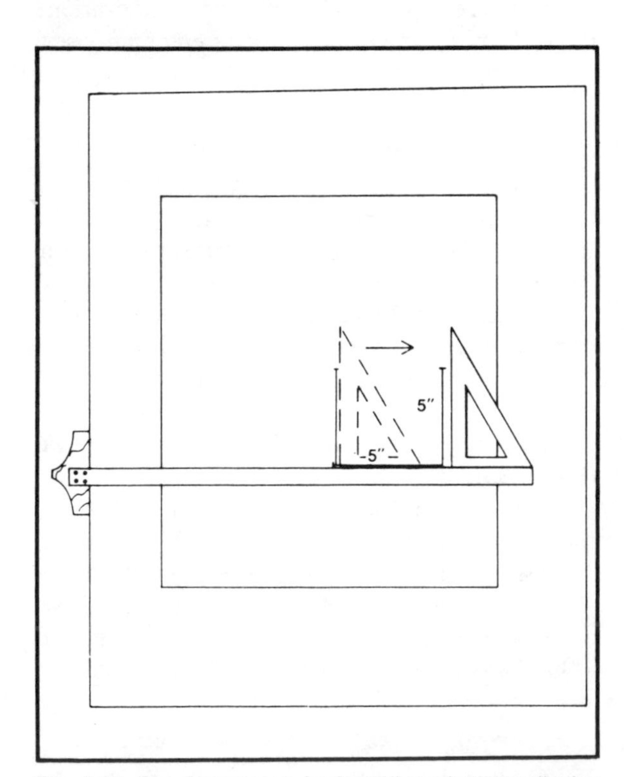

Fig. 4-35. Drawing square, horizontal, and perpendicular lines.

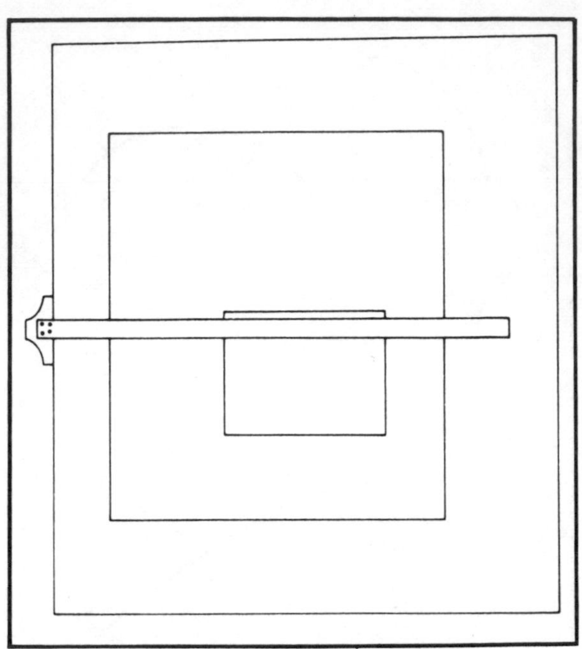

Fig. 4-36. Finishing the square.

Construct your rectangle following instructions for the square above, but using the measurements given for the rectangle.

Measure off 1 inch increments along the bottom line.

Measure up 2 inches from the bottom and draw a line across the rectangle (Fig. 4-37).

Measure off 1 inch increments on that line to match the bottom line.

One the remaining 2 inches of the vertical, measure off 1/2 inch increments.

Using your T square, lay in the horizontals along the 1/2-inch marks (as shown in Fig. 4-38).

Using your T square and triangle, lay in the verticals on the 1-inch marks.

Make sure that the lines meet exactly in the corners and other junction points. Erase any other lines. Your finished drawing should look like Fig. 4-39.

Various Kinds of Lines

When you draw lines, do them first horizontally using your T square. After you have some practice,

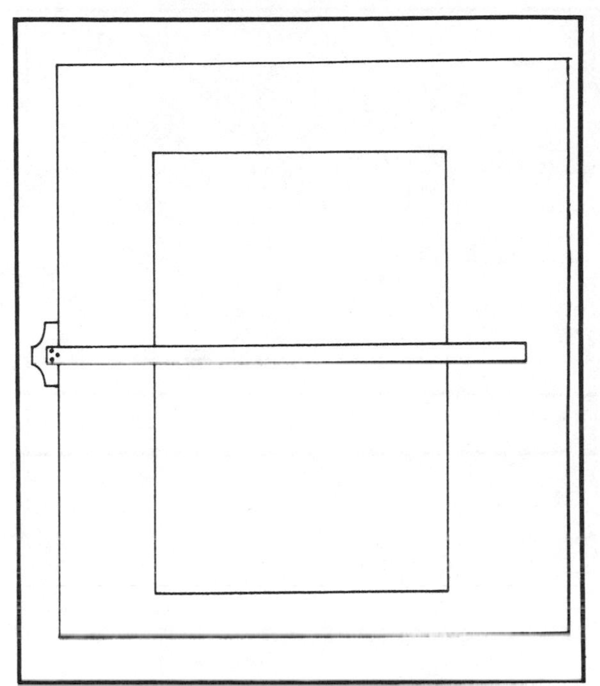

Fig. 4-37. Drawing a rectangle.

draw the same lines vertically using your T square and triangle together. Then try drawing the same lines at a slant following the hypotenuse of your triangle.

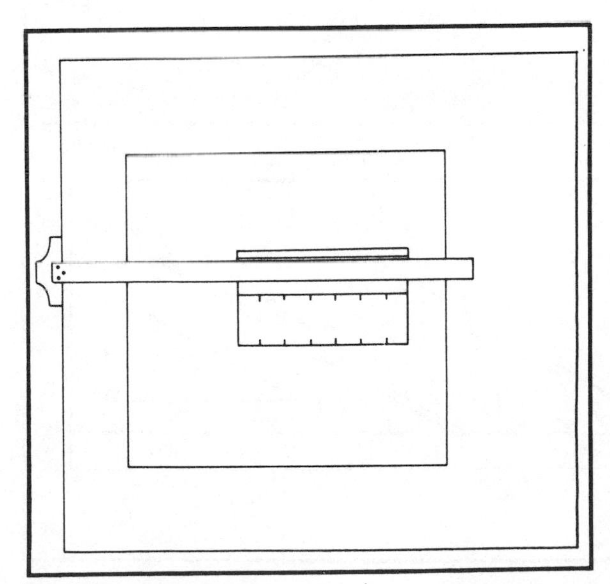

Fig. 4-38. Laying in horizontal lines with a T square.

Draw horizontal lines. Vary the lengths from 2 1/2 inches through 4, 5 1/2, and 7 1/4 inches (Fig. 4-40).

Repeat the exercise with vertical lines (Fig. 4-41).

Once more with feeling on the slant (Fig. 4-42).

Next is the broken or dashed line that is used to indicate things present, but not visible from the viewing angle. Try to keep the dashes the same length (Fig. 4-43).

Practice dashed lines on the horizontal, vertical, and slant.

Next comes the center line that is indicated in Fig. 4-44. Practice this one.

This is the dimension line that is easy so you'll only have to do it in one direction, your choice. See Fig. 4-44.

There are other lines that can be done and they are used for various purposes. There's the dotted line and the dot-dash line and any combination of the two you can use if you feel the need. They look like Fig. 4-45.

The convention that architects observe in regard to the thickness of a line, thick for outlines and short break lines is medium for hidden outlines (that's those dashes) and thin for center, extension, and dimension lines. But this is getting a bit too professional for us. We recommend that rather than going by the thickness you pay attention to the different kinds of line and use them.

Circles

Draw a series of circles with your compass. Start

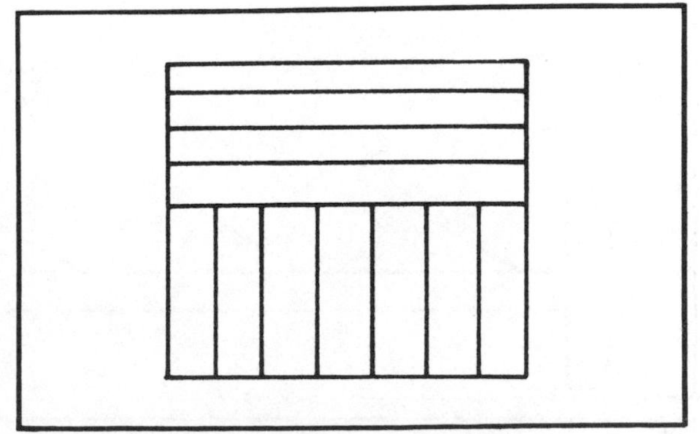

Fig. 4-39. A finished drawing of rectangle.

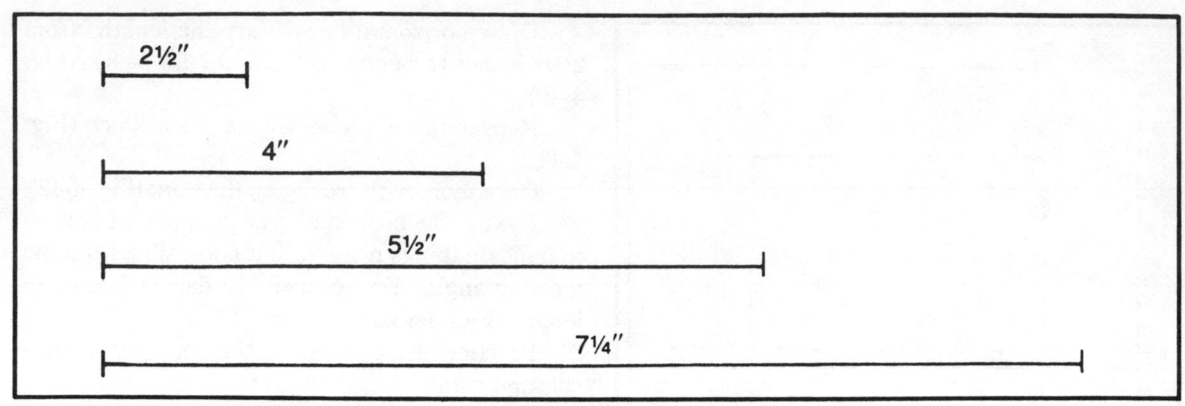

Fig. 4-40. Measuring and drawing horizontals with a T square and a ruler.

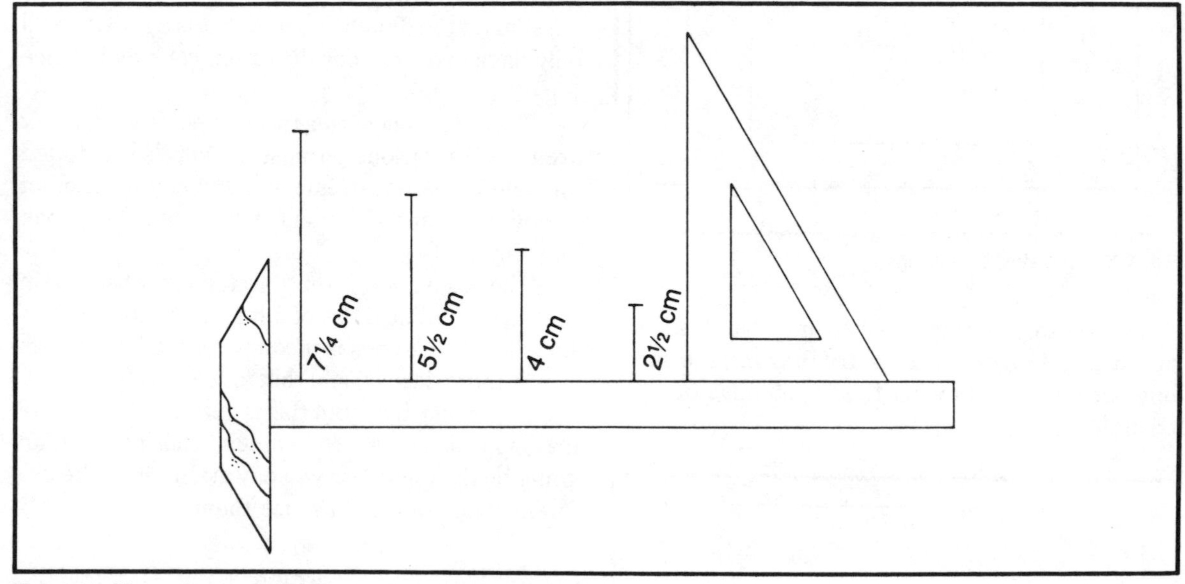

Fig. 4-41. Measuring and drawing verticals (perpendiculars).

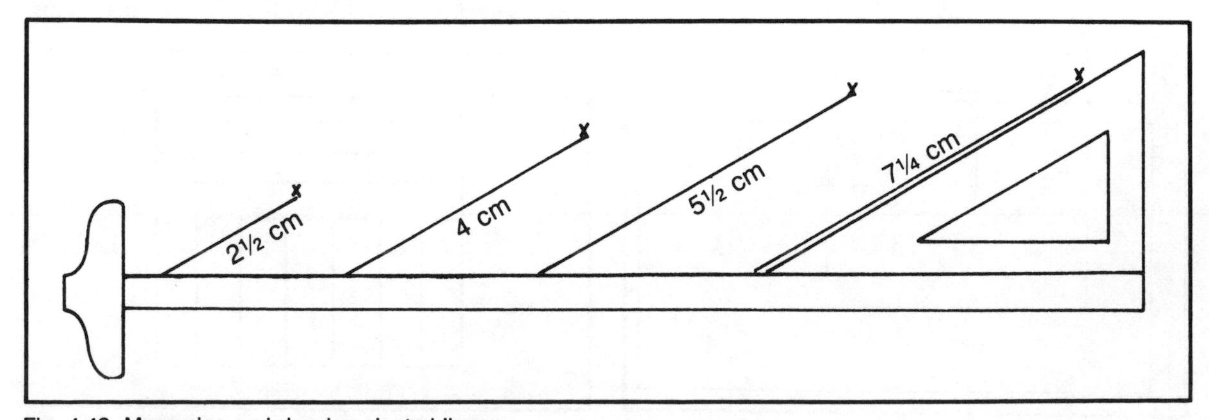

Fig. 4-42. Measuring and drawing slanted lines.

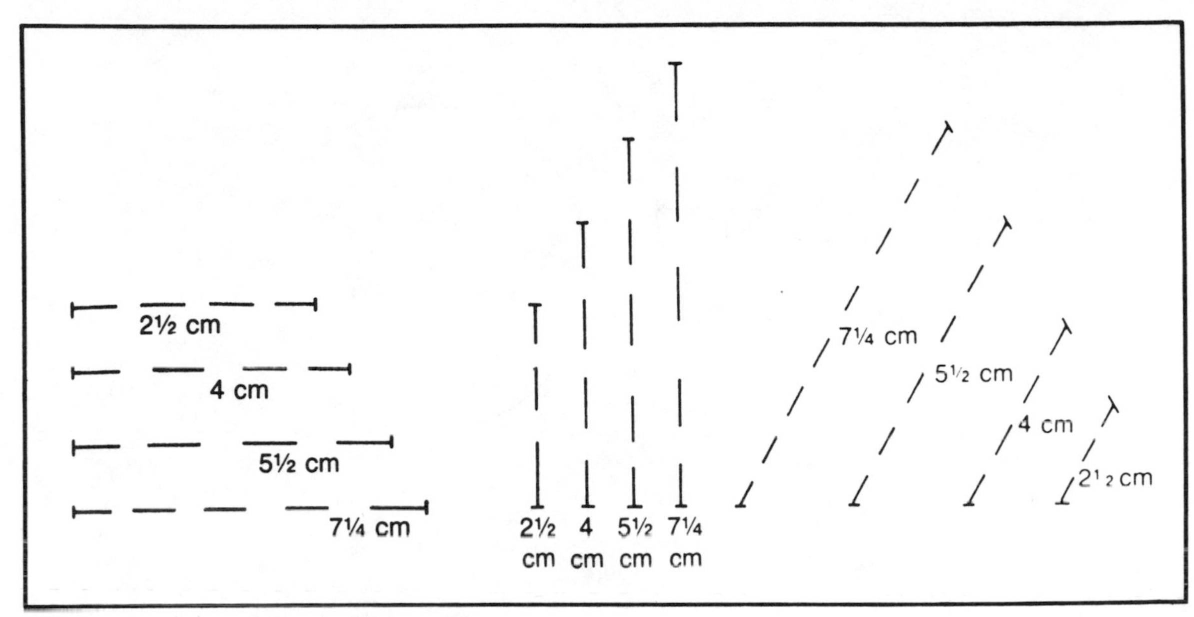

Fig. 4-43. Measuring and drawing broken lines.

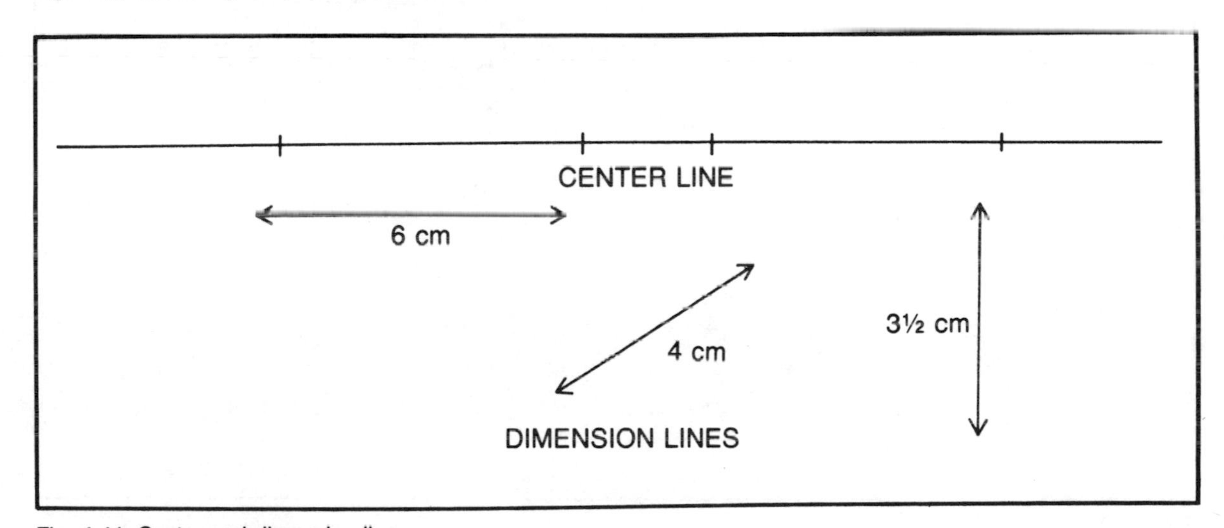

Fig. 4-44. Center and dimension lines.

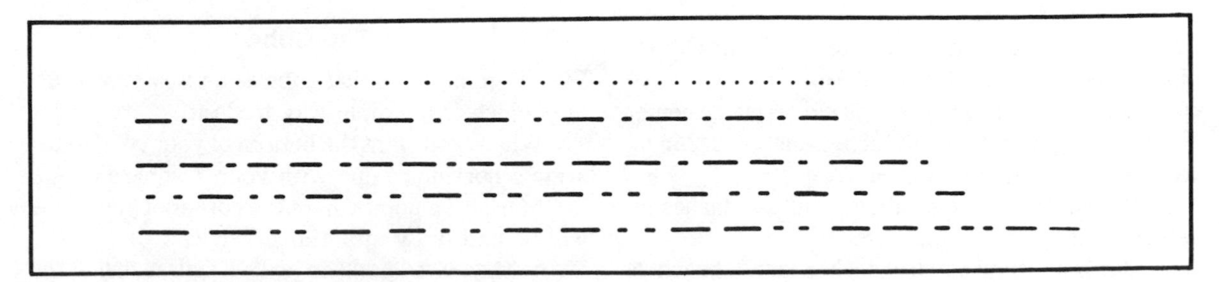

Fig. 4-45. Various kinds of lines used in plans.

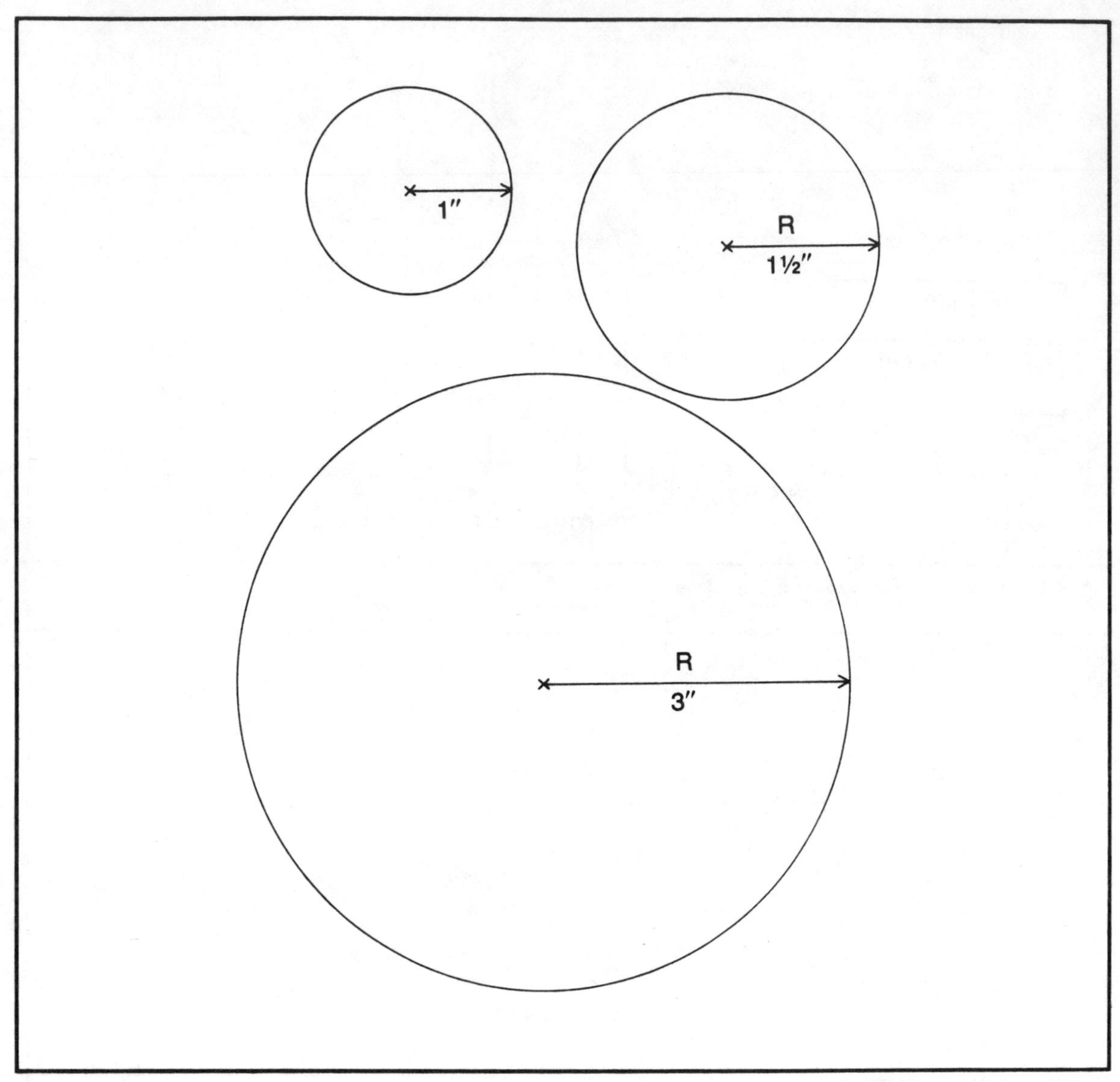

Fig. 4-46A. Drawing circles.

with a 1-inch radius and work up to 3 inches by 1/2 inch. See Figs. 4-46A and 4-46B.

Draw a series of concentric circles (same center point) using 1-inch, 1 1/2-inch, 2-inch, 2 1/2-inch, and 3-inch radii. See Fig. 4-47.

Repeat the exercise above, but use dashes instead of lines. See Fig. 4-48.

One more time and use dashes for all the whole numbers. See Fig. 4-49.

The Cube

We are going to do that cube as an *isometric drawing,* which is a simple way to do it.

Where you want the bottom of your cube to be, draw a horizontal line with your T square.

Mark off a point where the corner of your cube will be and draw in a 4-inch vertical.

With your triangle on your T square, using the 30-degree angle, draw a line to the right of your vertical and one to the left. See Fig. 4-50.

Measure off 4 inches along each of these lines and put in verticals at those points, also 4 inches long, by turning your triangle so you'll have the 90-degree angle. Use your T square and the horizontal line for a base. See Fig. 4-50.

With your triangle, connect the verticals parallel to the bottom lines. See Fig. 4-51.

Using the T square and the triangle, now draw parallel lines, one to each, forming a square on top. See Fig. 4-51. If you'd rather, you can use two triangles, but the T square is safer.

Put in your hidden lines, the fourth corner, and the bottom square. Congratulations! You've done it. See Fig. 4-51.

If you can, practice a bit more until handling the T square and triangles becomes easy and familiar. Calibrated triangles and T squares surely speed up things. Otherwise you'll need to keep your ruler handy. If you want to trace part of a drawing, your best bet is to use some tracing paper. You can get this in sheets, blocks, and pads.

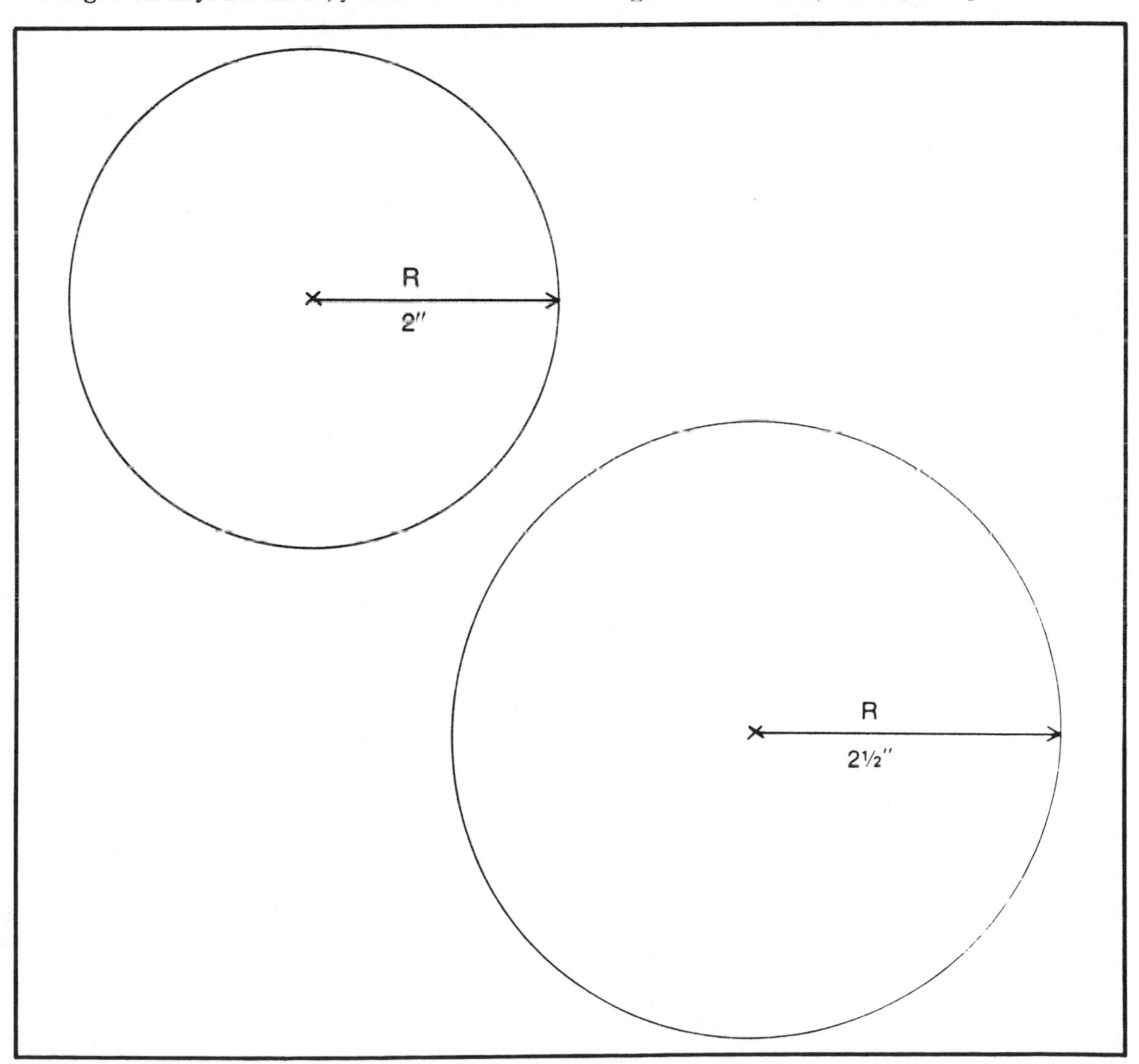

Fig. 4-46B. More circles.

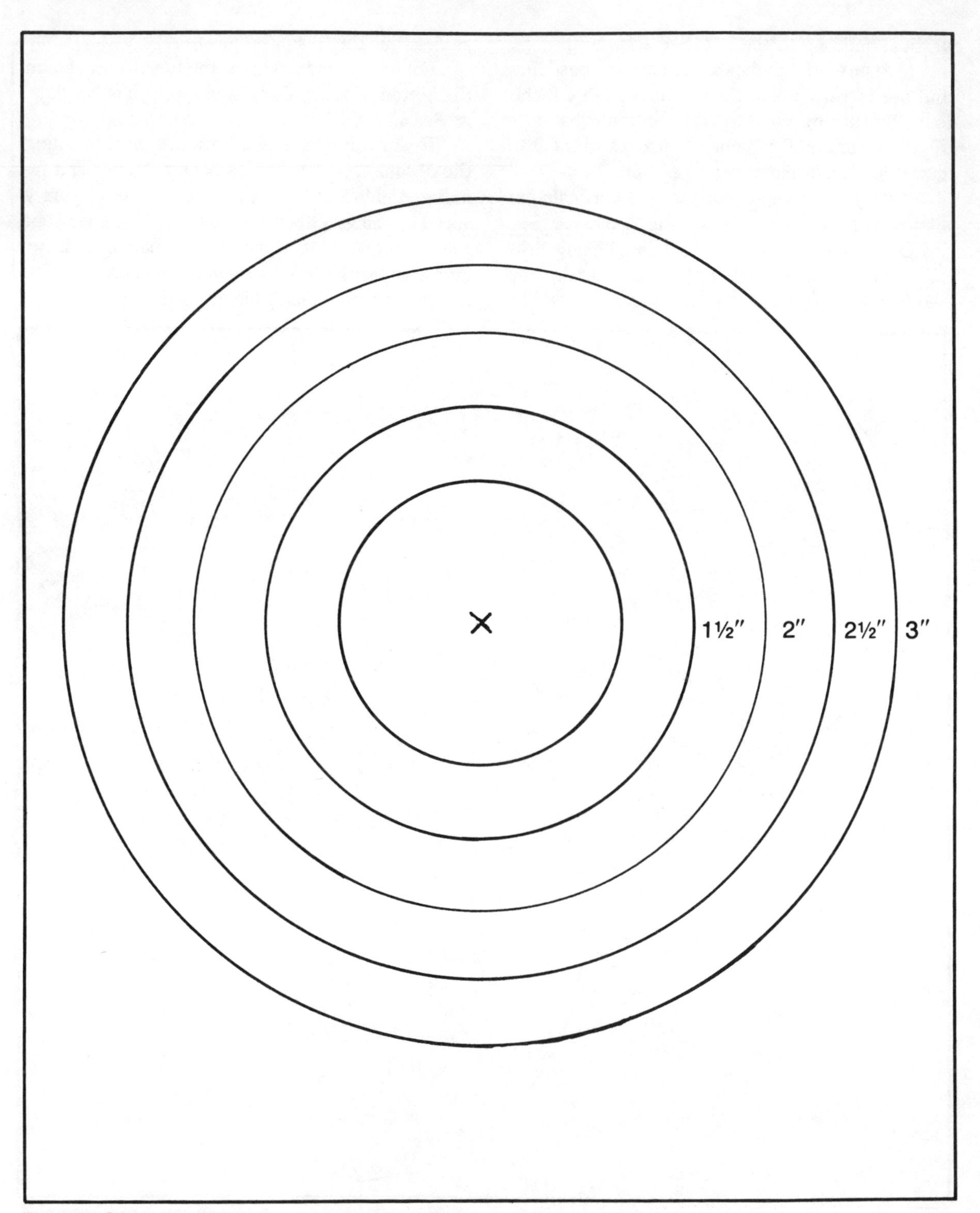

Fig. 4-47. Concentric circles.

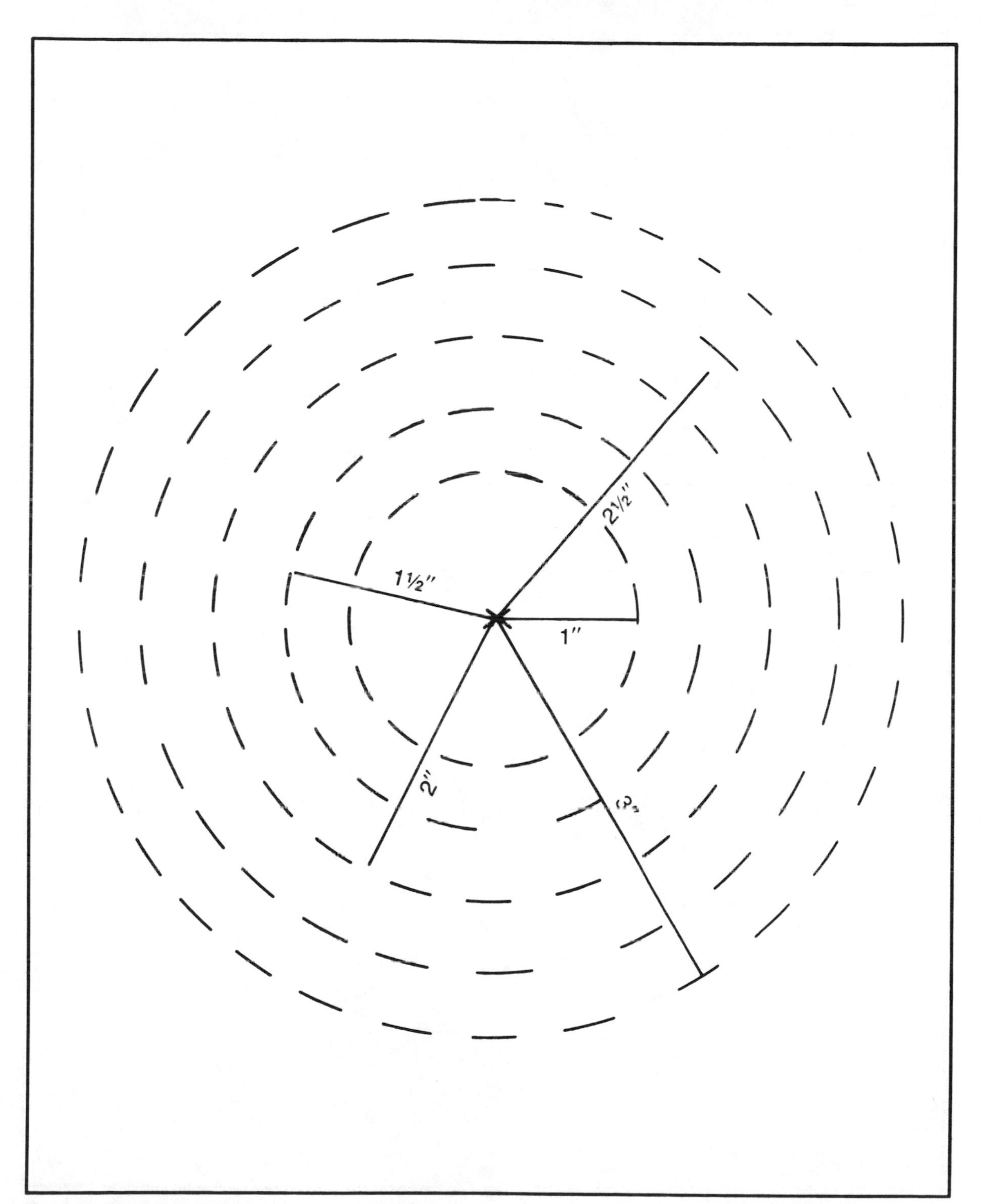

Fig. 4-48. Concentric circles drawn with broken lines.

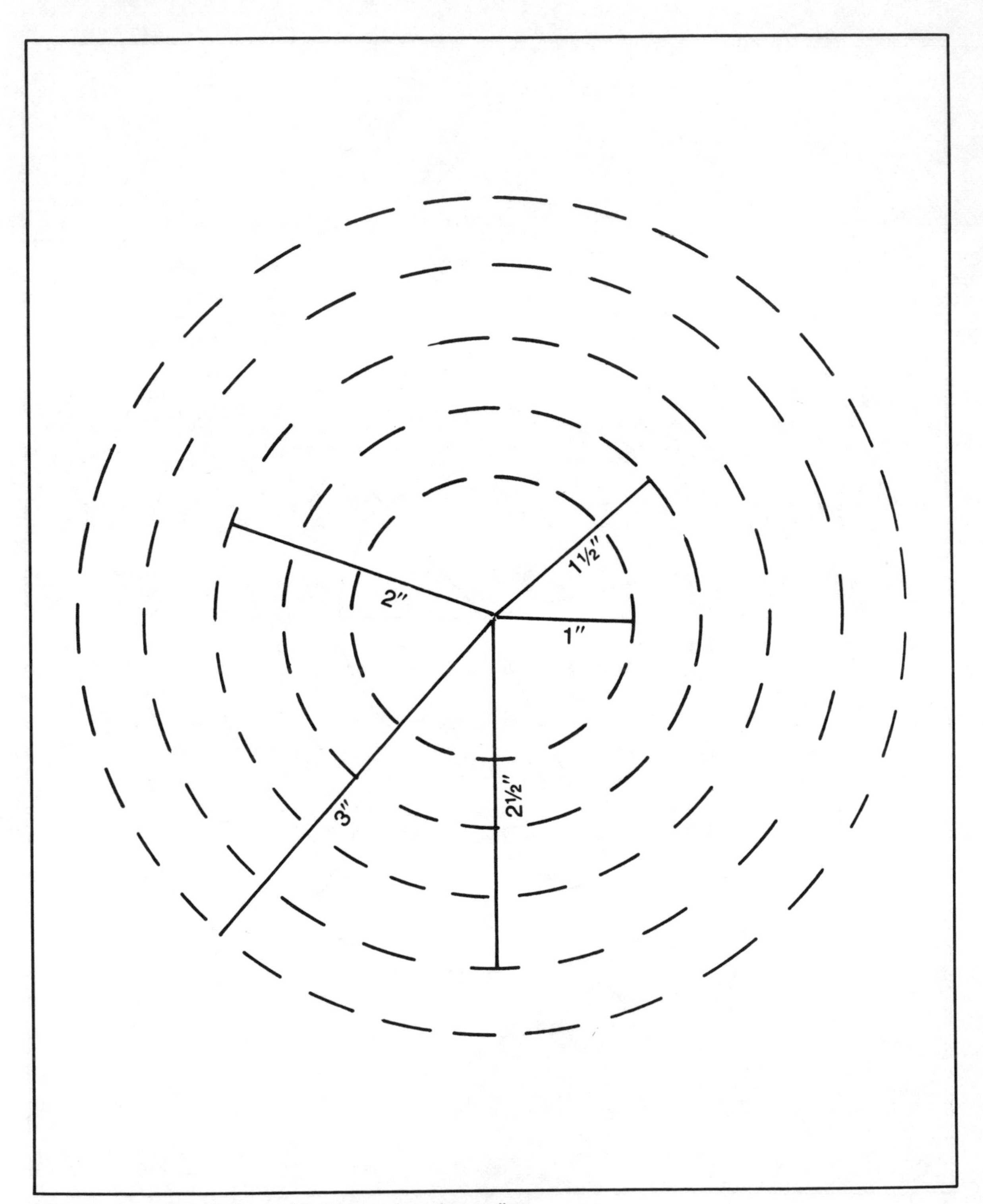

Fig. 4-49. Concentric circles, broken lines, and continuous lines.

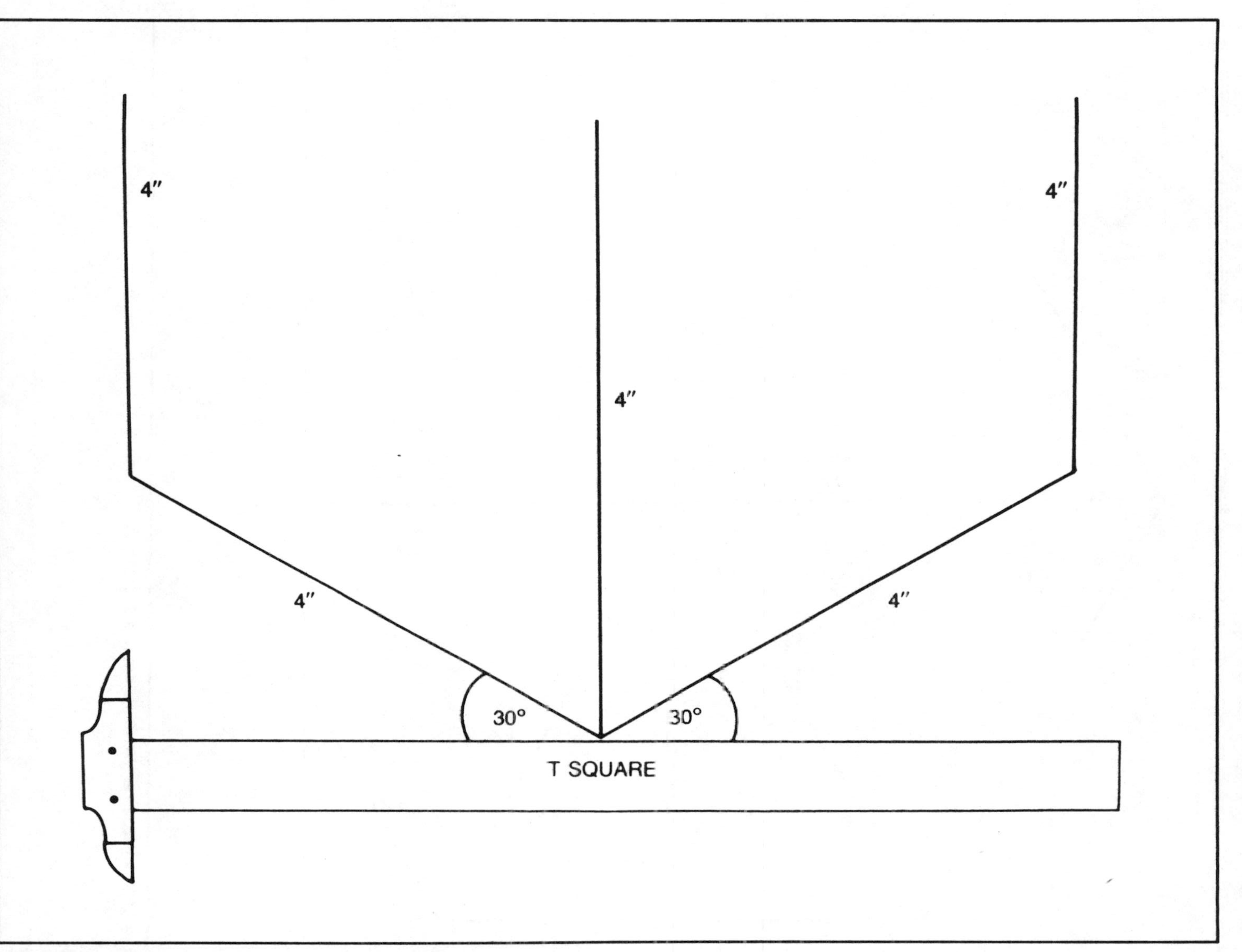

Fig. 4-50. First steps in the isometric drawing of a cube.

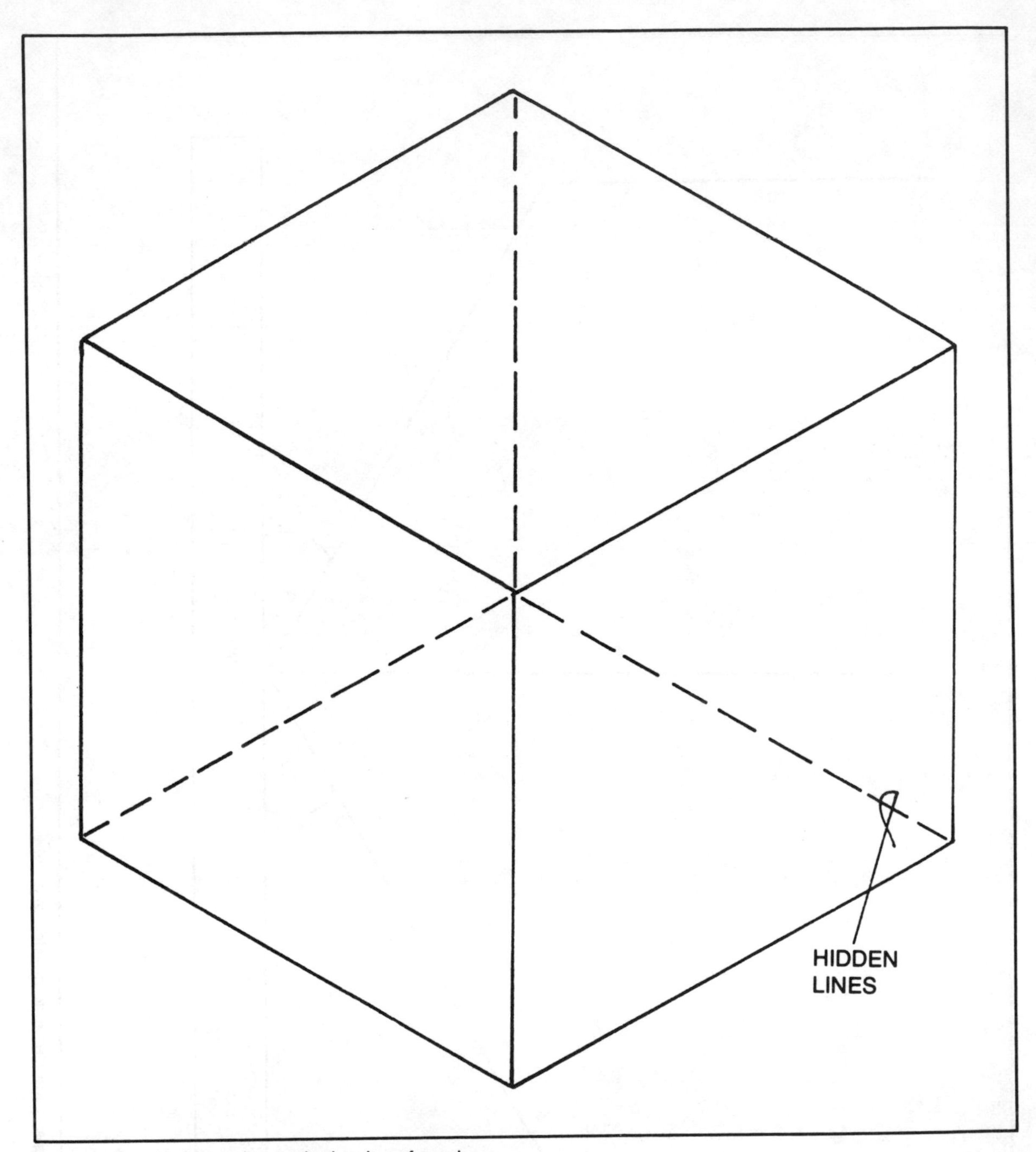

Fig. 4-51. Completing an isometric drawing of a cube.

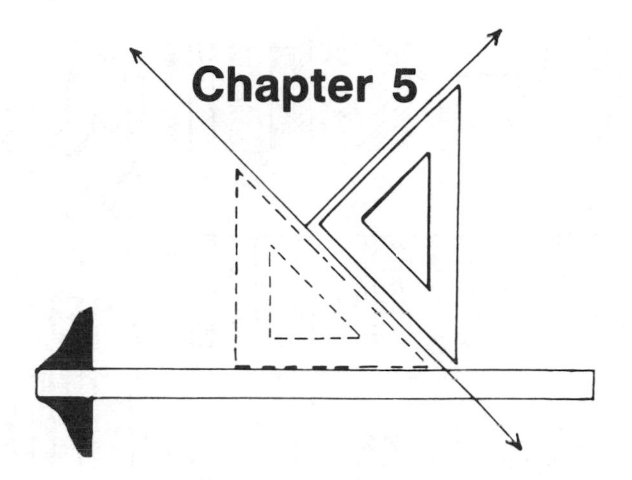

Symbols, Dimensions, and Lettering

As stated at the beginning of this book, the purpose of a house plan is to tell a builder, as completely and accurately as possible, how to build a certain structure.

Contractors, estimators, and workmen must rely on your drawings in preparing cost estimates, ordering materials, and executing the work. Thus the more complete the information on your drawings, the more likely you are to get the house you have planned. But his does not require that you show *every* minute detail, for competent builders are expected to know common and usual construction requirements.

Because of the nationwide distribution of house and building magazines as well as more technical publications, standard ways of showing the many materials of construction are widely used. These graphic ways of indicating construction materials are called *symbols*. The symbols you are likely to need when drawing house plans are shown in Fig. 5-1.

Dimension lines are those that indicate sizes on a house plan (Fig. 5-2). Outlines of walls, foundations, partitions, and the like should be darker than dimension lines. Usually dimensions are written immediately above dimension lines. Dimension lines should be unbroken.

Keep your dimension lines orderly and easy to read. Inside dimensions should be located where they are clearly distinct from construction lines.

Many planners dimension (i.e., draw dimension lines) inside partitions to the center of wall or floor plates (Fig. 5-3A). But particularly in frame construction, it is easier for workmen to mark partitions on the floor when the dimensions read from wall to wall, though this requires that the width of the floor plate be accounted for on the plan (Fig. 5-3B).

Dimensioning with coordinates is very helpful in accuracy, especially when there is more than one story and the first and second floor partitions must align because of plumbing and heating risers. The method is simple and practical. Start dimensioning with one lower corner of your plan considered as 0 feet 0 inches, both horizontally and vertically. As you proceed with the layout and dimensioning of

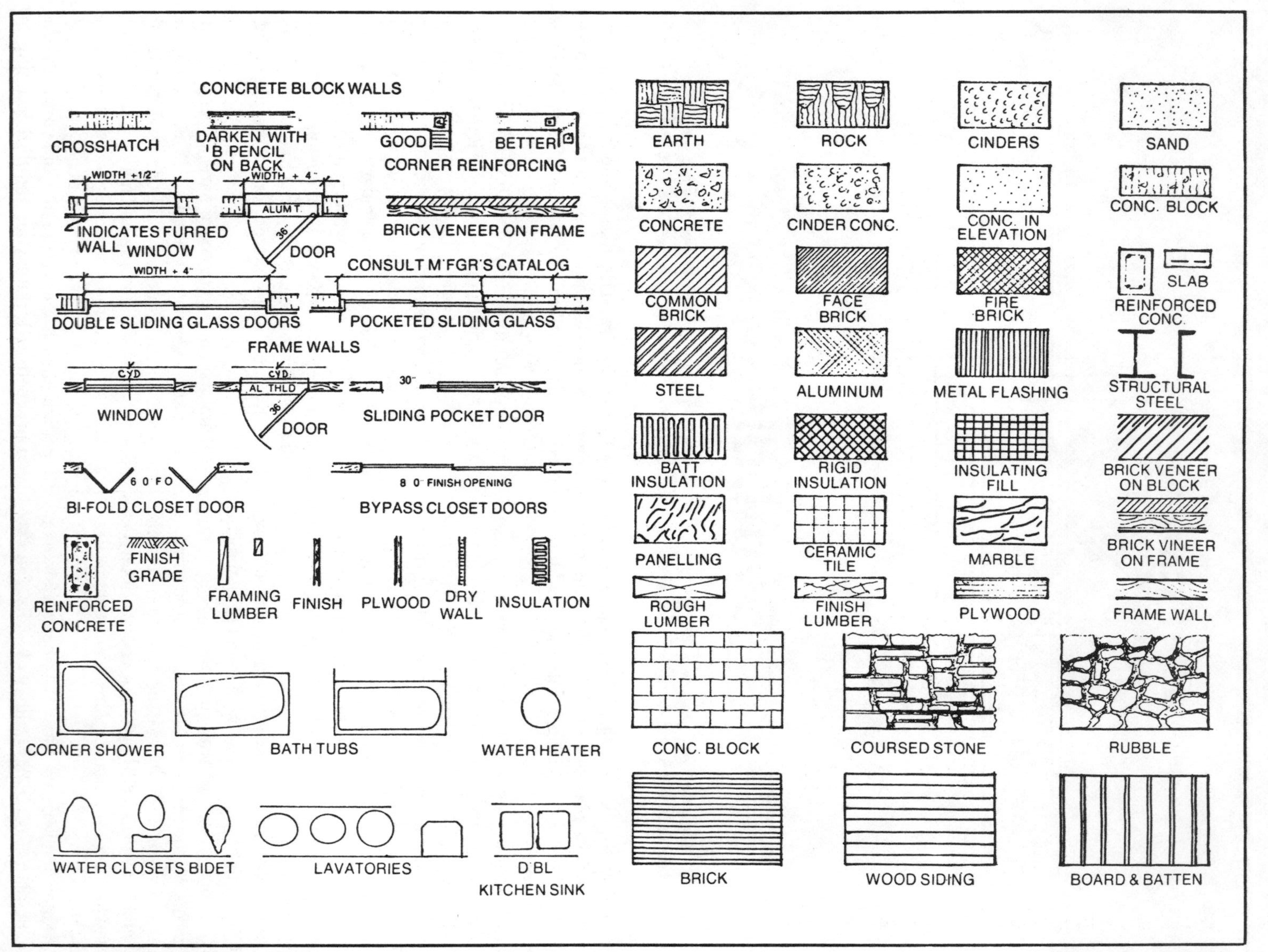

Fig. 5-1. Standard materials symbols.

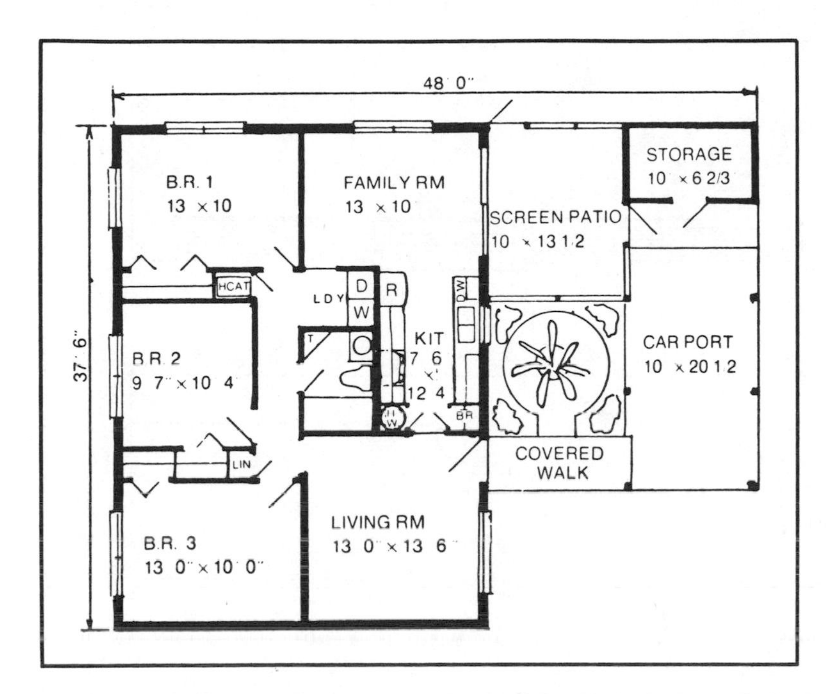

Fig. 5-2. Dimension lines indicate the size of the components of a drawing.

windows, doors, and partitions, note the distance from the zero point to each extension line (Fig. 5-4). Write in the distance above each extension line. These distances will be your coordinates, telling you exactly how far away the partitions, doors, or windows, are from point 0 feet 0 inches. For example, in Fig. 5-4 there are two windows shown in the drawing. The left edge of the left-hand window is exactly 4 feet away from the zero point; its right edge is 8 feet 1/2 inch away from the zero point. Such coordinates allow you to accurately position any object on your plans.

Take a look at Fig. 5-5. The size of the house is 28 feet 8 inches wide and 27 feet 8 inches deep. The intersection of the front wall and right-hand side wall is the zero point. Since the planner decided that the stairs should be 4 feet 8 inches beyond the front wall line, the stairs' vertical (from top of drawing to bottom) coordinate is + 4 feet 8 inches. (The plus sign distinguishes coordinates from other numbers on the drawing.) The length or run of the stairs is 10 feet 3 1/2 inches, therefore the coordinate at the second floor landing is + 14 feet 11 1/2 inches. The stairs landing must be at least 3 feet deep, so if dry wall (dividing the dining room and landing) is 1/2 inch thick, its coordinate will be 18 feet. And the coordinate for the 2 × 4 partition studs (3 1/2 inches thick) against the dry wall will be 18 feet 3 1/2 inches.

Note: In this and many other house plans, only frame walls are shown, that is, skeletal walls made of studs without dry wall or other coverings.

The coordinate for the rear wall of the house is + 27 feet 8 inches. Deduct the 3 1/2 inch thickness of this rear frame wall for a coordinate of + 27

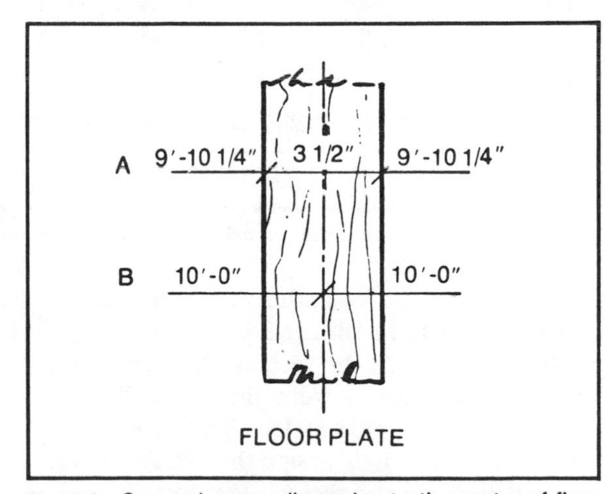

Fig. 5-3. Some planners dimension to the center of floor plates (A); others dimension from wall to wall (B).

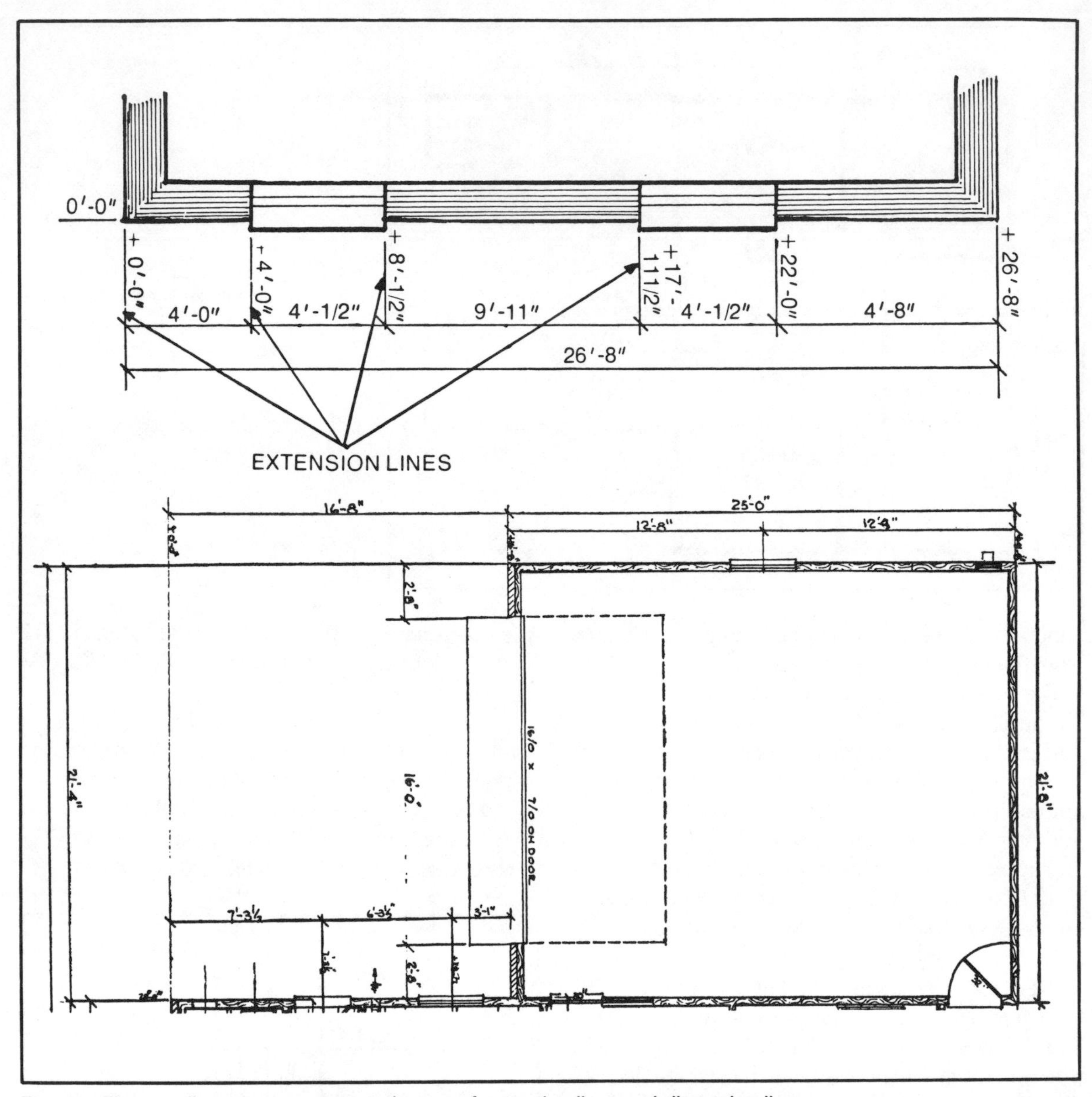

Fig. 5-4. The coordinate system must make use of extension lines and dimension lines.

feet 4 1/2 inches. Subtract the coordinate of the opposite frame wall (+ 18 feet 3 1/2 inches) and you get 9 feet 1 inch as the distance between frame walls. The distance between finished walls (after installing dry wall) will be 9 feet. These coordinates allow you to accurately assign the locations of certain partitions on the second floor too.

The same system of coordinates measuring can be used *horizontally* on a drawing, using the same zero point.

Using similar coordinates or exterior walls, a mason or carpenter could hook the end of his measuring tape on a building corner and accurately mark the locations of doors and windows.

A good example of the value of this coordinate system is seen in the complicated layout of angled

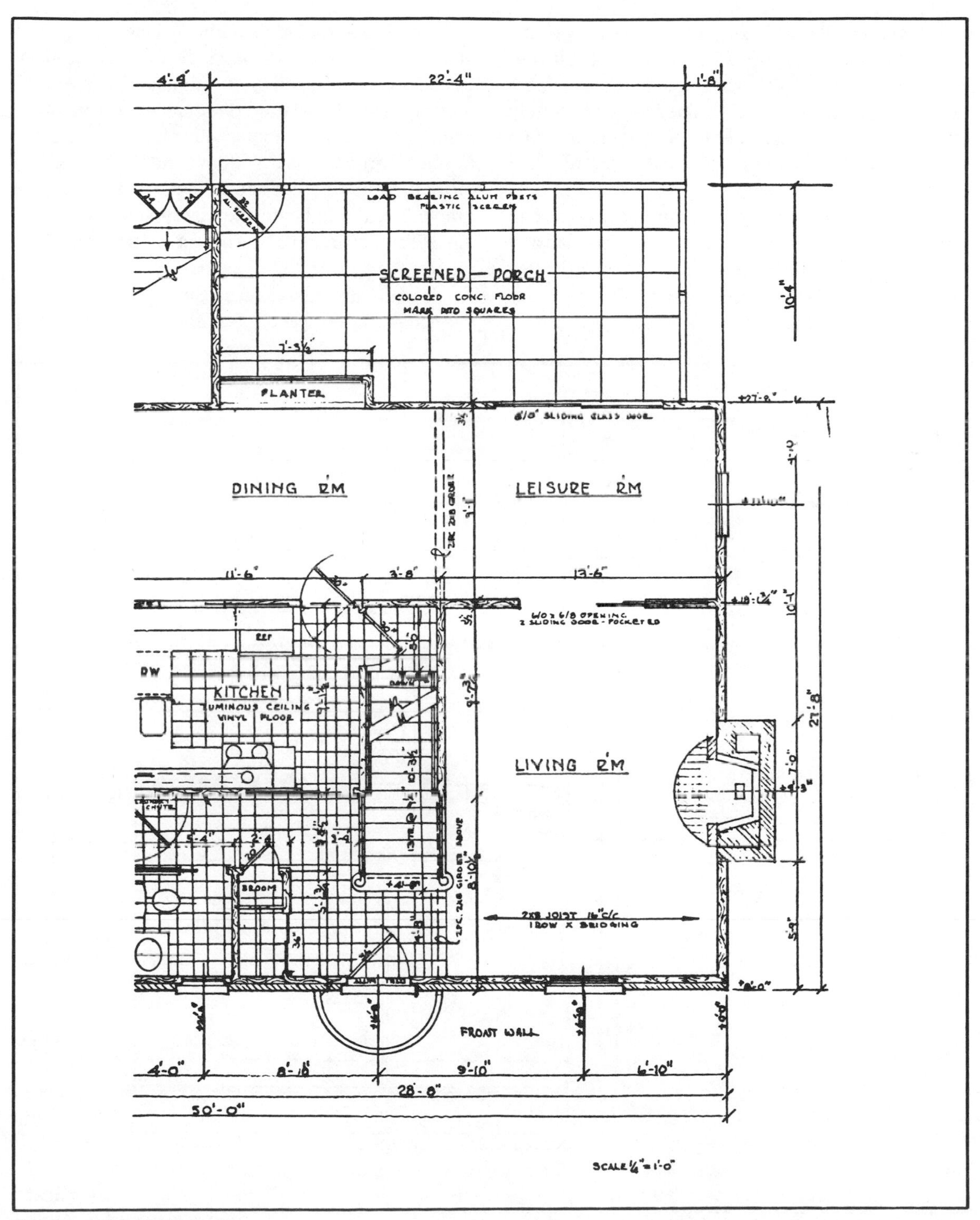

Fig. 5-5. A finished floor plan.

and straight walls at the second-floor hallway of the *same* two-story house (Fig. 5-6). After such a layout is drawn on the 1/4 inch scale second-floor plan, its final dimensions can be determined more accurately by drawing them to a larger scale so you can make certain that there will be sufficient space to place the rough framing.

All this should give you an idea of how you can work out final dimensions of a complicated plan and make sure that it can be built. With two coordinates given, one vertical, the other horizontal, any point on a plan can be definitely located.

Remember that *written* dimensions are the final authority for your drawings. If any part is positioned incorrectly on your plan and scales differently than the written dimensions, it can cause confusion and doubt about your plans. If you draw in some part such as a window and then decide to move it from where it should be (according to the drawing scale) indicate the discrepancy by lettering NTS just under the correct dimension. That means "not to scale."

Usually dimensions less than 1 foot are given in inches and fractions (eighths of an inch) for

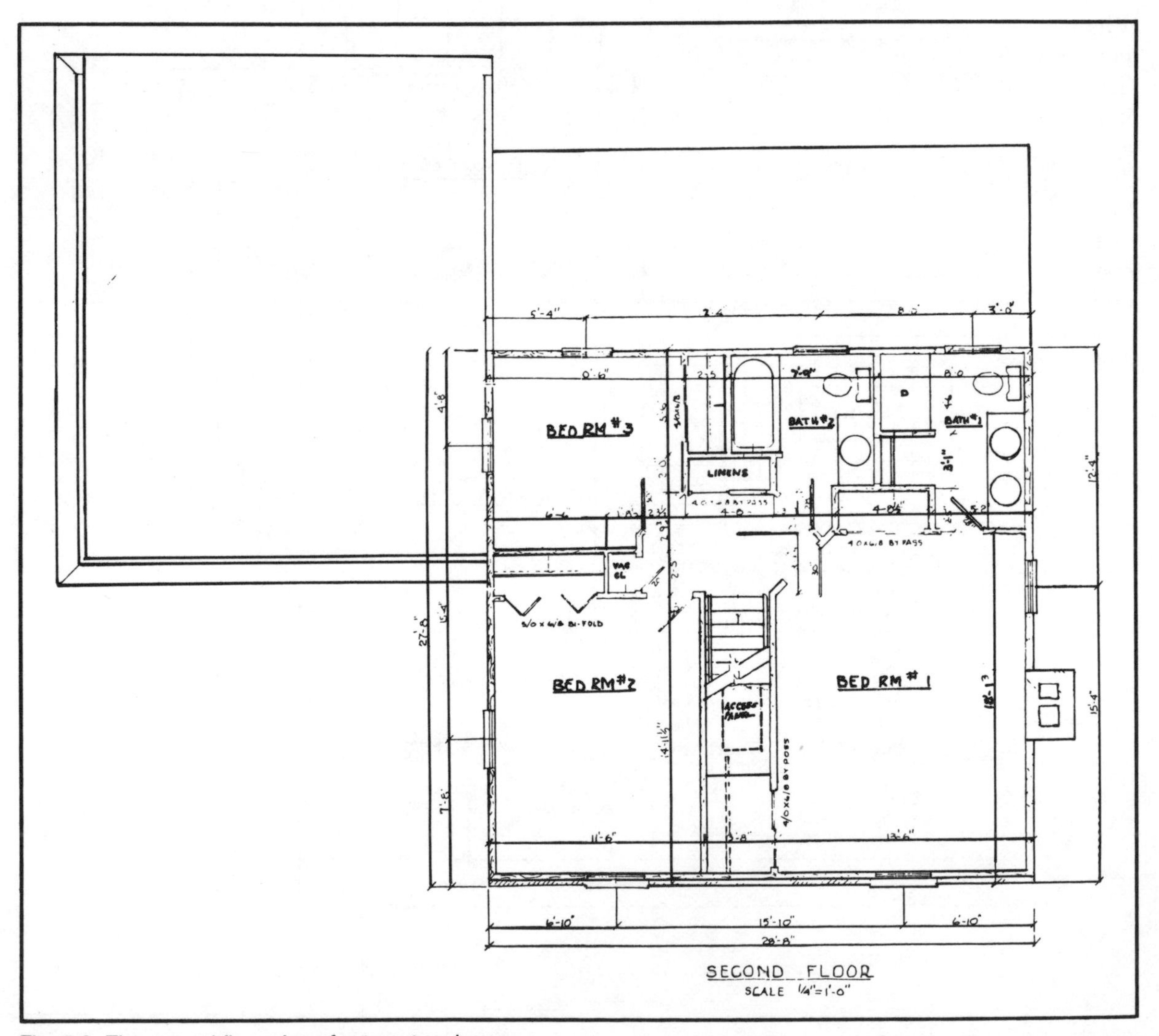

Fig. 5-6. The second-floor plan of a two-story house.

general house construction. Dimensions over 1 foot are given in feet, inches, and fractions (e.g. 14 feet 11 1/4 inches). Cabinet work is dimensioned *only* in inches, and an accuracy of 1/32 inch if called for. Dimensions for masonry and house framing are given to the nearest eighth of an inch. When working with moldings and the like, sixteenths may be used.

In masonry construction, window and wall openings must allow for side frames and other things. For example, for a 36 inch wide entrance door, the opening may have to be 3 feet 4 inches wide, which allows for 1 1/2 inch side frames and 1 inch for plumbing or vertical adjustment. Openings for metal windows must take into account 1/2 inch or 5/8 inch each side for nailing strips. Thus it is usual in catalogs of building components (such as windows, sliding glass doors, etc.) to see the dimensions for the masonry opening or M.O., and the rough framing opening, or R.O. If there is any question, you can dimension to the center of the opening and leave it to the builder to provide the proper opening; however, this may lead to generously wide openings which may require makeshift buildup and cause leakage problems.

The usual height of the *head* of door and window openings is 6 feet 8 inches above the floor, and doors are manufactured to that dimension. Windows should be located at that height, and the location of their bottom edge depends on the size of the window and is never dimensioned.

Lettering on a drawing gives a builder the dimensions and information he needs. Neat lettering makes a drawing easy to read and pleasant to look at.

If you find it impossible to do freehand lettering, there are lettering guides or templates, such as the Rapidesign, that you may find helpful. But good freehand lettering is not only preferred, it's faster.

Some professional planners use mechanical lettering devices to make a drawing more uniform in appearance.

Anyone who can write well can learn to letter if he will take the time, but there is no objection to using script if you can write more legibly than you can letter.

All draftsmen use *guidelines* to maintain even lines and uniform heights of letters. Place the point of a 4H pencil through a hole in your small lettering triangle, sliding the triangle back and forth along the edge of the T square to draw parallel guidelines (Fig. 5-7).

The guidelines should be light and should

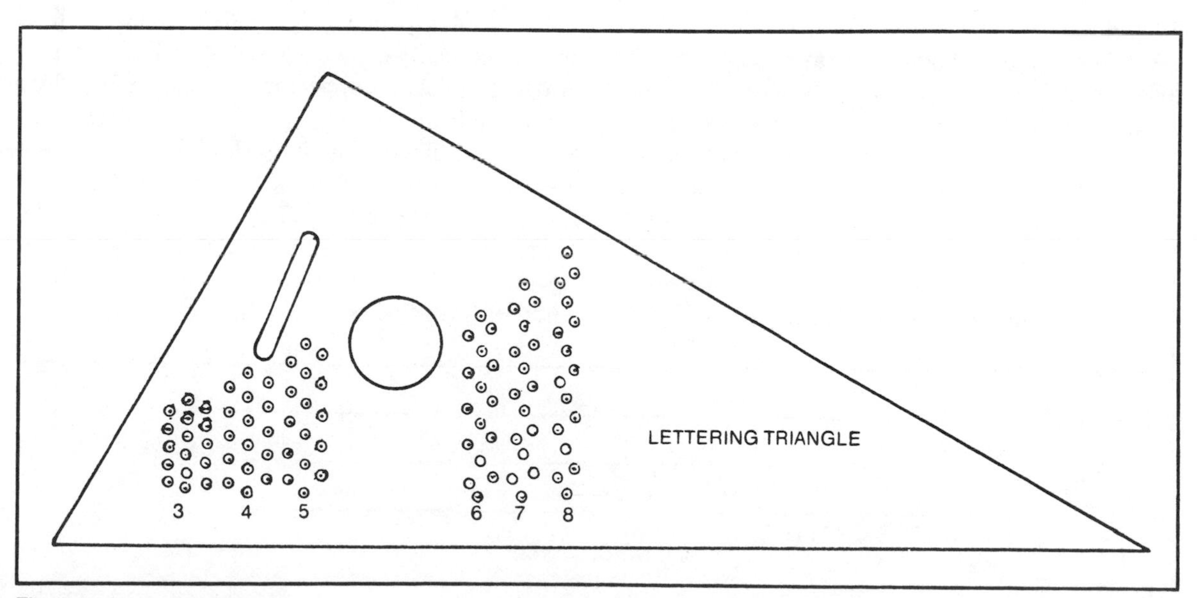

Fig. 5-7. A lettering triangle.

Fig. 5-8. Lettering is a lot easier if guidelines are drawn first. Letters should be formed consistently from word to word and plan to plan.(A); others dimension from wall to wall (B).

scarcely show on blueprints.

After the guidelines are drawn, you're ready to letter. Rest your lettering hand on the T square, which acts as a bottom stop. Only the fingers holding the pencil move. The letters are formed mostly with down strokes, as shown in Fig. 5-8. Maintain the same blackness of line throughout the drawing for best appearance. Use the softer H or HB pencil with a more blunt point for lettering titles.

Dimension and extension lines can be drawn with the 4H pencil (Fig. 5-9). Dimension lines ending in neatly drawn arrows (A of Fig. 5-9) give drawings a finished appearance, while crudely done arrows indicate sloppy work. Many professionals use the 45° diagonal endings (B of Fig. 5-9), which are easy to draw with the 45° triangle and are best for an amateur planner. They are approved by the Bureau of Standards but are less definitive than arrows: they can be confused with extension lines. Another variation is shown in C of Fig. 5-9.

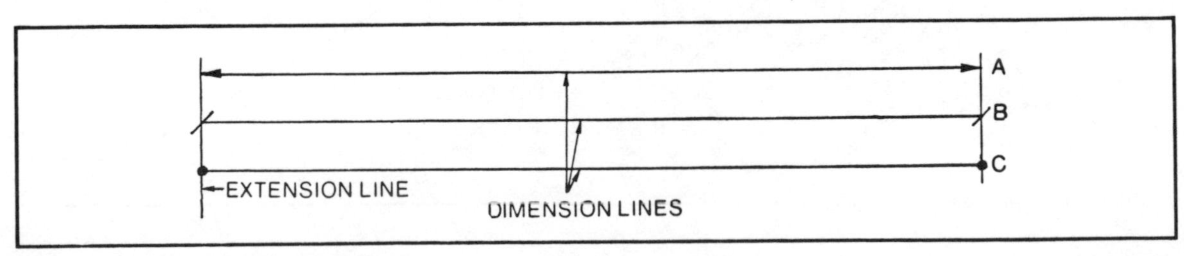

Fig. 5-9. Three kinds of dimension lines.

The title block, or identification, for each sheet of a plan should be uniform in size (Fig. 5-10). Draw a sample title block on a small sheet of paper and trace it onto each sheet of a set of plans. A title block usually contains the following information:

- ☐ Name or type of building.
- ☐ Name and location of owner.
- ☐ Name and address of the planner.
- ☐ Date the drawing was made.
- ☐ Date of revisions if any.
- ☐ Initials of draftsman.
- ☐ Initials of checker.
- ☐ Identification number of sheet.

In order to show all the work required for construction or remodeling of a structure, a whole series of drawings is necessary. These are called *working drawings* or *construction plans*. The series is known as a *set* that usually consists of at least the following.

Plans. These are drawings that describe what the designer has in mind, as seen from above, if you sliced off the top part of the structure. A *floor plan* will indicate sizes, shapes, and arrangements of rooms, kitchen equipment, bathroom fixtures, windows, doors, etc. A *survey* and *plot plan* will give information about the site on which a structure is to be built and show where it is to be located on the site. A *landscape plan* gives a view from above of all the plants and small structures that make up the landscape surrounding a structure.

Sections. Technically, a plan is a horizontal section, but the term is usually applied to drawings used to show the interior construction of various structural parts.

Elevations. Elevation and plan views are the two most important drawings in the set. Elevations show what the exterior of a structure is to look like. A separate drawing is required for each side of the structure. Most house plans have four such drawings. Elevation drawings always contain much information and many instructions about how a structure is to be built.

Details. Any information that cannot be shown on plans, sections, or elevations is put into detail views. These are drawings of the details of structural assembly, trim, special equipment, etc. Sometimes detailed information about electrical, heating, air-conditioning, and plumbing work is shown on a separate set of drawings called *mechanical detail* drawings.

Perspectives. This is a drawing that will look almost like a photograph of the finished structure. It is used to show people who cannot read the mechanical drawings what the designer visualizes as the finished and landscaped structure. Perspective drawings help builders and contractors visualize what their finished job is supposed to look like.

Because one set of drawings would not serve to communicate to all of the persons concerned with building a structure, the original set of drawings must be duplicated several times. These copies are usually black and white and they are called simply prints. They used to be called *blueprints*, after the blue-line method of printing them. Any color com-

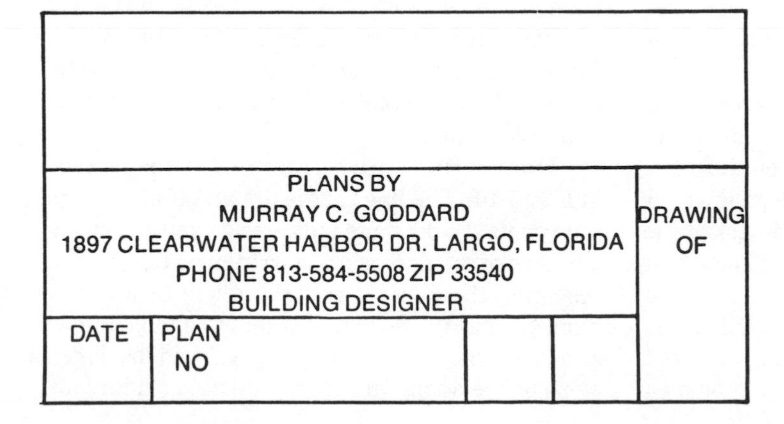

Fig. 5-10. A typical title block.

bination that shows contrast works fine. Often the term "blueprint" is used just to describe a copy of a set of drawings that might actually be copied in black and white.

When you have completed your set of drawings for your remodeling project or your new house, you will need copies of the complete set of drawings for the contractor, the builders (carpenters, masons, etc.), the estimators, and the financier of your project.

Architectural drawings will also include written *specifications* (or *specs* as they are called for short) for all phases of the work. The specs are put right on the drawings and they include instructions about all materials, methods of construction, standards of construction, and manner of conducting the work.

SYMBOLS USED IN HOUSE PLANS

Learning to read architectural drawings is mainly a matter of learning the language of conventions and symbols. These are the standardized methods of representing walls, windows, doors, stairs, fireplace and chimney, kitchen and bathroom fixtures, etc.

On a plan of a frame house, Fig. 5-11 shows exterior and interior walls drawn as two parallel lines. These lines are 6 inches apart on exterior walls and 4 inches apart on interior walls. This convention pictures walls as they are actually constructed.

Figure 5-12 shows a detail of the actual wall framing of a house at a corner and at a partition junction. Wallboards are nailed to the outside faces of the 2- × -4 inch wall studs. On the inside faces of the studs, plasterboard is nailed. The studs, wallboard, and plasterboard—together with exterior siding and interior paneling—will add up to approximately 6 inches. Partition walls, on the other hand, have only a 2- × -4 inch stud with paneling or plasterboard on each side and the thickness of the wall adds up to approximately 4 inches. (The 2- × -4 inch studs actually measure only 1 1/2 × 3 1/2 inches when dressed and installed.) The triple studs at the corners and where partitions meet are not drawn in on the floor plan because the floor plan

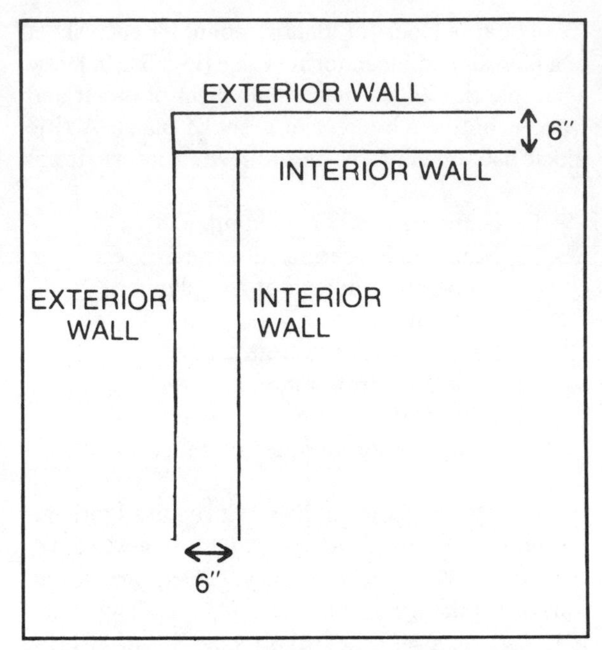

Fig. 5-11. Drawing a complete wall for a floor plan.

scale would be too small to clearly show the detail.

Windows. The convention or symbol for a window (double-hung) is shown in Fig. 5-13. It can be drawn exactly as wide as the window or it can be drawn as wide as the rough opening to accommodate the window, which will be slightly larger. We recommend drawing the window the width of the rough opening because this gives the builder the information he needs. Window manufacturers always supply the information on the size of the rough opening needed for their windows. This information is available to you when you pick out your windows. The location of the window in the wall is determined by a measurement from the outside face of the stud at the corner of the structure to the *middle* of the window. This is indicated by the dimension lines.

Doors. Doors are represented by the symbols in Fig. 5-14. The line that is drawn off at an angle represents the door and indicates its angle of swing. The size of the door is often written on the line representing the door. If not, there will be a circled number near the door that refers to a table known as a door schedule on which you will find the dimensions of the door. On exterior doors, an additional

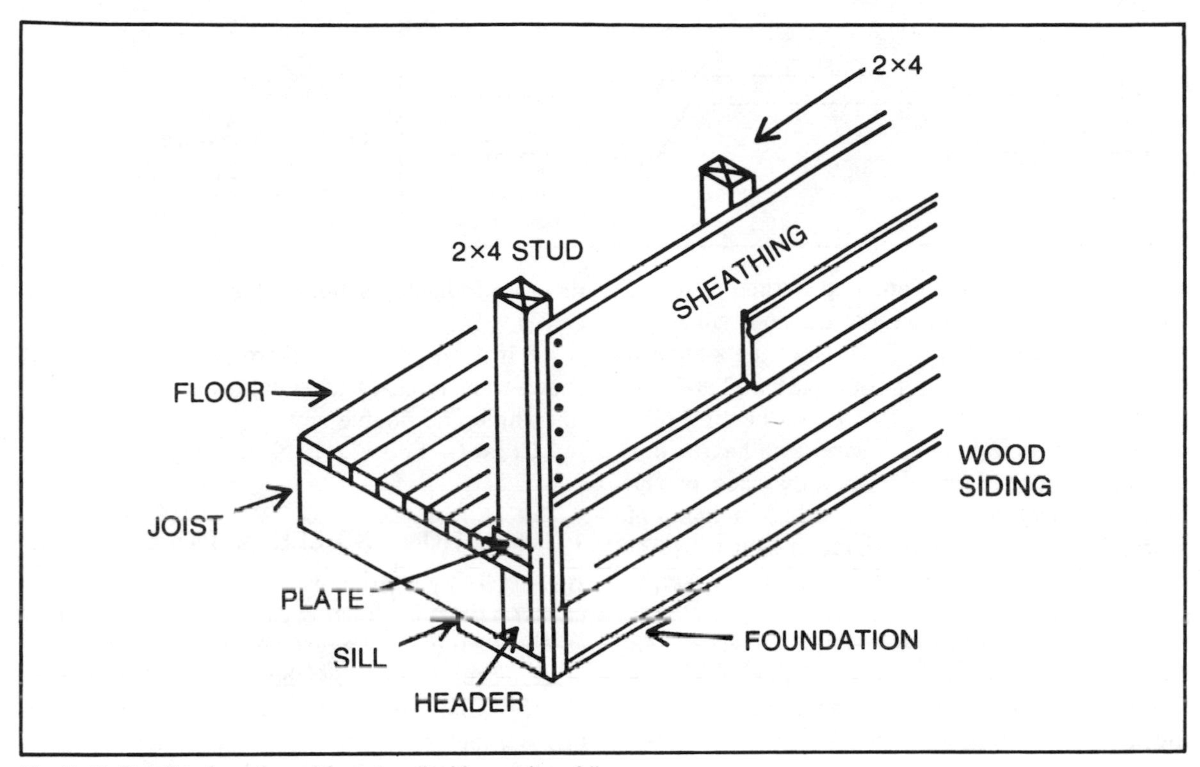

Fig. 5-12. Details of section of frame wall with wooden siding.

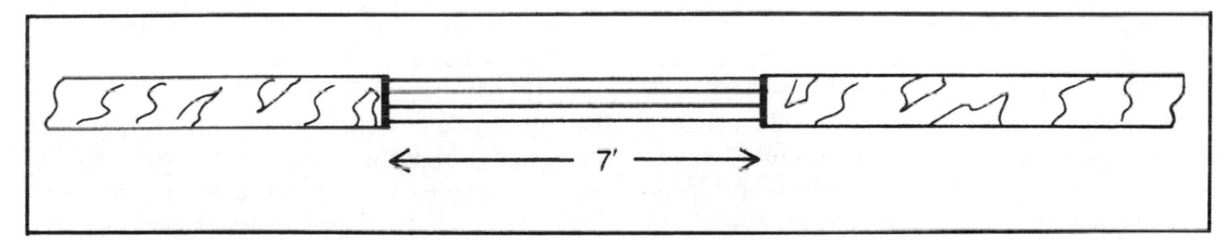

Fig. 5-13. Drawing in a double-hung window.

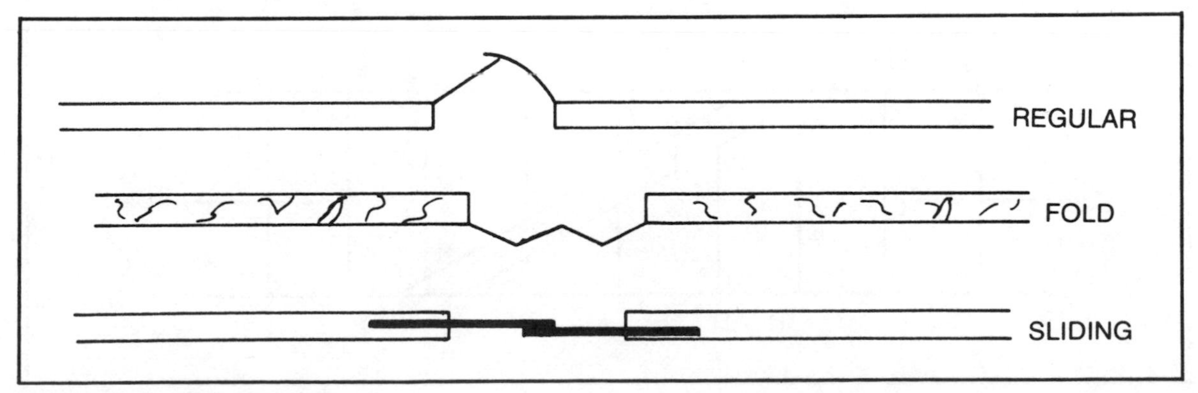

Fig. 5-14. Drawing door symbols.

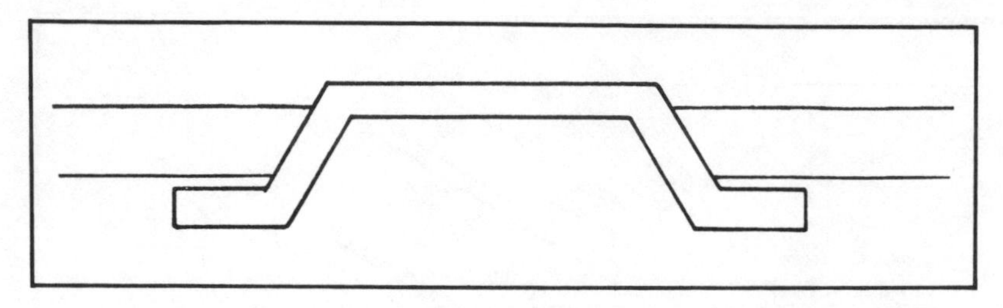
Fig. 5-15. A fireplace.

line is drawn across the opening on the outside of the door opening to represent the door sill.

Fireplaces. Figure 5-15 shows the standard convention for drawing in a fireplace. There also might be a flue venting the central heating plan drawn in as a circle or a square. The hearth line will be drawn 18 inches from the face of the fireplace. Firebrick is always used in the lining of a fireplace. Firebricks are made of heat-resistant material far superior to ordinary brick and usually of larger dimensions.

Stairs. Stairs are represented on scale floor plans as indicated in Fig. 5-16. An arrow drawn through the center of the treads indicates whether the stairs go up to the second floor or down to the basement. The number of risers is also given. If one stairway is on top of another as is often the case with basement stairs down and second story stairs up a line is drawn at an angle across the stairs to separate the up-stairs from the down-stairs.

Bathroom Fixture Symbols. Figure 5-17 shows some standard symbols for bathtubs, shower stalls, lavatories, toilets, and medicine cabinets. The diagonal lines drawn through the shower stall diagram indicates a floor pitched to drain in the center.

Kitchen Fixture Symbols. Figure 5-18 shows some typical kitchen-fixture symbols, including single and double sinks.

Closets. See Fig. 5-19.

FOUNDATION AND ELEVATION SYMBOLS

A concrete foundation wall 12 inches wide is indicated on the plan drawing shown in Fig. 5-20. The small dots represent aggregate or stone in the concrete. Figure 5-21 shows how some other materials are symbolized in drawings. If uncommon or unusual materials are used in the construction of a structure, the plans will include a *symbol key* to indicate how the material is symbolized.

It is important to the builder to have the proper representation of exterior wall treatment on the elevation drawings. Figure 5-22 shows horizontal lines drawn close together approximating the width of a brick to indicate a brick exterior wall. Asbestos roof shingles are also indicated by a series of horizontal lines, but these are drawn closer togeth-

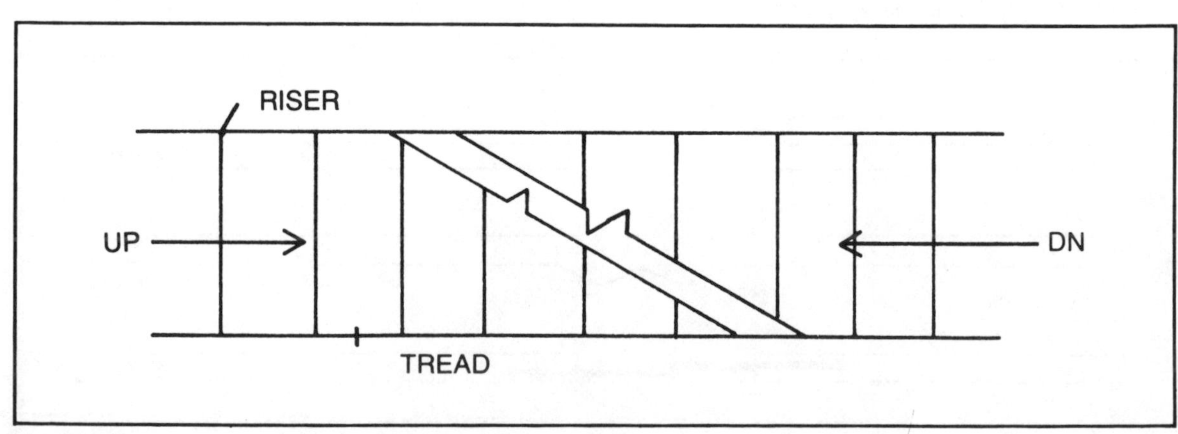

Fig. 5-16. Stairs.

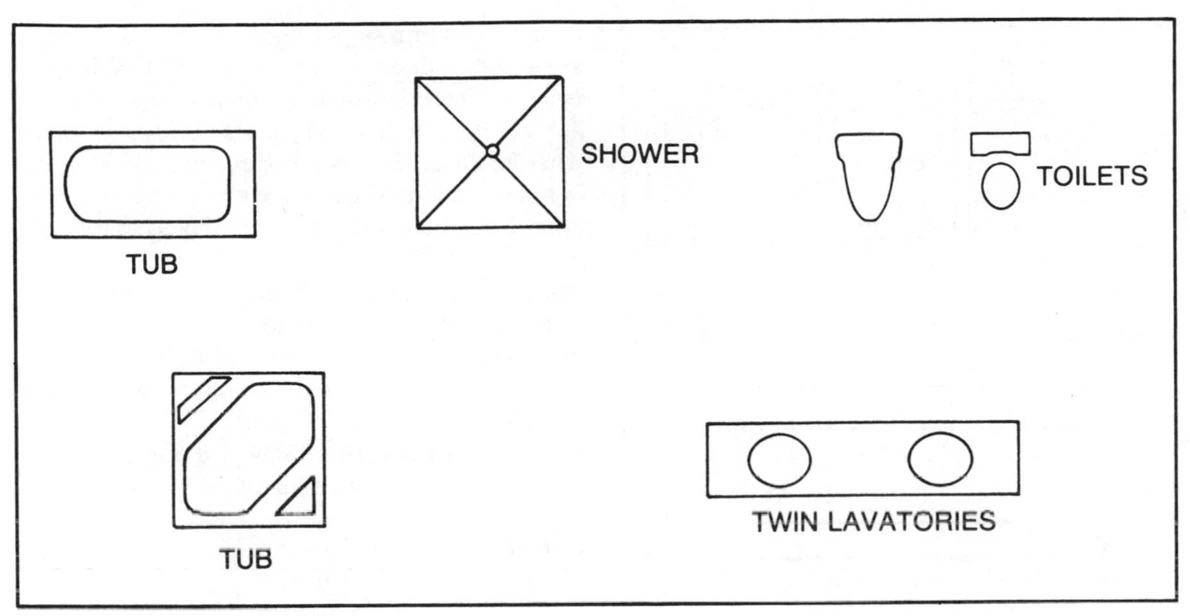

Fig. 5-17. Bathroom fixture symbols.

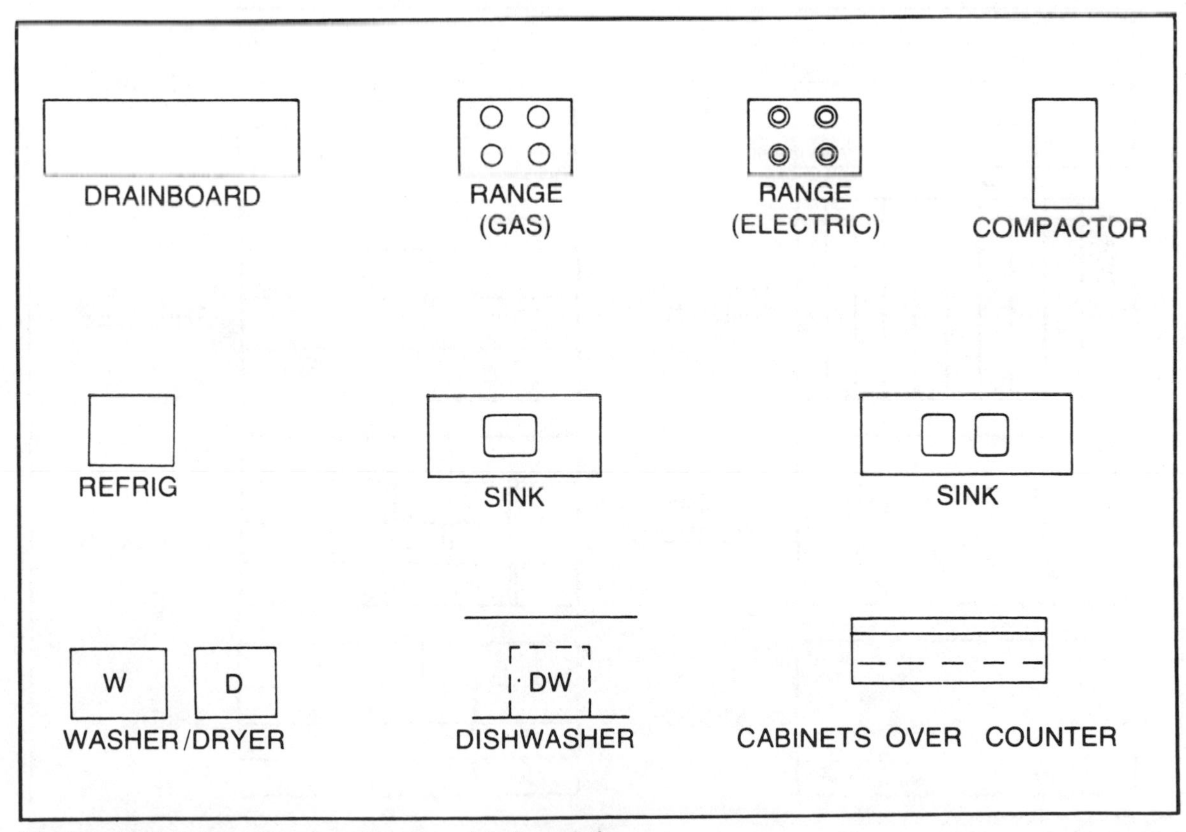

Fig. 5-18. Kitchen fixtures and appliances and their symbols.

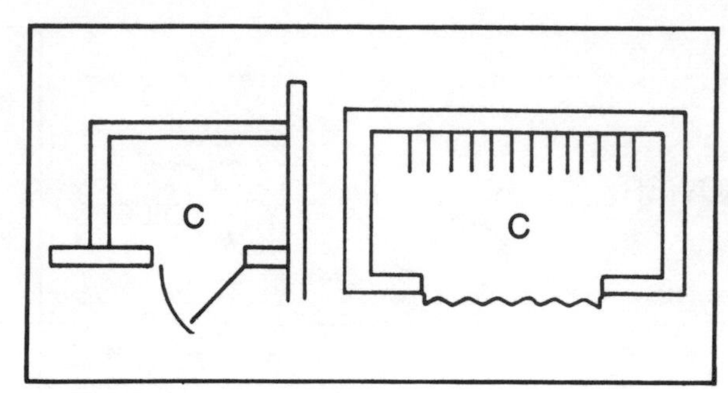

Fig. 5-19. The closets.

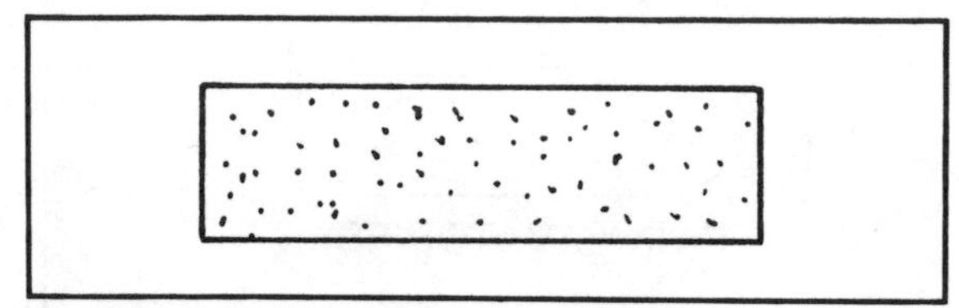
Fig. 5-20. A concrete foundation wall.

er and fade out near the ends. A lettered notation as part of the drawing further clarifies what is symbolized. These lettered notations have terms and abbreviations with which you should become somewhat familiar. Here is a list in alphabetical order of some of the most common terms. Table 5-1 lists typical abbreviations. Also see the Glossary.

apron—A finishing board placed beneath the base at the bottom of a window.

architrave—Another name for *trim.*

ashlar—The cut-stone facing on the outside of a wall.

awning window—A window that opens out from the top, as does an awning.

batten—A piece of wood used to fasten other pieces of wood together.

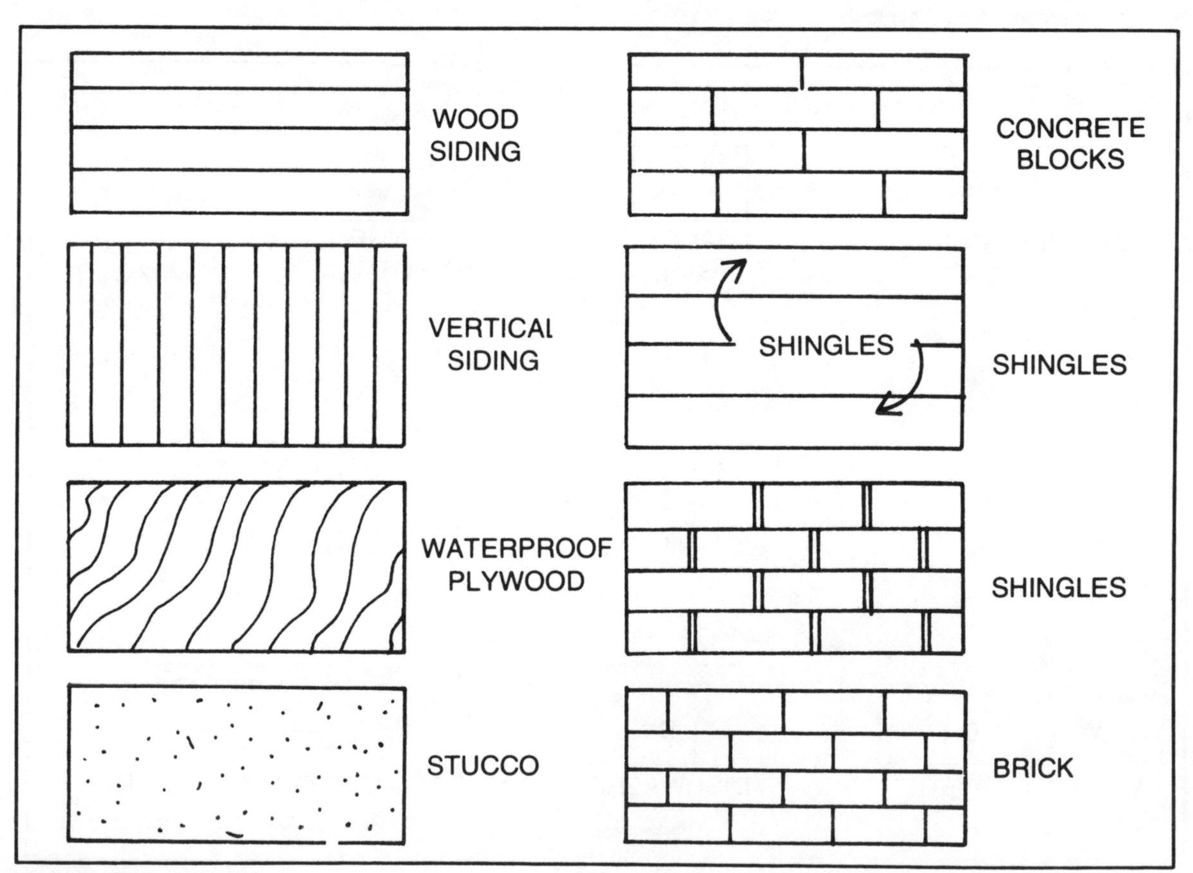

Fig. 5-21. Material symbols.

Table 5-1. House Plan Abbreviations.

Abbreviation	Meaning	Abbreviation	Meaning
Asph.	Asphalt	Ht.	Height
Bldg.	Building	L.	Center Line
B.R.	Bedroom	L.P.	Low Point
Br.	Brick	L.R.	Living Room
B.S.	Bevel Siding	Lt.	Light
Cem.	Cement	Mldg.	Molding
D.C.	Drip Cap	Mull.	Mullion
D.G.	Drawn Glass	No./#	Number
D.H.	Double Hung	Obs.	Obscure
Diag.	Diagonal	O.C.	On Center
Diam.	Diameter	O.C.	Outside Casing
Dim.	Dimension	O.S.	Outside
Div.	Divided	Pl.	Plaster
Dn.	Down	Pl.	Plate
Do.	Ditto	Cond.	Conduit
D.R.	Dining Room	Cop.	Copper
Dr.	Door	Corn.	Cornice
Dr.C.	Drop Cord	Csmt.	Casement
Drg.	Drawing	C.T.	Crock Tile
D.S.	Downspout	Crys.	Crystal
Ea.	Each	Pl. Ht.	Plate Height
El.	Elevation	R.	Radius
Ent.	Entrance	Rm.	Room
Ext.	Exterior	R.W.	Redwood
Fin.	Finish	Scr.	Screen
Fin. Ceil.	Finished Ceiling	Sdg.	Siding
Flash.	Flashing	Specs	Specifications
Fl.	Floor	T and G	Tongue & Grooved
C.I.	Cast Iron	T.C.	Terra Cotta
Clg.	Ceiling	Th.	Threshold
Clr.	Clear	Typ.	Typical
C.O.	Cased Opening	Ven.	Veneer
Conc.	Concrete	V.T.	Vertical tongued
Cond.	Conductor	and G.	and grooved
Ft.	Foot/Feet	W.	Wide
Ftg.	Footing	W.C.	Wood Casing
Gar.	Garage	Wd.	Wood
G.I.	Galvanized Iron	W.G.	Wire Glass
Gl.	Glass	W.I.	Wrought Iron
Gr.	Grade	Wp.	Waterproof
Gyp. Bd.	Gypsum Board	Yd.	Yard
H.P.	High Point		

board and batten—Wide boards used vertically as siding with battens over the cracks between boards.

casement—Any hinged window.

cement plaster—A mix of cement and sand used as fireproofing or waterproofing on the exterior of foundations.

clapboard—Long, thin siding that is graduated in thickness.

clerestory windows—Small windows in the walls between roof segments.

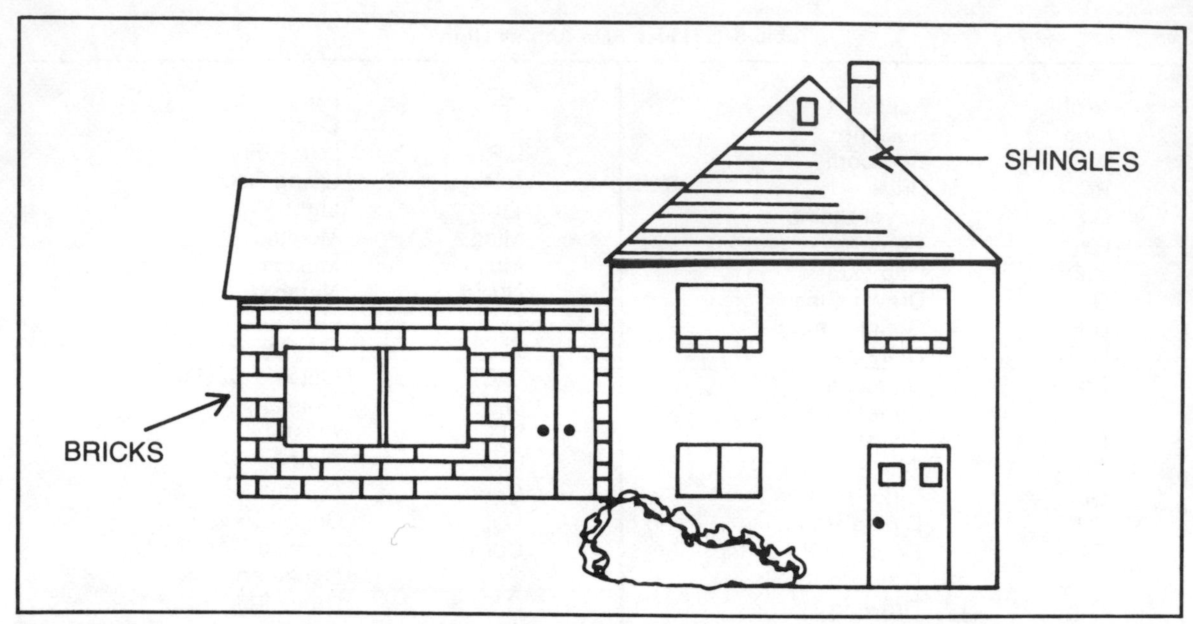

Fig. 5-22. Elevation with finishing materials indicated by symbols.

concrete block vent—A vent made in a concrete block wall by leaving a hole the exact size of a concrete block.

corbel—An arch formed by building out successive courses of masonry.

corner boards—Finish boards used to trim the outside corner of a frame house.

cornice—That part of a roof that projects beyond the wall below it.

course—One continuous row of bricks or other masonry material.

dormer—A structure projecting from a sloping roof to provide light and air under the roof.

drip—A projection over a window to drain water away from the window.

fascia—The outside, flat member of a cornice.

finish strip—Used with a crown molding as part of a cornice.

finished ceiling—Term used on elevation drawings to indicate ceiling height.

finished first floor—A term designating the first-floor level on elevation drawings.

fixed wood sash—Windows that do not open.

flashing—Sheet-metal strips installed to prevent water leaking into doors, windows, and roof openings.

flat roof—A roof with almost no pitch.

frieze—A piece of trim used just below the cornice.

gable—A triangle formed by a sloping roof in an end wall.

gambrel roof—A roof that slopes up at two different angles.

grade—The level of the ground around a building.

gutter—A trough, usually metal, for carrying water from roofs.

hip roof—A roof that slopes up from the corners of a structure.

hood—A small roof over a doorway.

jamb—The inside vertical face of a window frame.

lally column—A support for beams, usually a cylinder of steel, sometimes filled with concrete.

leader—A conduit, usually metal, used to carry water from gutter to the ground.

lintel—A beam used to support a wall over window and door openings.

louver—An opening for ventilation covered by angled slats to exclude wind and rain.

meeting rail—The horizontal center rails of the sash in a double hung window.
metal caps—Waterproof flashing over doors and windows.
millwork—Finished and partly assembled wooden parts.
miter—The beveled surface cut on the ends of a molding.
mullion—The vertical division of a window opening.
muntin—That strip of wood that separates the panes of glass in a window frame.

nosing—Overhang of a stair tread.

panel—A framed piece of wood.
parting strip—That strip in a double hung window frame that separates the upper and lower part of the sash.
pitch of roof—The amount of slope displayed by a roof.

rake board—That board that runs down the slope of a roof from the top of the gable. Also called a *verge board.*
ridge—The top edge of a roof where the slopes meet.
riser—The vertical portion of a stair step.
rowlock course—The course of brick that is set on edge under a window.
R.W. siding—Redwood siding.

saddle—Build-up metal around a chimney to throw water away from the chimney.
sash—The moveable part of a window in which the panes of glass are set.
sheathing—The rough boards nailed to the outside of the studs in an exterior wall over which is laid the siding, shingles, or brick finish.
sidelights—The fixed windows on each side of a doorway.
sill—The bottom member under a window or a door.
soffit—The underside of a cornice.
soldier course—Bricks laid on their ends with edges exposed.
stool—The support at the bottom of a window.
stringer—The supporting member at the sides of a staircase.
structural glass—Special glass, available in colors for use inside or outside.

T and G siding—Tongue and groove boards used as siding instead of flooring.
tread—The horizontal member of a stairway.
trim—The finish around openings or adjoining parts.

valley—The groove where two roof slopes intersect.
verge board—That board that runs down the slope of a roof from the top of the gable. Also called a *rake board.*

water table—A shelf of masonry or wood projecting from the top of a foundation to protect the foundation from rain water.
weather strip—Any strip of any material used to keep drafts and dirt from entering a building. Used to cover joints and cracks.
window wall—Fixed glass windows with large panes that form one wall or nearly all of one wall of a structure.
window water table—A *drip* over the top of a window to throw rain water away from the window.

PLUMBING SYMBOLS

Typical plumbing symbols are shown in Figs. 5-23 and 5-24. You should become as familiar with these symbols as with the words that describe plumbing fixtures and accessories. Plumbing symbols are drawn accurately to scale so that the builder will allow sufficient room for their installation.

HEATING AND VENTILATING SYMBOLS

Heating and ventilating symbols, like plumbing symbols are drawn to scale so that the builder will know exactly how much space to allow for the equipment. Figure 5-25 shows the most-used symbols.

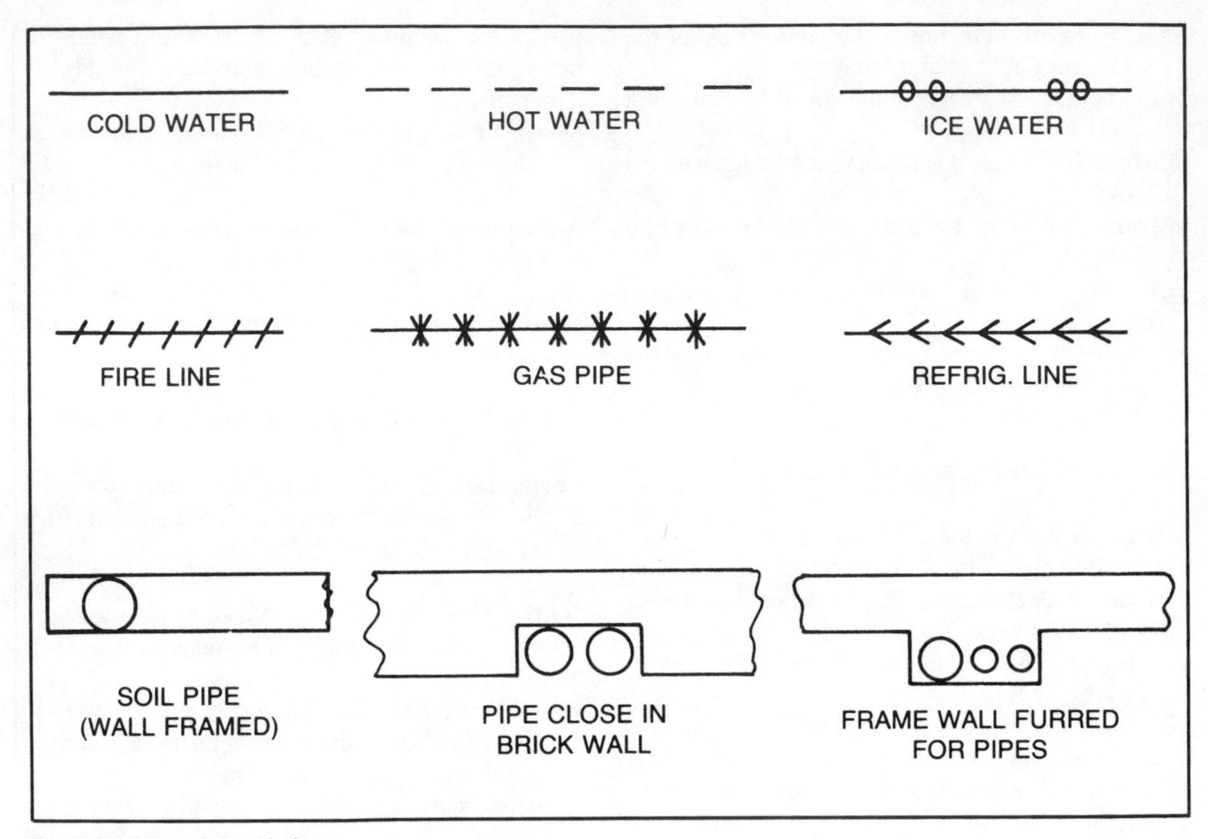

Fig. 5-23. Plumbing symbols.

SOLAR AND WIND GENERATOR SYMBOLS

New innovations in energy generation and distribution have brought about whole new designs relating to heating, air-conditioning, and electrical work described on house plans. Some of the solar energy extractors also have a great deal of plumbing connected with their construction. Figure 5-26 shows some of these new symbols as they are generally

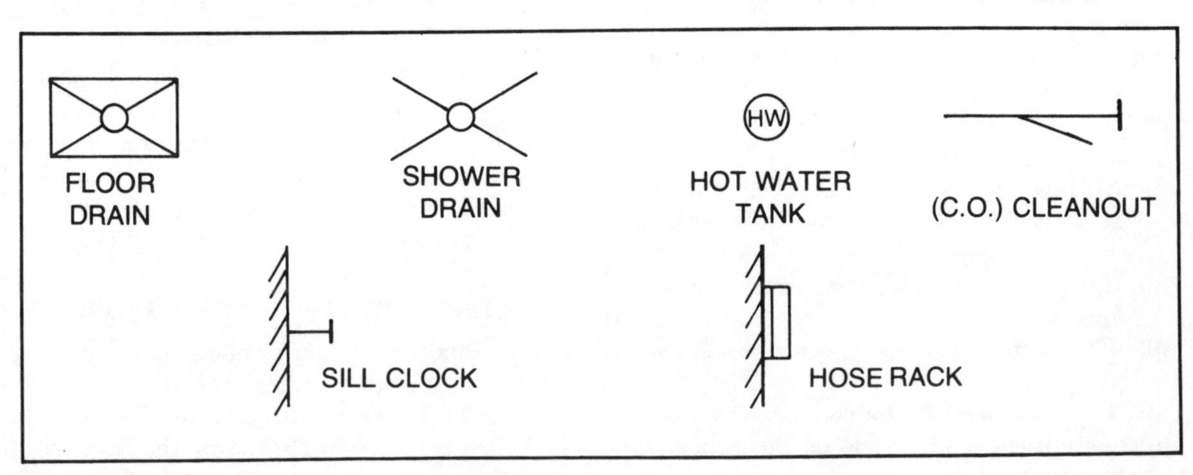

Fig. 5-24. Plumbing symbols.

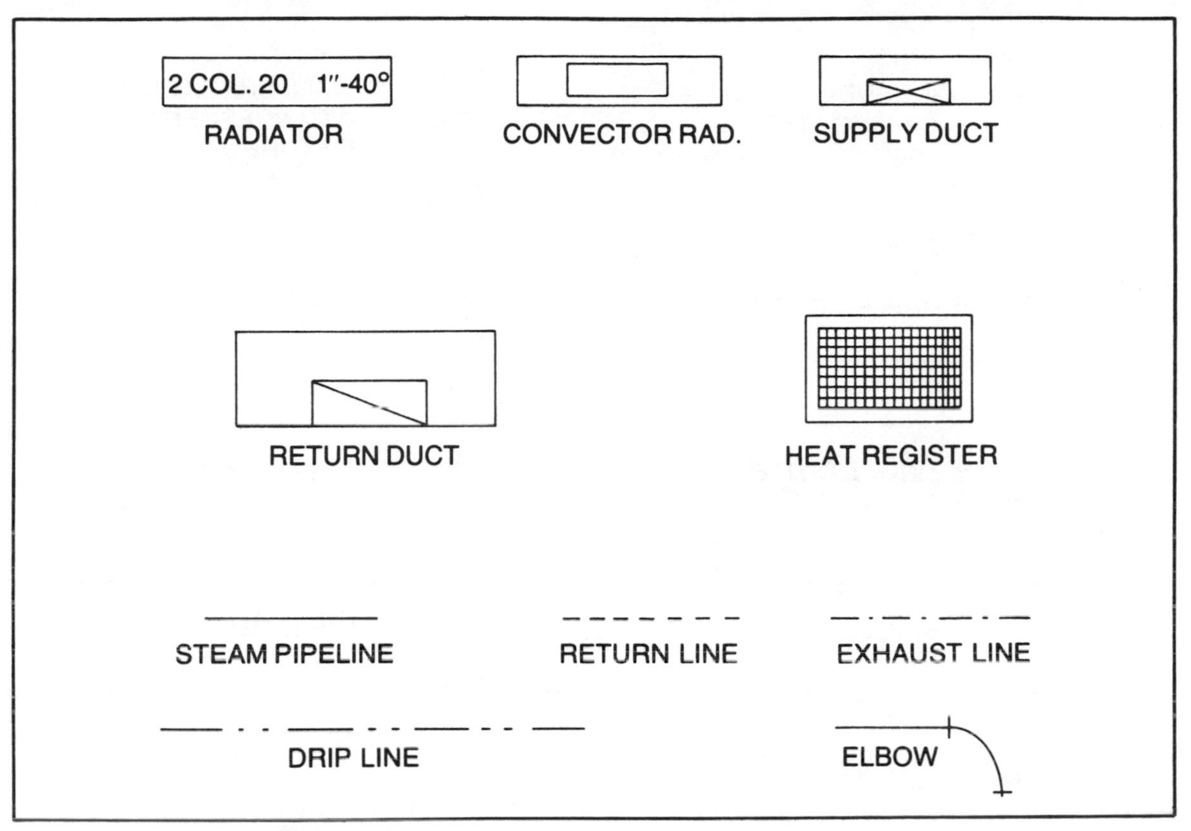

Fig. 5-25. Heating and ventilating symbols.

represented on plans being drawn to include these new ideas in the capture and conservation of energy.

ELECTRICAL SYMBOLS

Electrical outlets are always indicated on floor plans in their approximate locations. Fixture type is indicated by an uppercase letter in the symbol. Lowercase letters designate switch control. Numbers identify the circuit. In the switch symbol, there will be a letter indicating the device that switch controls. The electrical symbols shown in Fig. 5-27 are typical of the variety of symbols used by architects. Once you learn to recognize these, you will be able to recognize variations that will appear on plans from time to time.

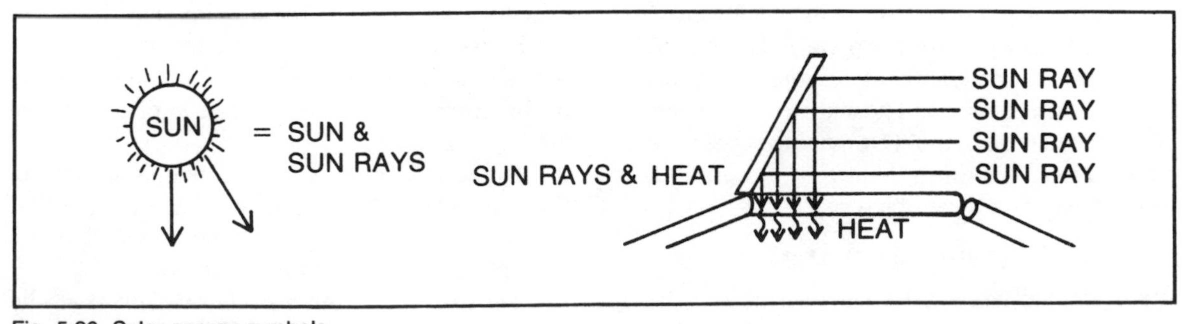

Fig. 5-26. Solar energy symbols.

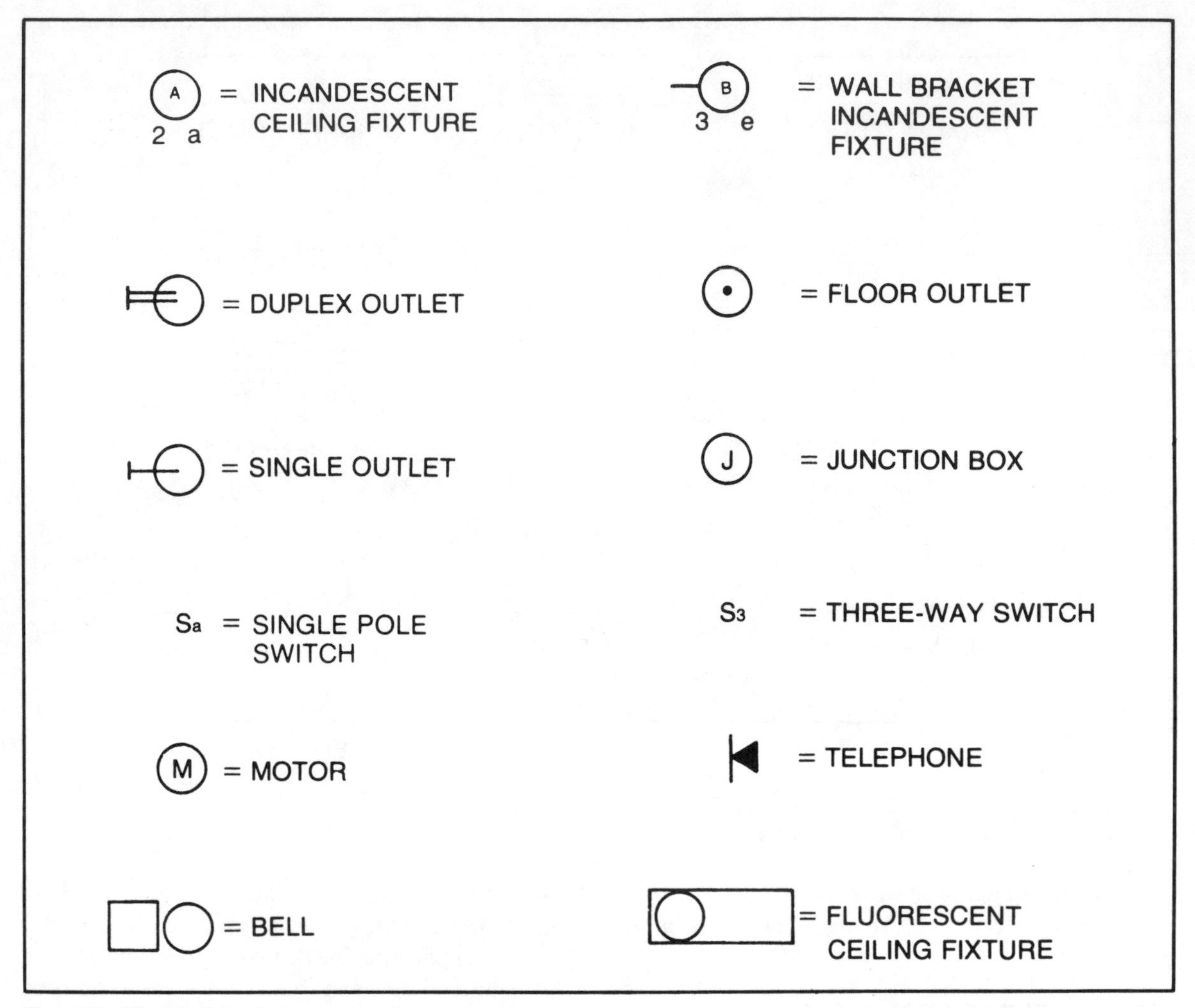

Fig. 5-27. Electrical symbols.

Be sure to pay careful attention to all notations you find on drawings. Most good drawings contain printed instructions relative to many of the details of construction. These notes will give information directly or refer you to other parts of the set of plans where you can find more information. When drawing your own plans, don't hesitate to explain details of what you have in mind by making notes on your plans. There are times, believe it or not, when a word is worth a dozen pictures!

LINES ON PLANS

When you are learning to read construction drawings—and later on in your own drawing of plans—you will need to pay close attention to the variety of special types of lines used for specific purposes on all architectural drawings. There are four types of lines with which to become familiar.

Invisible Lines. This is a line made up of a series of short dashes. It is used to indicate parts of the drawing that cannot be seen or to indicate hidden edges.

Broken Lines. A broken line is to indicate that parts of the drawing have been left out or that the full length of something is not shown. This line has an uneven break in it at intervals.

Section or Reference Line. This is a solid line with an arrow head at each end pointing in the direction of a detail drawing on another sheet of the set of plans. Reference letters or numbers may be shown at the ends of the lines.

Center Lines. Dash-and-dot lines are drawn lightly. They are used to indicate the centers of things.

Figure 5-28 should be studied to familiarize yourself with the various kinds of architectural drawing lines and what they indicate.

SECTION VIEWS AND DETAIL DRAWINGS

While plan and elevation views are the basic and most important drawings in a set of construction plans, they often cannot show enough information to enable the builders to see exactly how to build or assemble some parts of a structure. For example, the plan and elevation drawings might show that the exterior walls of a house are to be constructed using 2-×-4 inch studs, with plasterboard on the inside and sheathing, covered with building paper and clapboard, on the outside. And that's about all the information you get. No indication about the location of the studs or what type of sill construction or method of bracing is to be used. This information will appear in *section views* and *details*.

Section views and details are more picture-like drawings than are plans and elevations. They usually contain more information in the form of symbols, abbreviations, and terms. Details and section views also show how things are assembled, how they relate to one another, and how they relate to the structure as a whole.

In general, section views and detail drawings are drawn to a much larger scale than plan and elevation drawings. This allows for a "close up" look. Sections and details are also often drawn isometrically. This makes them more picture-like and dimensional than the flat, two-dimensional plan and elevation drawings. This added dimension means that much more information can be included.

Staircases and fireplaces are good examples of structures that need additional drawings to show carpenters and masons how many stringers to use, and their size, or how brickwork over the fireplace opening is to be supported. Section views are used primarily to indicate the interiors and arrangements of individual or related structural parts. Detail drawings are those section views that show blow ups of items too small to be represented by the scale used for plan and elevation drawings. We will treat them as section drawings from now on and refer to both as *sections*. These sections are drawn carefully to scale (a larger scale) just like all the rest of the construction drawing set.

When you visualize a section, you imagine that the structure or part of the structure, has been cut through *vertically* as shown in Fig. 5-29. This is a section of footing and foundation wall; it shows how the two are tied together. Just imagine that you can use a saw and cut the header, sill, and footing on the line *xy*. The dashed line *fg* shows that the saw cuts the footing and foundation wall vertically. Now, imagine that once we get the saw all the way through the footing, we can swing the *a* and *b* corners backward.

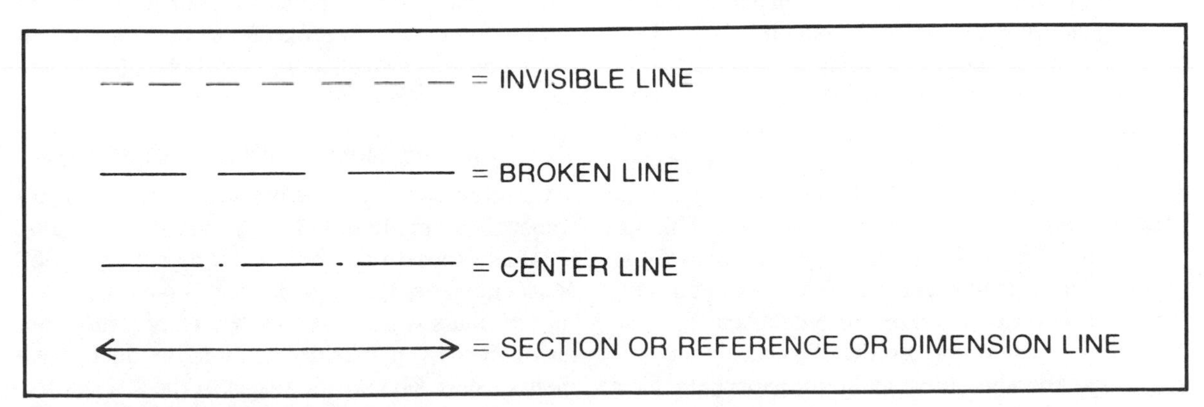

Fig. 5-28. Some architectural lines.

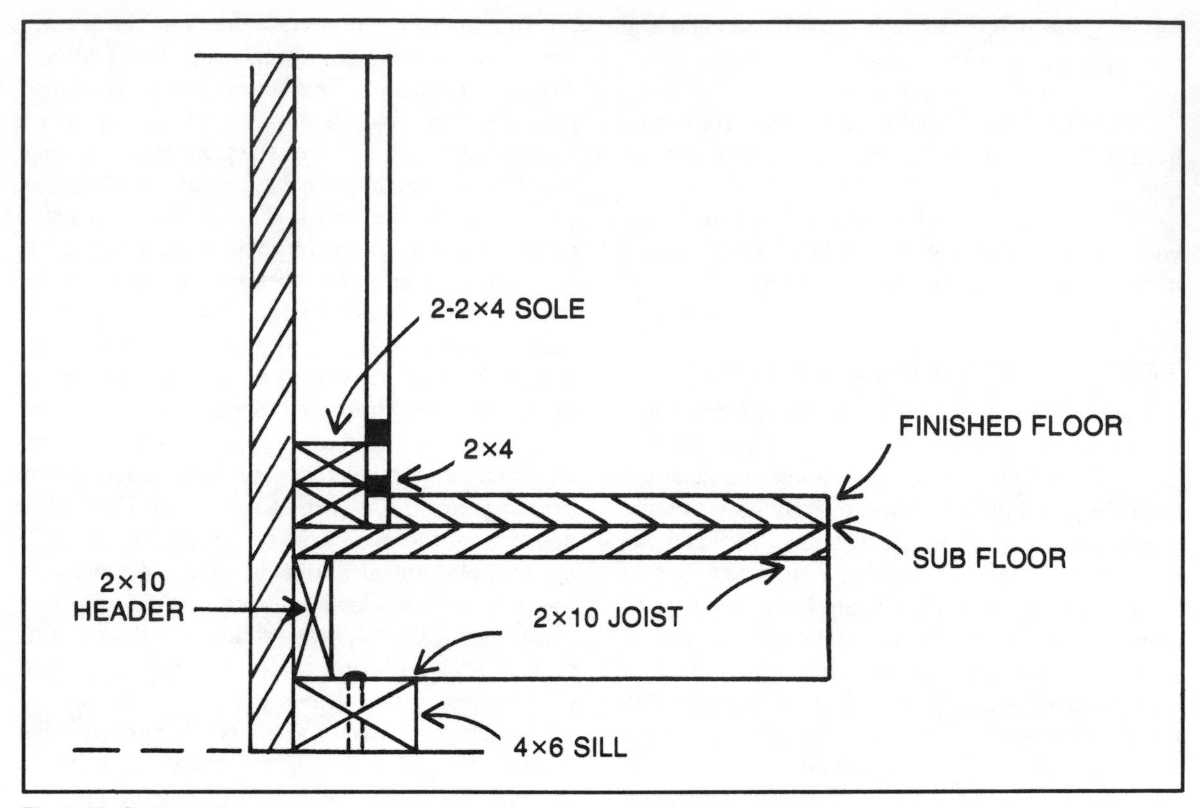

Fig. 5-29. Section of a wall.

We can see "inside" the anchoring arrangement between footing and foundation and visualize just exactly how it all goes together. The subfloor is nailed to the joists that are at right angles to the header and the sill. The sill is anchored to the foundation wall that is resting on top of the footing. The header and sill are continuous members running completely around the building. The joists are usually spaced 16 inches on centers at right angles to the header/sill combination.

If you continue to study the sections, you will discover the information needed to construct and finish the walls. Figure 5-30 shows how the sole plate relates to the subfloor in Fig. 5-19. This in turn receives the 2- × -4 inch wall studs. Figure 5-31 shows a frame wall covered with a veneer of brick. The section drawings show such details as the extra width of the foundation wall to accommodate the brick, the sheathing, building paper, etc.

Figure 5-32 continues the wall section up to the second floor to show the studs resting on the second floor sole plate and the relationship between those elements of construction and the joists, top plate, and first-floor studs below.

Topping off the structure with a section (Fig. 5-33) that carries you to the roof, you can see how the construction is to be carried out in relation to the top plate on the second floor, the rafters, and ceiling beams, as well as some detail on the finishing of the roof.

Concrete and Masonry Wall Section Views

There are two basic types of concrete and masonry walls: those made of concrete cast in place (Fig. 5-34) and those made of precast units (Fig. 5-35). Most engineers recommend cast-in-place concrete for buildings where external water pressures are encountered, on hillside locations or deep basements where severe side thrust of the soil is a factor, and where load requirements are severe—as in

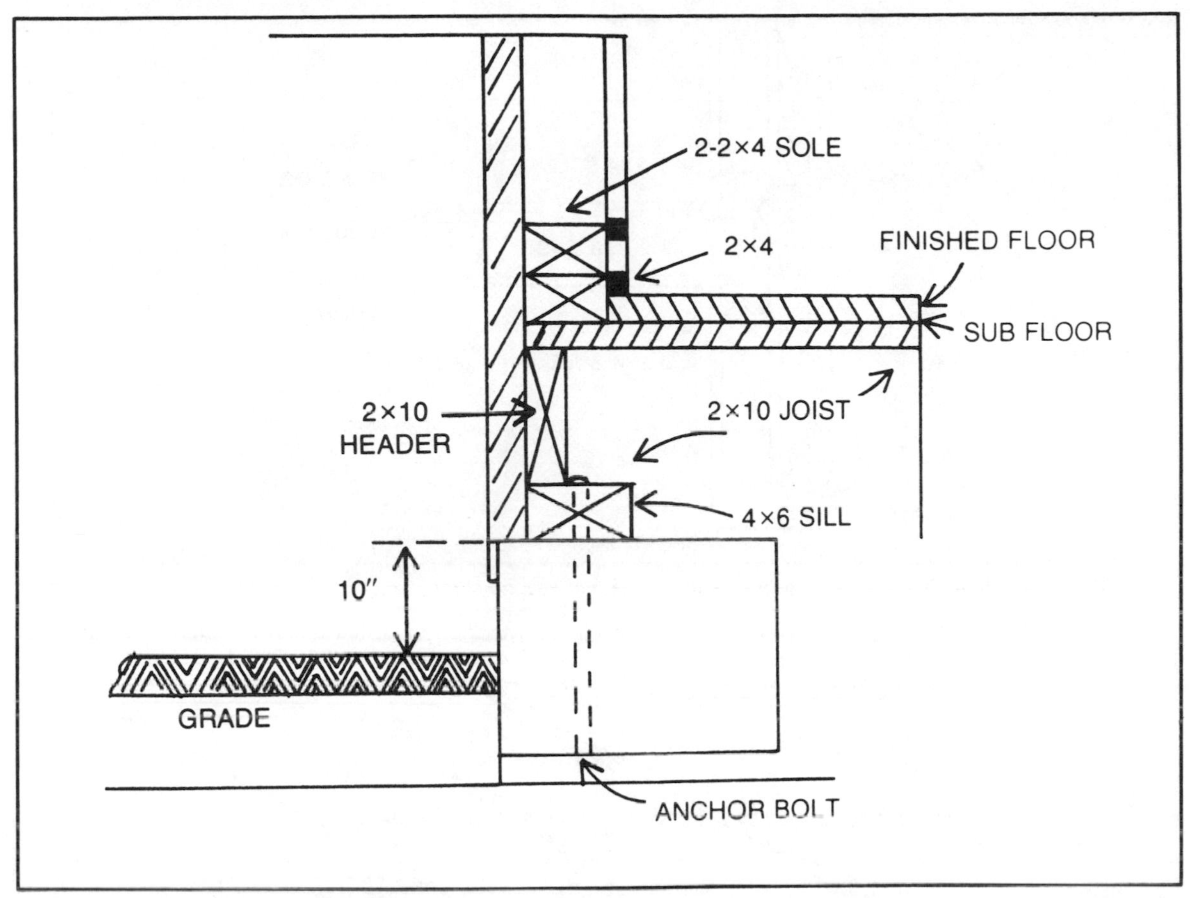

Fig. 5-30. Section of a wall and footings.

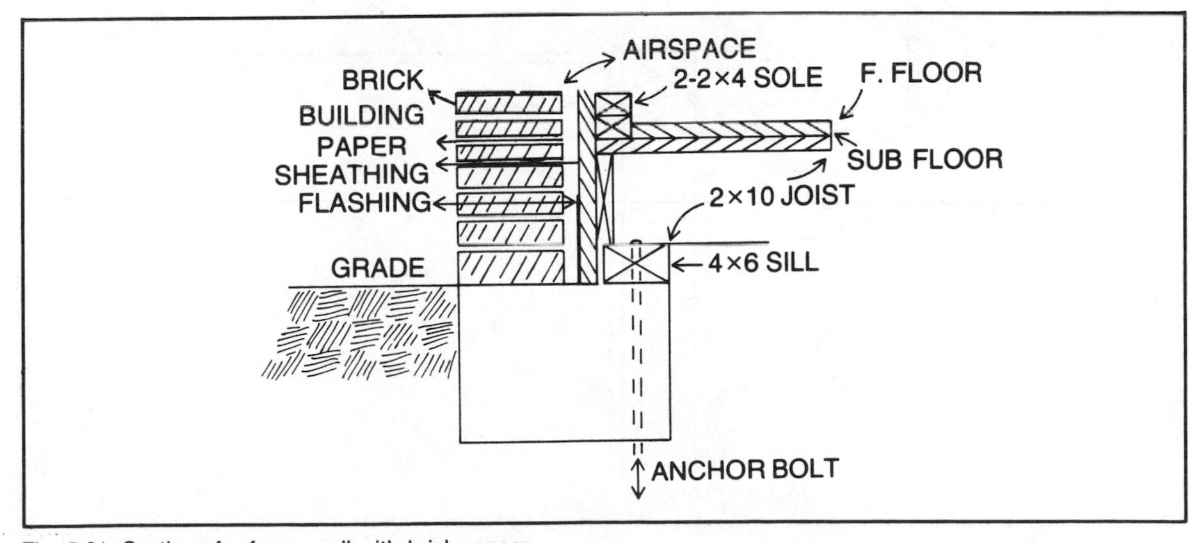

Fig. 5-31. Section of a frame wall with brick veneer.

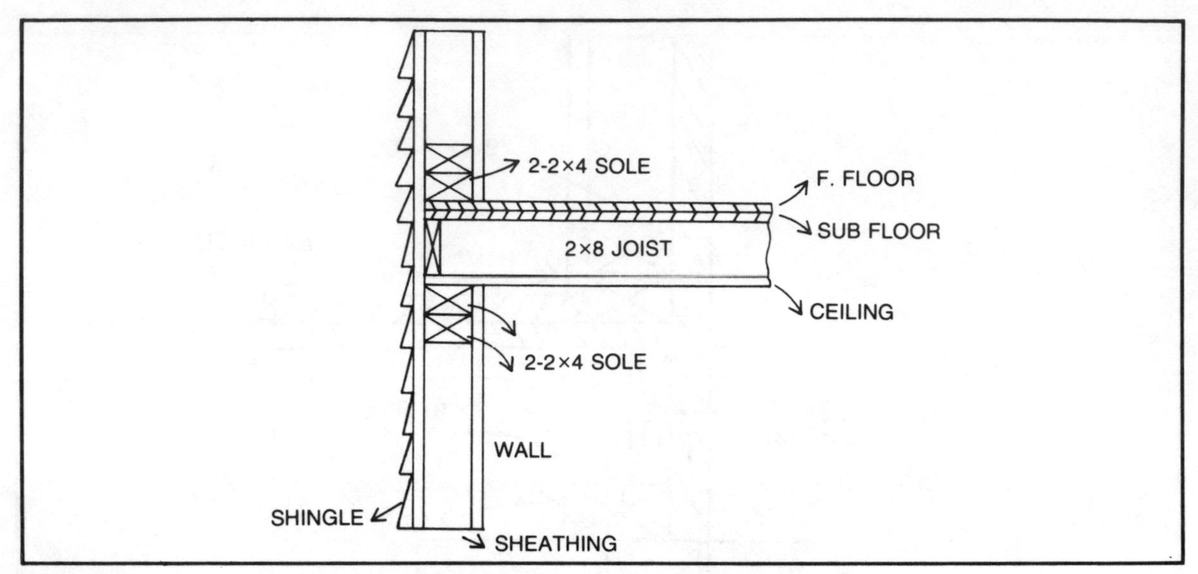

Fig. 5-32. Wall section showing relationship between first and second story.

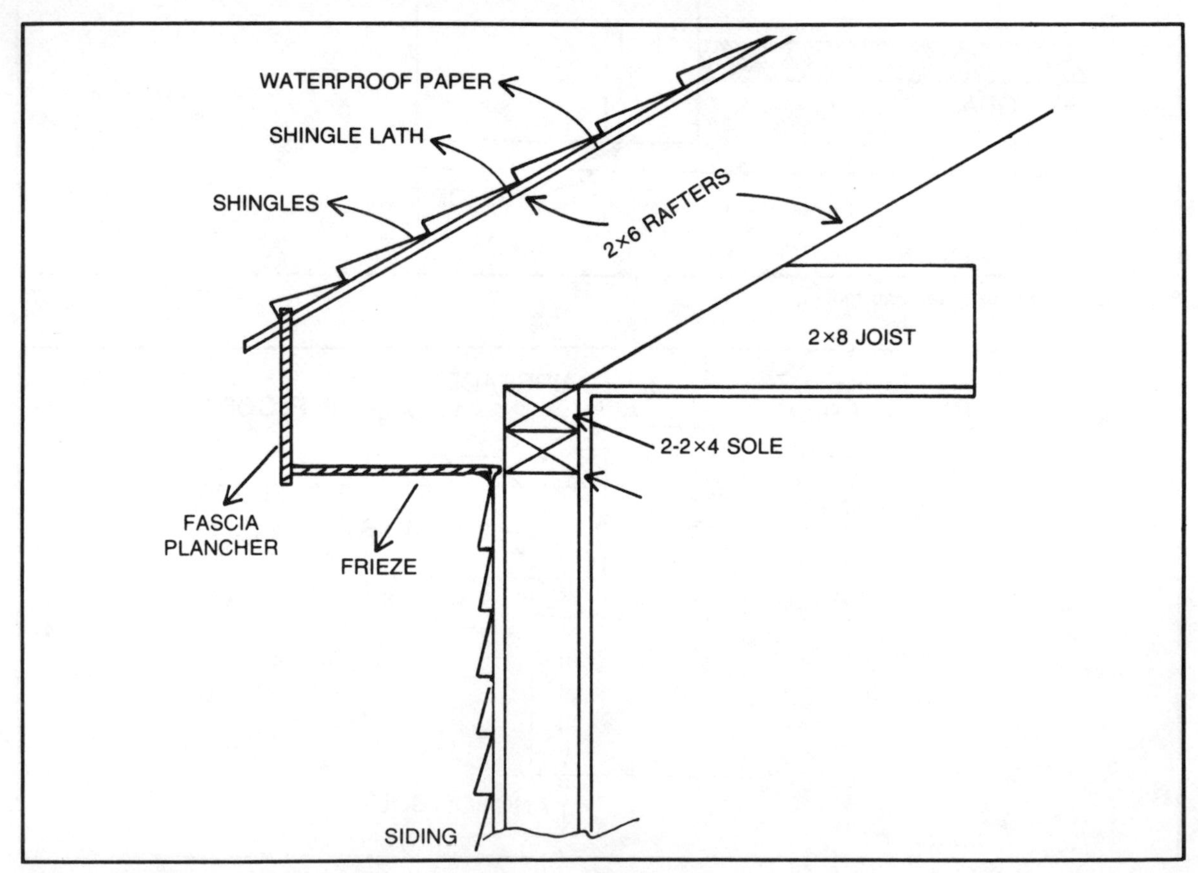

Fig. 5-33. Detail of roof construction.

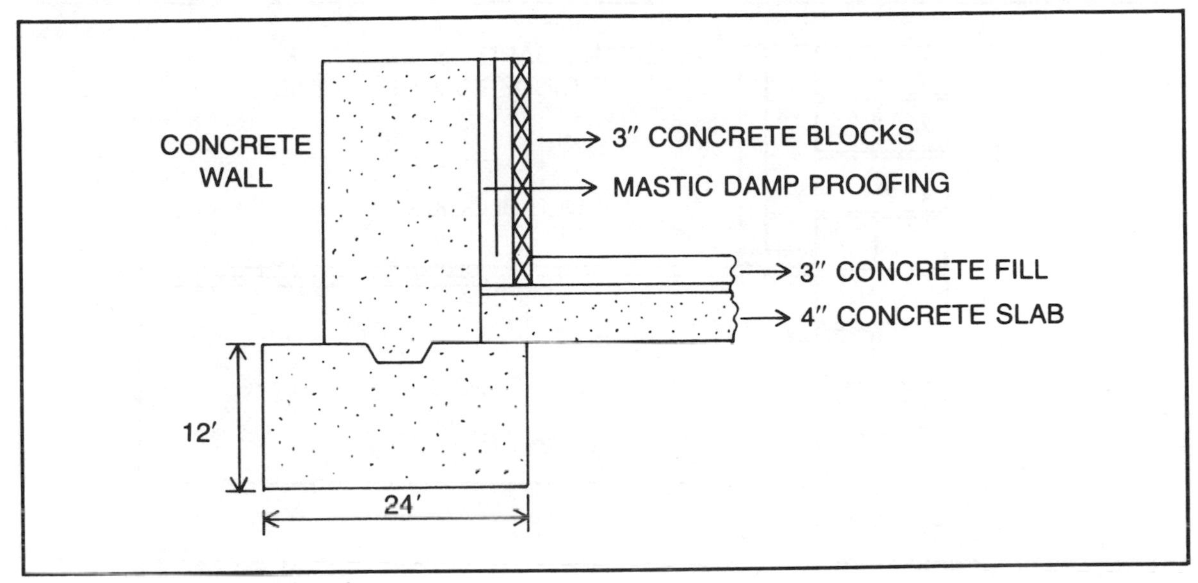

Fig. 5-34. Cast-in-place wall section.

tall buildings. All types of foundation walls should rest on concrete footings to provide an even surface on which to base the wall proper.

Other construction details you can pick up from concrete and masonry wall section views is that concrete walls below grade level are usually 12 inches thick, going to 8 inches above grade. A brick veneer is often applied to the face of the concrete above grade to improve the appearance of the face of the wall. Notice, too, that the floor joists rest on brick rather than on the concrete to ensure an even alignment of joists. Be sure to learn the symbol for concrete block and for brick and tile.

A tile wall that can be used in light structures, hollow building tiles come in a variety of sizes. Perhaps the most common size is 12 inches long by 8 inches wide by 5 inches deep. In this type of wall, fill with concrete any tiles on which floor joists rest (as noted in the detail drawing). Good engineering practice recommends inserting steel reinforcing rods into the tiles before packing with concrete. This construction detail will be shown on the drawing.

Brick walls are bonded every fourth course to the concrete block behind them. Concrete slab floors need to be from 4 to 6 inches thick depending upon the load to be supported and on the pressure of ground water. If soil at the construction site is water-logged, a 4- to 6-inch fill of cinders or gravel needs to be specified to provide a dry base for the concrete slab. The slab is usually placed directly on the earth. Where the slab meets the wall, a little space needs to be provided for expansion.

Study the construction details shown in Figs. 5-36 and 5-37. These detail drawings show a solid brick wall rising from a poured concrete foundation supported by a keyed footing. A 3-inch cinder block is used as furring with air space provided between it and the wall. The basement and first floor are of 4-inch concrete with a 3-inch cinder block on top. The fabric flashing between the foundation and the brick wall prevents moisture from seeping into the structure.

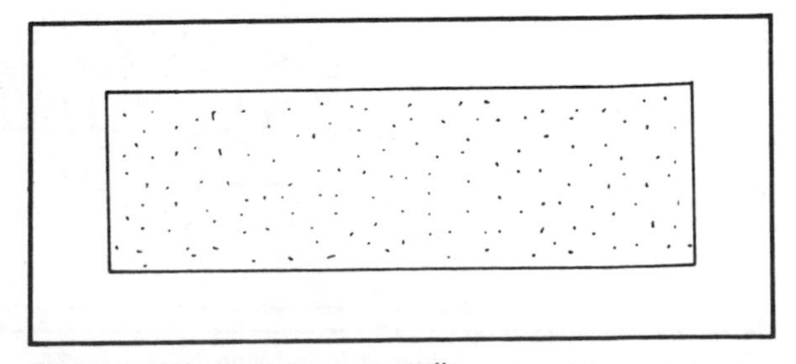

Fig. 5-35. Pre-cast concrete wall.

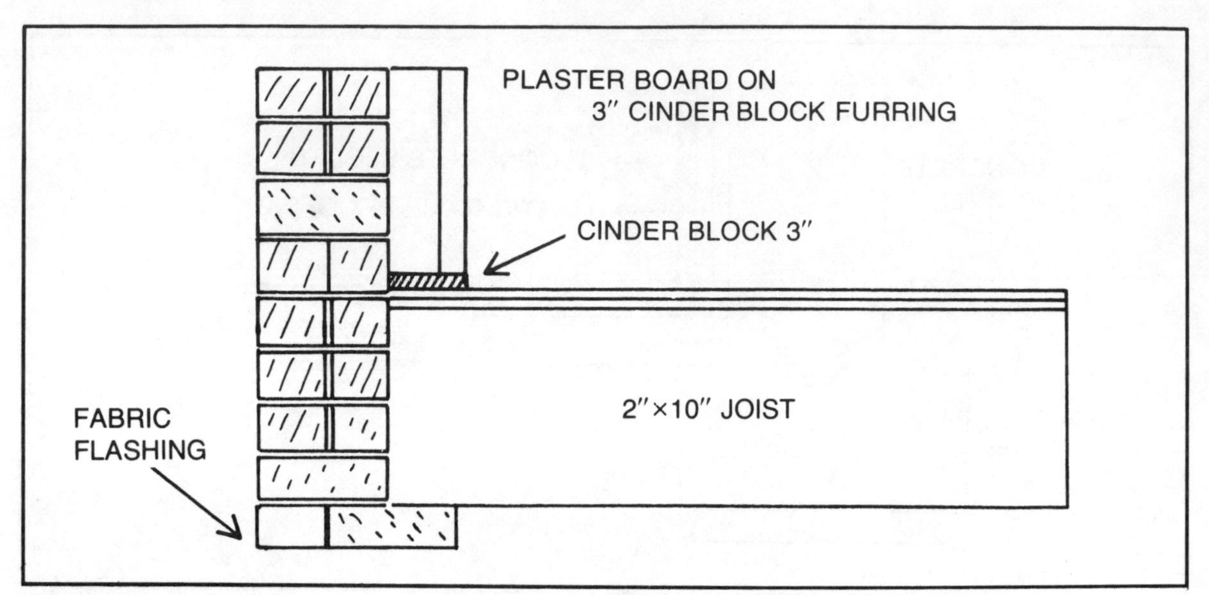

Fig. 5-36. Details of a wall.

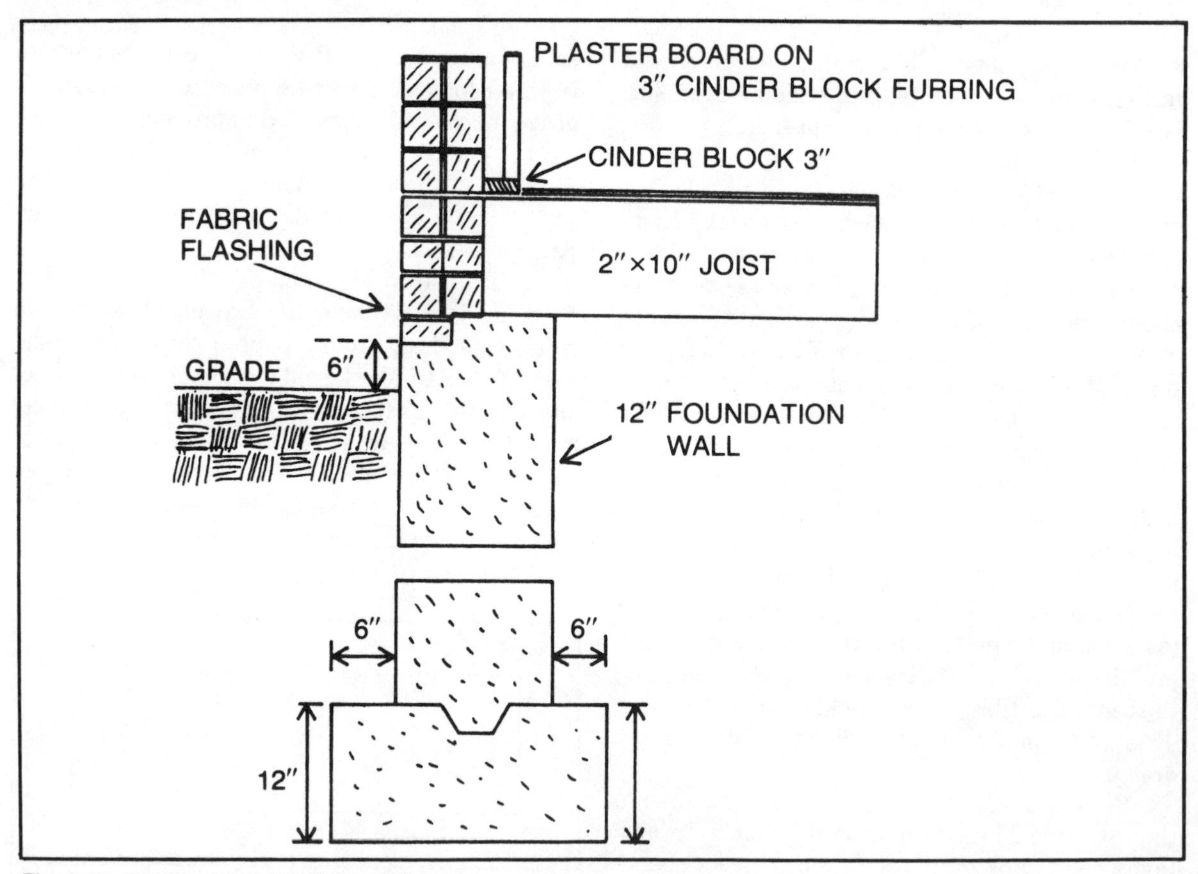

Fig. 5-37. Details of a wall and foundation.

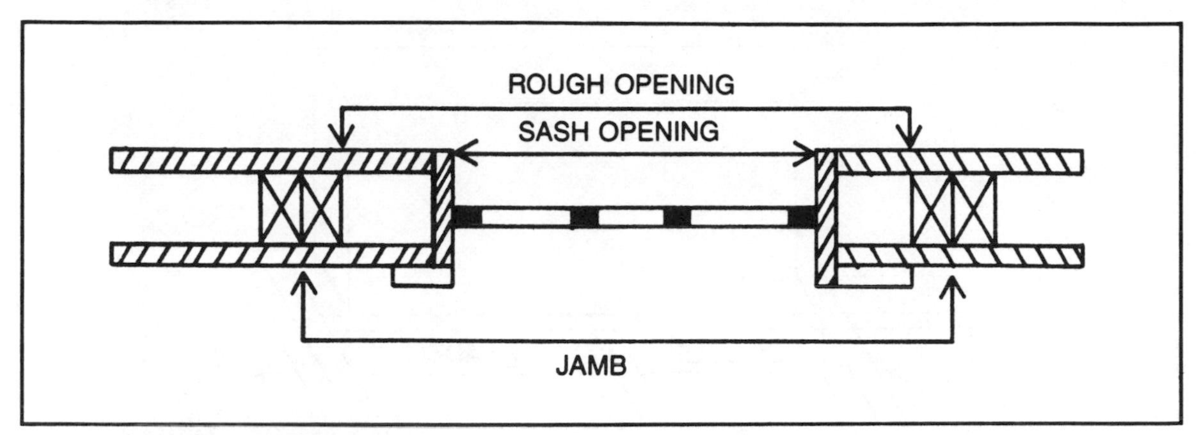

Fig. 5-38. Details of a double-hung window.

Large-Scale Details

Figure 5-38 shows the details of an ordinary wood double-hung window set into a frame wall. The drawing projects the three distinct components that make up the window construction. The window sash goes into the window frame; in turn, this goes into the rough opening in the wall. Figure 5-39 is a large-scale representation of a small section of the window frame and sash installed in the rough opening. All parts of the window are shown clearly along with the outside wall finish around the window.

Framing Plans

A complete set of house plans does not have to include detailed plans for framing the structure to be built. Most builders are quite familiar with where to place framing members and how they go together. Although you do not need to draw framing plans, it is best to be able to read these drawings and to understand what they illustrate.

Framing members consist of *joists*, *headers*, *tail beams*, *trimmers*, and *bridges*.

Joists. The term joists applies to both floor and ceiling beams as shown in Fig. 5-32. Spacing of joists in both floor and ceiling is usually 16 inches from the center of one joist to the center of the next.

Headers. These are beams used around openings for windows, doors, stairwells, etc. Fireplace and chimney openings also require headers. Headers are placed at right angles to joists and are used to carry the ends of joists that are cut. Because headers carry extra loads, they are usually doubled or tripled. Figure 5-40 shows construction details of headers as well as tail beams and trimmers.

Tail Beams. Tail beams are simply cut joists butting against headers around openings. Tail beams carry their loads to the header. Sometimes they are supported by metal joist hangers.

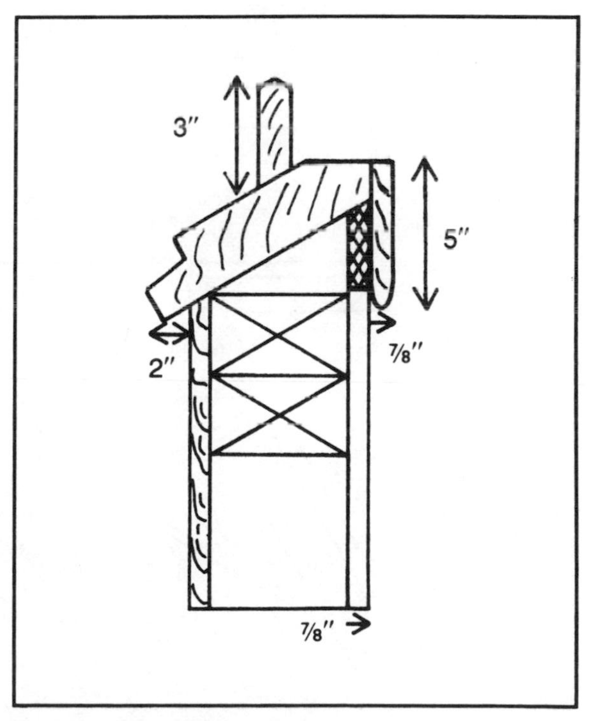

Fig. 5-39. Sill detail of a window.

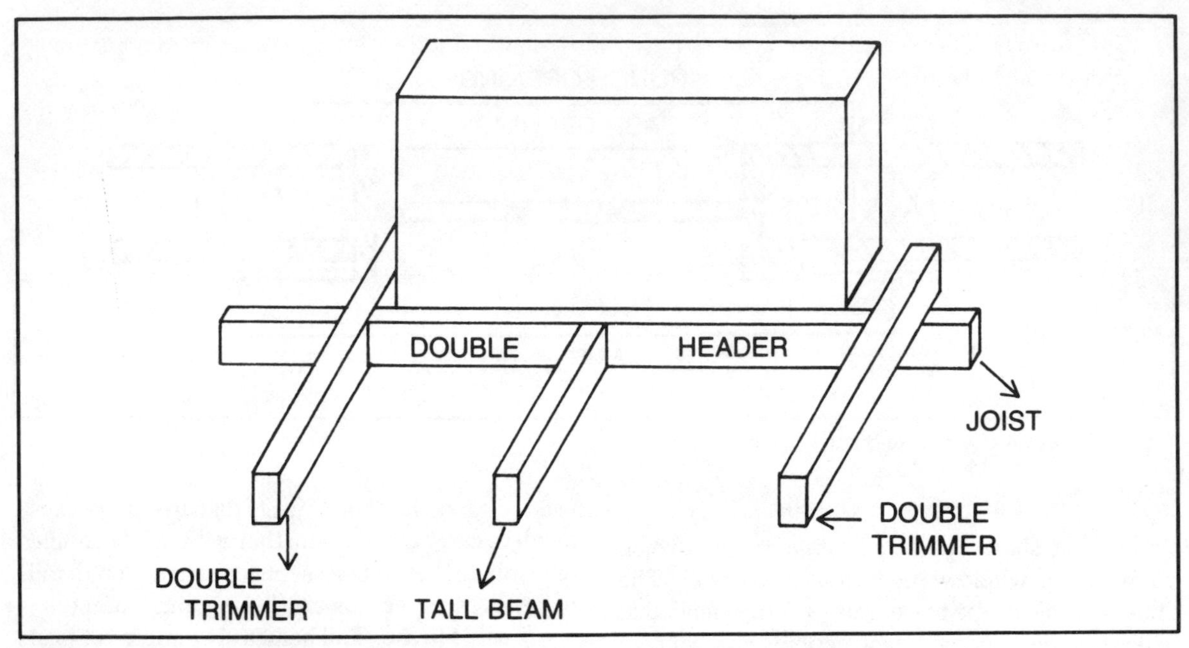

Fig. 5-40. Construction details showing trimmers, joist, and a tail beam.

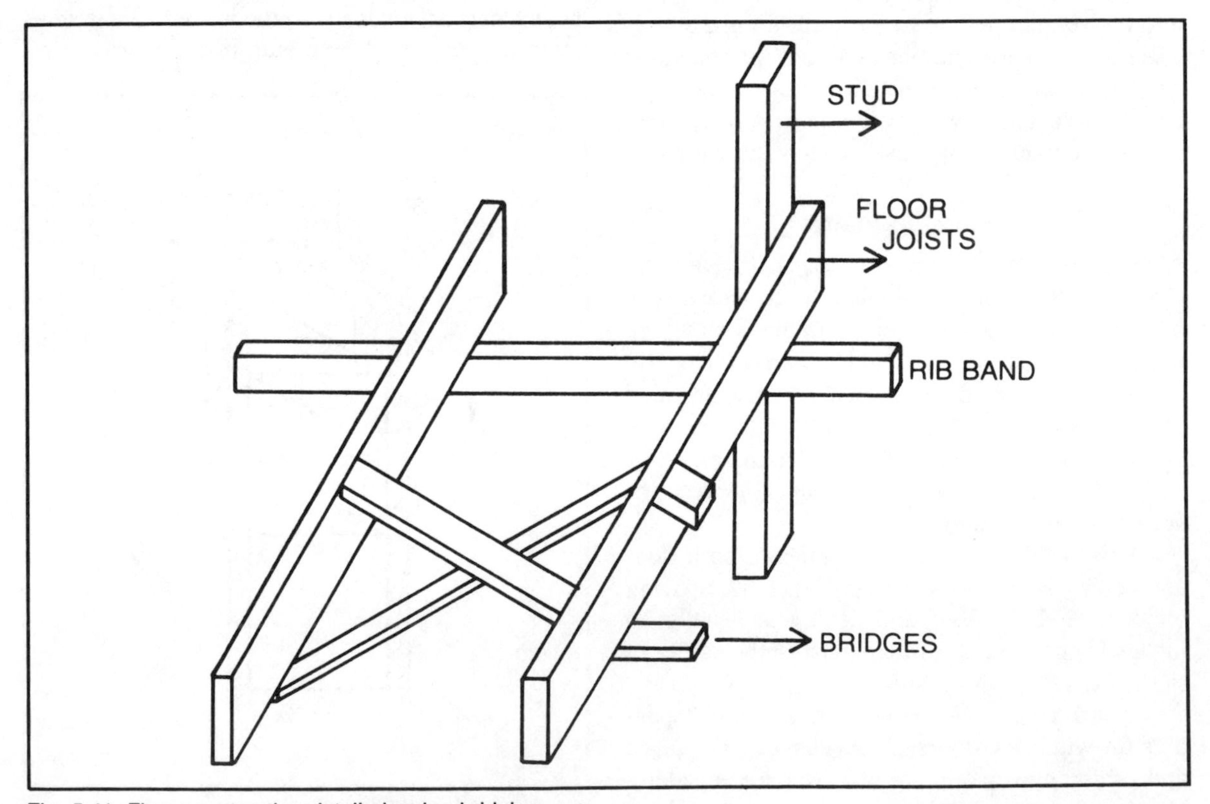

Fig. 5-41. Floor construction detail showing bridging.

Trimmers. These are floor joists that are doubled, run at right angles to the header, and are fastened to it.

Bridging. This is the 1-×-3 inch pieces of wood nailed crosswise between joists, tail beams, and trimmers to keep them perfectly aligned with one another. One row of bridging is needed for every 8 feet of joist span. Good builders will use lots of extra bridging to ensure sturdy, rigid construction. Figure 5-41 shows cross bridging underneath a floor.

Complete floor and roof framing plans are sometimes included in construction plans. This is especially true if the designer wants to deviate from conventional construction methods. Otherwise, these plans can be deleted from sets of plans for remodeling, build-ons, or complete new construction.

Chapter 6

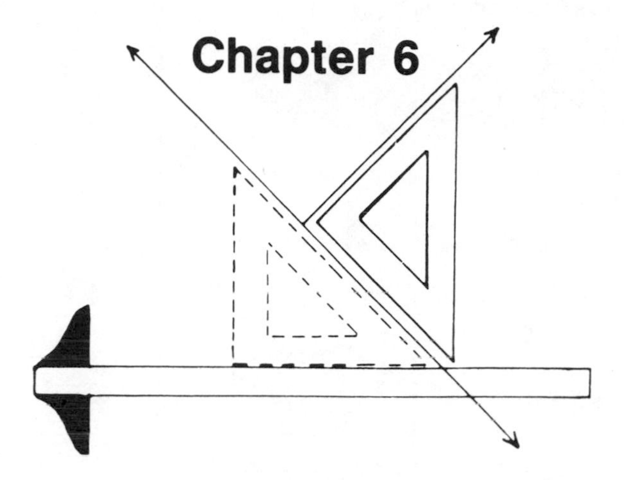

Construction Materials

To draw good plans, you have to know something about construction materials. And it's not just a matter of knowing what's available. It's mostly a matter of understanding how construction materials are used so you can draw plans that work.

STOCK MATERIALS

Plan to use construction materials that are commonly found in your local supply yards. Stock materials usually cost less than those that must be specially ordered.

Some building supply yards can give you a list of available stock as well as catalogs and other information that will be helpful. Mail-order catalogs offer quite a variety of construction materials and are a useful source of information as to sizes and prices.

There are wholesale firms, such as Morgan-Wightman, which issue extensive building material catalogs and have warehouses in many cities. Such books are for established builders and you should expect to pay for such a catalog for they are costly to print and mail. The Morgan-Wightman catalog, "Building Materials," is available free to professional builders and contractors. For all others there is a charge of $3, which is refunded on a first order of $350 or more. Send request for catalog to Morgan-Wightman, Mail Order Dept. H, Main P.O. Box 1, St. Louis, MO 63166.

Using stock sizes helps you eliminate waste and keep costs at a minimum. As far as possible, dimension your plans so you can use *standard sizes* without cutting.

Figure 6-1 shows standard sizes of concrete blocks. An 8 × 8 × 16 inch concrete block, much used in foundations and exterior walls, actually measures 7 5/8 × 7 5/8 × 15 5/8 inches, but when laid in a wall with 3/8 inch mortar joints, it has a face measure of 8 × 16 inches. And when these blocks are stacked 10 high (with 3/8 inch mortar joints), which is the height of a door or the head of a window opening. Because blocks can be had in half lengths (7 5/8 inches), lay out your floor plan in increments of 8 inches.

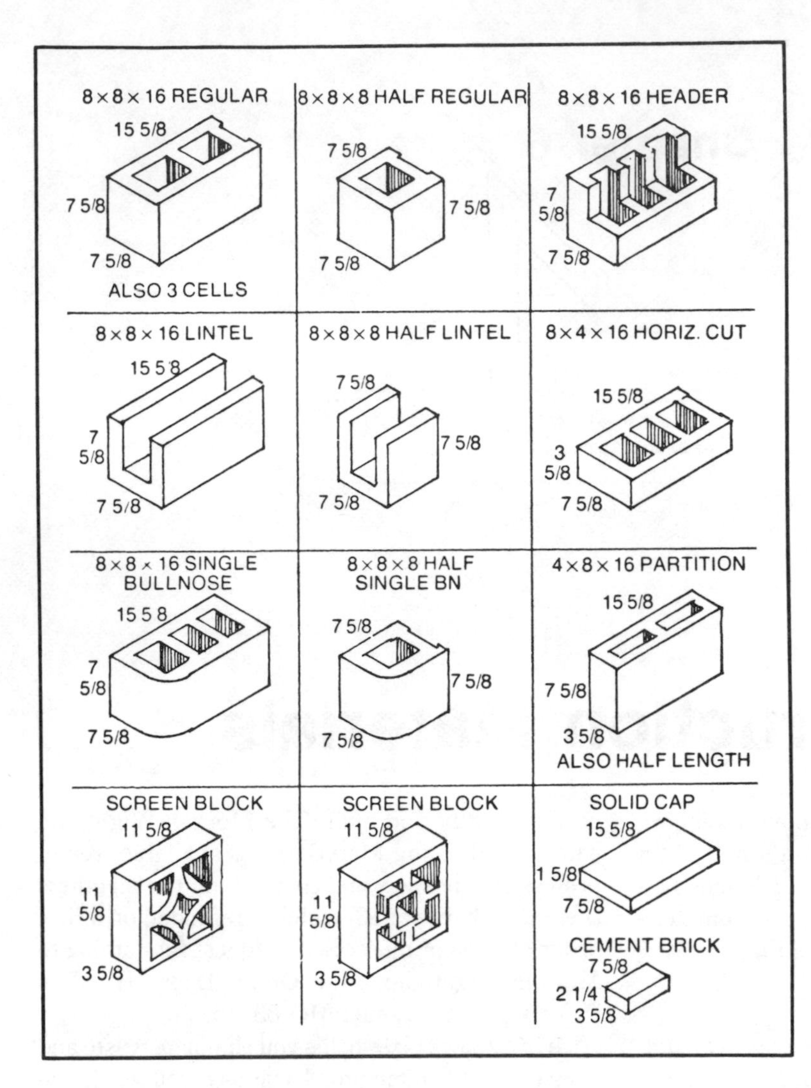

Fig. 6-1. Typical concrete block units.

Framing lumber is well standardized too—in lengths from 8 feet or less to 24 feet. Both rough and finished lumber are named for its presawn size. Thus a 2 × 4 when *kiln dried (DK)* and planed to a smooth finish actually measures 1 1/2 × 3 1/2 inches. More information about wood will be found in later pages.

Most frame walls use 2 × 4s and 2 × 6s for vertical framing (studs), and the studs will be spaced 12 inches, 16 inches, or 24 inches on centers. Such spacing means that 48 inches wide sheathing, sheathing plywood, gypsum dry-wall, and other materials can be nailed to the studs without cutting or fitting.

MODULAR PLANNING

As you become better acquainted with building materials, you will realize that their sizes tend to be multiples of 8 inches, which corresponds to the 8 inch and 16 inch blocks used in the foundation. Thus the closer you can fit the dimensions of your plan, both horizontally and vertically, to 8 inch, 16 inch, and even 4 inch increments, the less waste you will have. This is called *modular planning*.

Keeping your plan to an 8 inch module will be easier when you first start rough freehand sketches if you use the 1/4 inch square cross-ruled pad. Each square can be counted as 2 feet and the plan will be to 1/8 inch = 1 foot scale.

REINFORCING MATERIALS

When the foundation footing trenches of a house have been dug, steel *reinforcing rods* are placed in them and held in position on wire "chairs" about 2 inches above the bottom. These bars come in 20 foot lengths and can be bent around corners. For most house construction, No. 3, 4, and 5 bars are used.

Reinforcing bars are rippled, that is, bumps are rolled into the steel when they are made so they are less likely to pull out of hardened concrete when subject to strain. The No. 5 bar is commonly used in footings and bond beams. Your building code probably tells how many must be used.

Reinforcing materials for slabs are usually welded wire meshes (WWM), sometimes called just *mesh*. Mesh is usually #10 gauge wire crossed into 6 inch squares (6 × 6 10/10); it comes in rolls 60 inches wide and 150 feet long.

CONCRETE

Concrete is a mixture of portland cement, sand, and graded crushed stone or gravel. When properly mixed with water, it can set to stonelike hardness. Its strength varies with the amount of cement in the mix and with the amount of water. A dry mix, properly compacted, gains strength faster than a wet sloppy mix, though a certain degree of wetness is necessary for concrete to be placed in a trench or form.

Concrete blocks are made of the same materials as concrete except that the blocks use a finer aggregate. The blocks are made by pressing a dry concrete into steel forms. They may be cured in open air but are often steam cured in cold weather.

Concrete blocks can also be made of powerhouse cinders, blast furnace slag, or shale rock that has ben heated to lava and expanded by dumping into water. Such blocks are light and have varying degrees of insulating value. For a concrete block house, such insulating block can be worth its additional cost because of the long term savings in heating costs. The insulating value of all types of concrete block can be increased if the cores are filled with granulated material such as Zonolite (an expanded micalike rock) or a chemical foam.

BRICKS

Bricks are of many types and sizes. Some are made of concrete; others of burned clay; others of a sand-lime mixture. Common bricks may vary in strength and ability to withstand weather. Face bricks can be had in different colors and surface finishes and may be used for decorative purposes.

A standard brick is 8 inches long, 2 1/4 inches high, and 3 3/4 inches wide. It is considered non-modular. *Modular bricks* have been designed so that by adjusting the size of mortar joints, they can lie in 8 inch lengths and three stacked bricks (with joints) can be 8 inches high.

One thickness of brick can be veneered to a frame wall with a saving in cost and weight. Such a brick veneer will afford a weather resistant facing that requires very little maintenance (Fig. 6-2). This brickwork requires skilled labor, but its cost may be worth it in the long run.

STONE

Stone can make a pleasing and long lasting exterior finish. It can be costly, but there are many areas where local stone is available at a fairly low price. For veneer, stone can be had in a nominal 4 inch

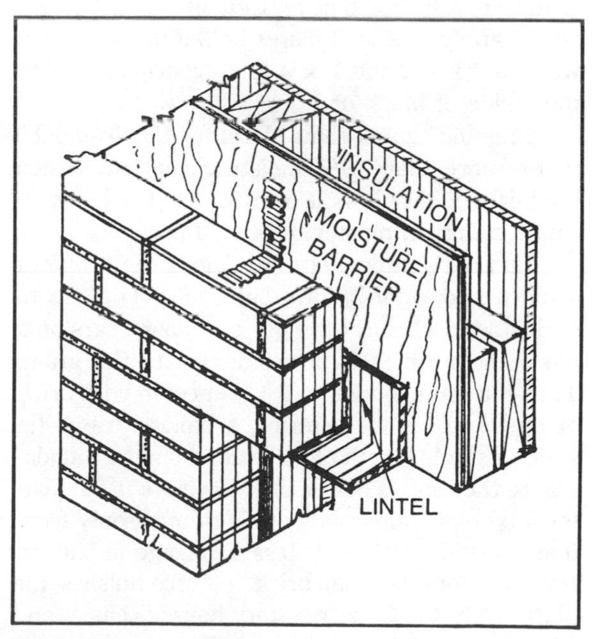

Fig. 6-2. Brick veneer.

thickness. For solid walls, stone may be used in selected natural sizes and shapes, or it can be had in slabs of different thickness that are broken to size for laying in a wall.

LUMBER

Lumber is called *yard lumber* when used for rough building work, and *finish lumber* when its surface is to be left exposed. Boards which are nominally 1 inch are actually 3/4 inch thick and vary in width from 3 inches to 12 inches (actually 1/2 inch less). The strength of lumber used in house framing is important, and it is usually grade marked for its fiber strength in bending. The common grade for house framing is 1200f, the old #2 grade. The stronger and more costly grade is 1450f or 1500f and may sometimes be used for beams and girders because of greater strength. Only when you need lumber stronger than 1200f grade is it necessary to specify it on your drawings.

Lumber is priced at so much per thousand *board feet* (BM for Broad Measure). A board foot is nominally 12 × 12 × 1 inch. Thus a 2 × 6 12 feet long will have 12 BM.

If lumber is to come in contact with concrete or masonry, it should be *pressure treated* (PT). This would apply to 2 × 8 plates bolted to foundation walls and 1 × 2 and 1 × 4 furring strips nailed to the inside of block or brick walls.

Framing lumber should be air or *kiln dried* (KD) to no more than 19% moisture content. Green lumber will shrink as it dries out and is likely to cause cracks in plaster or other finish.

Lumber shrinks very little along the *length* of a piece (along the grain), but even KD tends to shrink a bit when the weight of a floor bears on a flat section which is perpendicular to the grain. Thus if a frame house is to be veneered with brick or stuccoed on the exterior, a balloon frame (in which a single length of stud runs from the foundation to the roof) is better than the more usual platform type of frame (with studs running only from floor to ceiling); there is less shrinkage in balloon framing. For other than brick or stucco finishes, the platform frame for a two-story house is easier and less costly to build than the balloon type.

Both boards and plywood are used to cover rough walls, floors, and roofs. Both board lumber and plywood come in different grades, as does dimension lumber. Boards come with different edges; there are square edges, tongue and groove (T & G) edges, dressed and matched (D & M) edges, shiplap edges, and others.

Sheathing boards made of recycled wood fiber (old newspapers) can be had in 4 by 8 foot sheets and even 10 foot, 12 foot, and 14 foot lengths. It is treated to be moisture resistant and is more mildew- and termite-proof than most species of natural lumber. It can be used directly exposed for both exterior and inside finishes and can be painted or may be had with a plastic surface.

Factor-built wood roof trusses, spaced on 24 inch centers, have come into common use in recent years. They use less lumber than older framing methods and can be installed as soon as the outside walls of a house are up. They can be quickly decked (covered with plywood or boards) and weatherproofed, permitting all inside construction to proceed whatever the weather. Check with your local lumber dealer about availability.

Whether your roof is conventionally framed or is of truss construction, it must be decked with boards or plywood, just as the walls must be sheathed. The roof is weatherproofed by a layer of black asphalt felt held in place by thin strips of wood such as lath or by tin tags and galvanized roofing nails. The finish roofing is placed *after* the exterior wood trim is installed around the edges. This trim should be redwood, cypress, or red cedar; these woods resist rot better than other woods.

ROOFING MATERIALS

For flat or nearly flat roofs, a built-up roofing of several layers of asphalt felt is mopped in hot asphalt onto the original weatherproofing layer. This built-up roofing is finished with a wearing surface of mineral chips set in a flood coat of asphalt. White marble chips make a white reflective finish, which is particularly desirable in warm sunny climates. Such roofing is rated by the number of layers or plies. Properly built 4-ply roofing should last 20 years or more. Even flat roofs should have sufficient slope so rain will drain off; a pitch of 1/8 inch per foot is satisfactory.

Shingles may be used on slopes from 2 1/2 inches per foot and steeper. Wood shingles were once in common use but are often barred by code because of their fire hazard. Asphalt shingles are relatively low cost and commonly come in 26 inch long strips and weights varying from 235 pounds per square (100 square feet) to 300 pounds or more. All asphalt shingles have self-sealing tabs so the wind will not blow them up and break them. They usually have a 5 inch exposure which is of colored mineral granules. The 250 pound shingles are a good buy. Check with your local supply dealer.

Other shingles may be of natural slate, cement or asbestos.

Cement roofing tiles, either the flat shingle type or the barrel type, are common in many southern areas, while burned-clay tiles are traditional in the Southwest. Cement tiles are usually laid in mortar on a layer of 90 pound mineral-coated roll roofing which has been hot mopped onto the original weatherproofing layer.

WINDOWS AND DOORS

After the roof is on and weatherproofed, or "dried in," windows and exterior doors must be installed. Windows may be had in a wide variety of types and sizes, both in wood and metal (usually aluminum). Wood is preferred where temperatures below freezing are common. Double-hung windows are common; they have upper and lower sashes that slide vertically in separate grooves. With single-hung windows, one sash is fixed. Casement windows are hinged at the side. In awning types, the sash is pivoted at the top and swings out to give some weather protection when open (Fig. 6-3). Some windows have horizontally sliding sashes (Fig. 6-4). Your building supply dealer can give you a catalog of most types of windows.

DOORS

Exterior doors are usually 1 3/4 inches thick and have solid cores for greater strength and weather resistance. Hollow-core doors are used inside and are 1 3/8 inches thick. Exterior doors should be 36

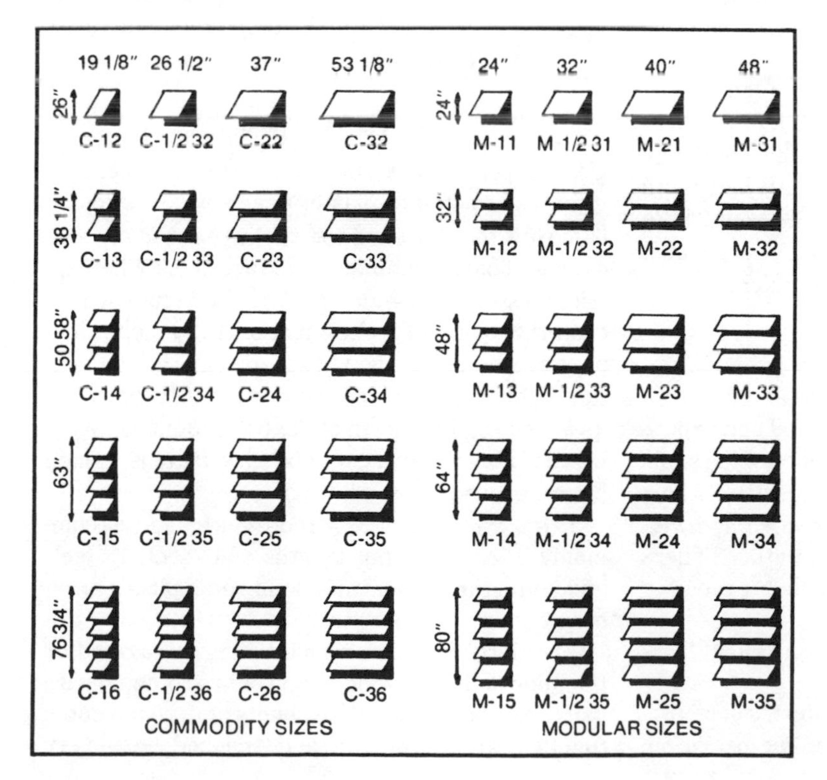

Fig. 6-3. Aluminum awning windows.

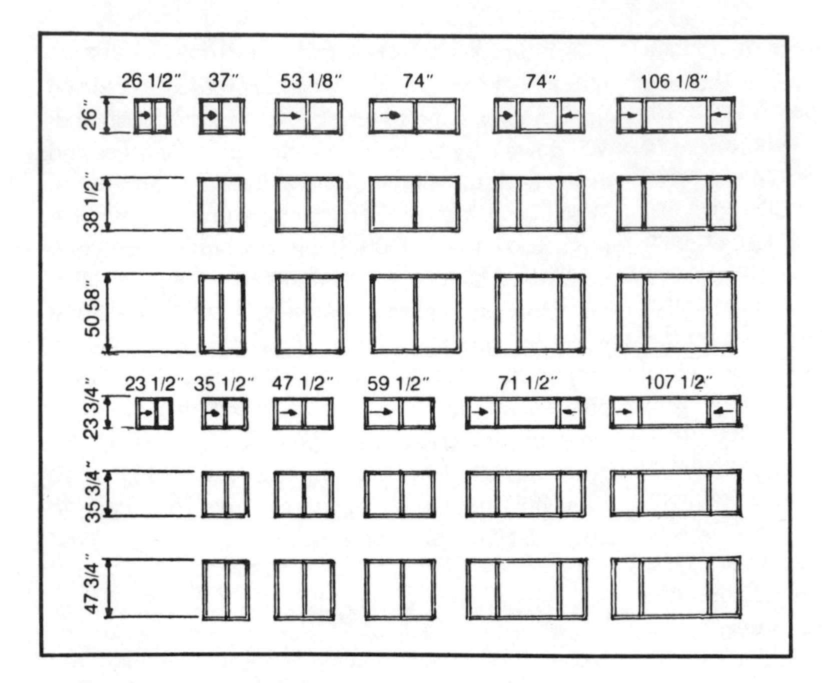

Fig. 6-4. Aluminum sliding windows.

inches wide for bedrooms but can be 28 inches for baths. All house doors are nominally 6 feet 8 inches high; frames and trim fit these sizes. Exterior doors should have aluminum thresholds with suitable weather strips.

Interior closet doors can be the hinged type or the bi-fold type (for openings from 2 feet to 6 feet wide). They are usually 1 1/8 inches thick and may be the flush type, panelled or louvered. Where there is limited space, bypass doors can be used. There are wood or vinyl accordion folding doors which may be used for closets or room dividers.

INSULATION

Exterior frame walls should be insulated with a blanket of mineral wool (wrapped in foil and paper). This blanket is stapled to the inside face of the studs with the moistureproof foil facing *inside*. This prevents the condensation of water vapor and wetting of the insulation during cold weather. There is also an insulation foam for walls, but it requires special handling.

Block walls are also insulated by a kind of blanket. It is stapled to the furring.

Ceilings, where heat loss is greatest, can be protected with double 3 inch thick blanket insulation, or mineral wool can be blown in (to at least 6 inches thick) after the ceiling has been placed. Do-it-yourself ceiling insulation, which comes in large 4 CF sacks, can be installed by hand and is especially valuable when improving an older house.

WALL COVERINGS

Until recent years, interior walls and ceilings usually consisted of plaster over wood or gypsum lath. Now the usual and less costly treatment is with gypsum board, commonly 1/2 inch thick when applied to studs on 24 inch centers (i.e., studs whose centerlines are 24 inches apart). A 3/8 inch thickness can be used when studs are on 16 inch centers, but the small additional cost for 1/2 inch makes its added rigidity worth the extra amount. Gypsum board is also called dry wall because there is no long period of drying as with plaster.

Dry wall comes in 4 foot widths and is commonly 8 feet long, but lengths of 10 feet, 12 feet, and 14 feet may be available in some places or on order.

Dry wall must be applied with corrugated or threaded nails that will not pop out as common wire nails do. The joints where sheets of dry wall come together are depressed and finished with very strong paper tape covered over with cement. This cement is sanded to a smooth seamless surface ready for paint or wallpaper.

If plaster is used 3/8 inch thick sheets of gypsum lath (16 feet × 48 inches) are nailed to the studs. Outside corners are protected with a metal bead; inside corners with metal lath cornerite. Short strips of metal lath are placed diagonally above openings. Wood screed strips are placed at the bottom. All this is done *before* the usual two coats of plaster are applied. When finished, plaster is usually 3/4 inch thick and requires wider door frames than dry wall.

Prefinished wood panelling is often used on inside walls and can be had in many species and colors. Its 4 × 8 foot sheets vary from 1/8 inch thick to 1/4 inch or more. The thin panelling is best applied over 3/8 inch dry wall. The heavier grade is nailed and glued directly to the studs. Matching moldings are available to go with the panelling.

FLOORS

Until recent years, a terrazzo finish over a concrete slab floor was common and low cost. The advent of carpet and easily applied vinyl have taken precedence though terrazzo is still a good choice.

For terrazzo, the concrete slab floor is given a rough broom finish. Then a soupy mixture of white cement (or gray), white marble chips, and water is spread onto the floor and leveled to an even thickness. Colored marble chips are then scattered over the surface and rolled with heavy iron rollers until the cement gets its initial set. Then within a day or so, the floor is ground smooth so that a pattern of white and colored stone is exposed. After the glassy smooth surface is given a liquid seal coat, it is ready for use.

For other than a terrazzo finish, the concrete slab is finished to a smooth surface by either hand or machine trowelling. Sheet or tile vinyl may then be glued to it, or carpet and padding can be used. Thin hardwood tiles can also be glued to the surface for a wood parquet floor.

For floors in frame construction, the initial rough floor of plywood or boards is a base for tile or carpet. Or thicker dressed and matched (D & M) hardwood may be blind-nailed to the base material.

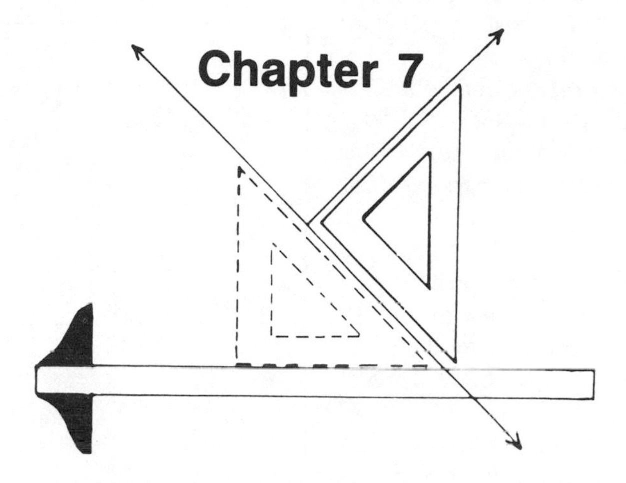

The Money Factor

For the amateur planner, planning a house means taking a hard, cold look at building costs. Unless you're independently wealthy, building costs will influence every planning decision you make.

GETTING THE MONEY

Home loan agencies have a rule of thumb: A family can afford a home costing from 2 1/2 to 3 times their annual income. This is a good basis to start from.

If you already own or are buying a site, you know that part of the cost. Most people must borrow some part of the building money. The value you have in a lot can often be part or all of the down payment of your building loan.

In granting a building loan, loaning agencies consider other factors besides income. If a wife is working, is her income likely to be interrupted by the birth of a child? What are the job prospects for the principal earner? For instance, a construction worker though well paid when working is likely to face layoffs in bad weather. The bank will weigh such factors. If you are a veteran you may get special consideration and a guaranteed loan. How well do you pay your bills? Do you have adequate life insurance? Do you drive a Cadillac on a VW income? If you are already buying a lot and have a record of steady payments, that's something in your favor.

COST ESTIMATES

For planning purposes, rough cost estimates can be based on the total floor area of the house. If you know the total building cost of a house that's similar to the one you want to build, you can divide the total cost by the total (adjusted) floor area to get the *unit cost*, that is, the cost per square foot. This will give you a general idea of your own unit cost for your own proposed house.

Check the advertisements for new homes that show pictures and floor plans. Often the plans are dimensioned, or the floor areas are given. If prices are also given you can figure unit cost. Sometimes you can determine the approximate area by using the 5 foot bathtub (shown on most plans) as a unit to measure and calculate the total floor area.

Because different parts of a house vary in cost, you must adjust the estimated floor areas. For living rooms, kitchens, dining rooms, and bedrooms use 100% °s ÷ mn$ of their total floor area. Use 50% °s ÷ mn$ of the estimated open porch floor area. The sum of these is the total *adjusted floor area*.

Look for a house that is similar to the one you hope to build. If it is advertised for a certain price built on your lot, deduct 10% °s ÷ mn$ for the advertiser's profit and selling cost and divide the balance by the adjusted floor area to obtain the cost per square foot.

For example, let's say you want to build a single-story house like the one shown in Fig. 7-1. Let's also assume that an advertiser claims that such a house can be built on your lot for $36,500. To figure unit cost, your computations should be as follows:

House priced at $36,500 on your lot

1305 sq. feet.	(living area) at 100%	= 1305 sq. feet.
578 sq. feet.	(garage) at 50%	289 sq. feet.
378 sq. feet.	(open porches) at 40%	= 151 sq feet.
2261 sq. feet.	(total estimated floor area)	1745 sq. feet. (total adjusted floor area)

$36,500	(price)
− $ 3,650	(10% of price for profit and costs)
$32,850	(probable cost)

$32,850 (probable cost) / 1745 sq. feet. (total adj. floor area) = $18.82 per sq. feet. (unit cost)

Applying this unit cost to your own plan will give you an educated guess to use in determining what size house you can afford. The larger the number of advertised houses you price, the more accurate the estimated unit cost will be.

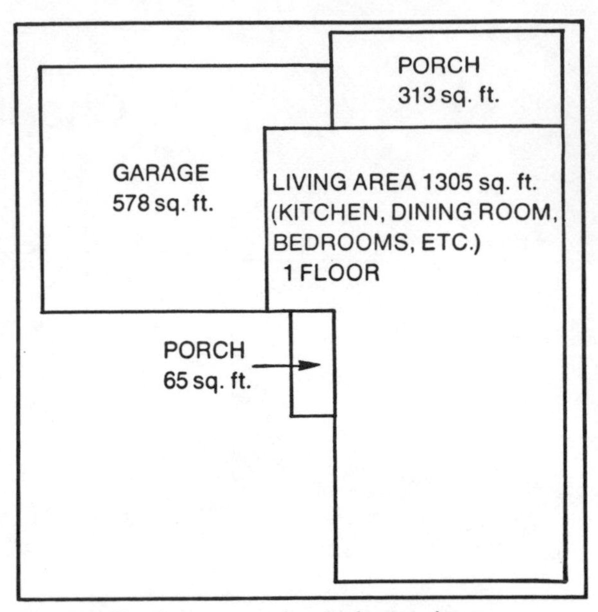

Fig. 7-1. The floor areas of a single-story house.

If the price of the advertised house includes the lot, deduct 1/3 as cost of land and profit and use the remaining 2/3 as the price on which to figure probable unit costs.

If you are planning a two-story house, with or without a basement, a *cubic foot unit cost* will be more accurate than a square foot unit cost but requires more figuring. Figure 7-2 shows the dimen-

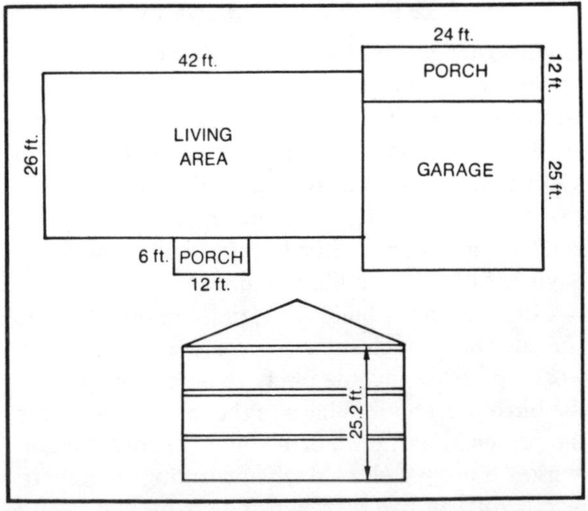

Fig. 7-2. The dimensions of a two-story house with a basement.

sions of a two-story house with a basement. The following computations would be necessary to determine the cubic foot unit cost for such a house:

House priced at $72,500 (including lot)

42 feet × 26 feet	= 1092 sq. feet	× 25.2 feet	=	27,518 cu. feet	(volume of living area)
12 feet × 24 feet	= 288 sq. feet	× 9 feet	=	2,592 cu. feet	(volume of porch)
24 feet × 25 feet	= 600 sq. feet	× 9.5 feet	=	5,700 cu. feet	(volume of garage)
6 feet × 12 feet	= 72 sq. feet	× 9 feet	=	648 cu. feet	(volume of porch)
	2052 sq. feet			36,458 cu. feet	

27,518 cu. feet	at 100%	=	27,518 cu. feet
2,592 cu. feet	at 40%	=	1,037 cu. feet
5,700 cu. feet	at 50%	=	2,850 cu. feet
648 cu. feet	at 40%	=	259 cu. feet
			31,664 cu. feet (total adjusted volume)

$72,500 (price including lot)
− $24,167 (1/3 of price for land cost and profit)
$48,333 (probable cost of house)

$$\frac{\$48,333 \text{ (probable cost of house)}}{31,664 \text{ cu. feet (total adjusted volume)}} = \$1.53 \text{ per cu. foot (unit cost)}$$

For most purposes the square foot method will give useful unit cost figures, though a second floor will probably cost about 10% less than a lower floor and basement.

When the working drawings of your house are completed, you might ask your loaning agency to appraise the house, for they have elaborate charts and tables by which they can calculate your total building costs.

Sometimes a builder will tell you his unit cost for the type of house you want. Often the local building supply yard can give you such information.

From your working drawings, it is possible to compile a *takeoff*, or materials list. Using the takeoff you can estimate the cost of all materials to be used in the building of your house. Then you can add labor costs for their installation. Your building supply dealer will price the materials. And the electrical, plumbing, and heating contractors can quote their prices. The total is the probable cost of your house.

SUBCONTRACTING

Many building contractors, both large and small, do not actually work on the site. They subcontract all the different jobs, add their profit, and sign a contract for the complete building. They may buy the materials and contract the labor or they may contract for both labor and materials. Thus a mason starts the job, then the carpenter builds the frame and installs windows, doors, trim, etc. Dry wall or plastering is another contract. Roofing another. Plumbers, electricians and heating men perform their part, and finally comes the painter and decorator with a final contract for cleanup.

An owner can act as his own general contractor and get a good job by subcontracting all the work—if he has the time, knowledge, and drive to see that each job is properly done in coordination with all other jobs. The advantage is that you can save the 10% or 15% charged by a general contractor. Subcontracting is well suited to an owner who can do some parts of the work himself and be on the job when other work is proceeding.

Chapter 8

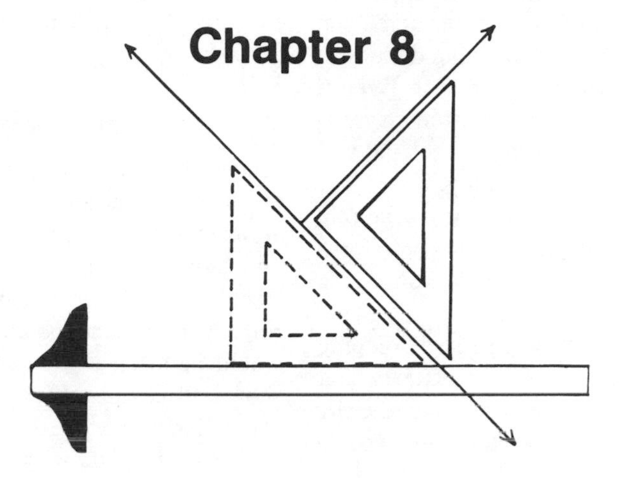

House Design

Driving along a street in a newly developed neighborhood, you may see a house that stops you for a longer look. Usually it is no single thing that catches your eye but a combination of things: perhaps an inviting entrance, the symmetry of features, the coloring, or the feeling that the house *belongs* in that landscape. Such design elements are the result of careful thought, of an awareness of desired effects.

STYLES

Styles in houses vary with the times and the region. Early New Englanders, after the first scramble for the shelter of a log cabin, were able to make use of the abundant timber of their area to copy, to some extent, the brick and stone houses of their homeland. We see the results in the white clapboard homes and attached outbuildings in many small New England villages.

The Cape Cod saltbox style must have evolved from the weather-beaten fisherman's shanty.

The Victorian age brought the turreted house with its gingerbread scrollwork—a style that was an imitation of the stone castles that were the homes of the feudal lords.

At the turn of the century, Frank Lloyd Wright influenced housing design with the wide overhangs and long windows of his prairie houses.

The Georgian styles and the Greek Revival gave us Jefferson's Monticello and the pillared plantation houses of the South. These styles still have great influence in many of today's finer homes, both outside and in.

The Spanish in the Southwest brought the Mediterranean style of plaster walls, small windows, and tile roofs.

The Japanese influence, with its simplicity of natural exposed wood, has resulted in what can be called the California style.

DISCOVERING THE RIGHT DESIGN

How can an amateur planner synthesize all such precedents and gain an instinctive sense of rightness that can be expressed in a beautiful home?

Build yourself a library of books and clippings.

The home betterment magazines are filled with beautiful advertisements asking you to copy them. If you have some artistic ability, carry a sketch pad and soft pencil. When you see a building detail you like, take the time to make a sketch of it. Even if you have never tried sketching before, you may surprise yourself. Just trying to draw it freehand will tend to fix the detail in your memory.

A $25 Polaroid camera can be a great help in compiling a library file of pleasing houses, for a single color shot can capture a whole house or a pleasing detail.

When working on the drawing board, keeping the windows and other features symmetrical about a center entrance is good. However, a necessarily off center entrance may often be balanced by some other feature.

Fit your plan to the conditions of your location. In the sunny South, wide eaves shadowing the windows against the summer sun are common. Open the house to the prevailing hot weather breeze with low interior partitions that permit a sweep of air through the house.

You will be happier if you don't fight the neighborhood. Who wants a dome house in a neighborhood full of conventional homes? Besides, you might not be able to get financing. Banks want to grant a loan on a house that can be sold readily in case you fail to keep up your payments.

Undoubtedly, the ever increasing costs of energy will affect the plans of homes in the future. A recent Florida law makes it mandatory to provide connections for solar water heaters. And it is probable that solar house heating will affect the appearance of many houses in the future—an entire roof may be part of a heating system.

For the amateur planner, it will probably be better to imitate designs that have endured through a considerable period of time. Many homes built 200 years ago remain in good taste today.

SECTIONS AND ELEVATIONS

Section and elevation drawings are necessary to show contractors and workmen how all the structural and architectural parts of a building relate to one another, and to show the expected finished appearance of relating those parts. Generally, sections show relationships and elevations show appearances. Elevations are usually drawn by projecting sections.

CROSS SECTIONS

Section is architectural jargon for *cross section*. A cross section is any view you get when you cut across a structure and look at the "inside" of it. If you have ever looked down on the stump of a freshly cut tree and counted the annual growth rings, you have looked at a section of the tree. That would be a horizontal section because the cut to show it would have been made horizontally. A vertical section of the tree would have a cut up and down the tree and show a different version of its structure.

Technically, a floor plan is a section drawing. It is what you would see if you cut horizontally across the bottom of a house just a few inches above floor level. The difference is that a horizontal section drawing will show, in addition to the floor plan, details of materials used in walls and foundations—structural details. For this reason, it is usually referred to as a *structural section*.

Drawings are made of structural sections that will best illustrate the relationships of all parts of the structure and can be taken through the structure horizontally or vertically. Often small sections are drawn with "cuts" made at any angle. Whatever best shows the desired construction details of stairways, fireplaces, doors, windows, etc., is appropriate.

Structural sections must show all dimensions and such important details as plate height and roof pitch. And they must be drawn to the same scale as the rest of your house plans. It is acceptable, and sometimes necessary, to draw small details on the same sheet as a section to a larger scale, but be sure to note the scale of the detail if it is different from the main section.

Sections should also show points of connection between structural parts and notations about the size and material of structural members. Usually one or two sections is all that is required for plans for a simple addition to a house. Three or four will suffice for a small residence of conventional construction, but anytime there is a change in shape or method of construction a new sectional drawing will be necessary.

Most structural sections are drawn on separate sheets of paper. If your project is not complicated you can combine them with other drawings. Do try to use a fairly large scale so that details can be shown clearly without confusion; 1/2 inch = 1 foot is a good scale to use.

Instructions in structural systems, represented by sectional drawings, should be applicable anywhere in spite of regional differences in materials and construction practices. For this reason, these construction instructions are usually broken down into five parts corresponding to the way construction usually proceeds:

—Foundation.
—Floor.
—Wall.
—Ceiling.
—Roof system.

You can usually specify any combination of materials you want for any one of the five categories, but you would not usually mix materials within a category. In other words, while you can use masonry materials for your foundation and wood for your floors, you might be in serious trouble trying to use both masonry and wood in your foundation.

There are certain obvious combinations you would not ordinarily use, such as a masonry roof over wood walls. To draw really good, workable section plans, you need to go out and look at the materials and construction practices in your locality and stick to making plans around what is most common in your particular locality. The following information will be helpful to you in your design work and plan drawing wherever you live. It will also help you to recognize what is going on when you visit home construction sites near your home.

DESIGN CONSIDERATIONS

Footings. The purpose of the footing of any building is the same as the purpose of your own feet: to distribute the weight of the building over the ground. The size of the footing varies with the weight of the building and the kind of soil that must support it. In areas where severe winter weather sends frost deep into the ground, footings must go quite deep. The rule is to always go below the frost line. In mild climates, where there is little or no frost in winter, footings can set right on top of the soil.

Footings and foundations are usually made from concrete and masonry. Those are often the *only* materials approved by building codes in most areas. Building codes usually specify the use of reinforcing steel in footings and foundations. We recommend that you show reinforcing rods in the footings of any structure you plan.

Foundation Walls. Foundation walls carry the load of the building to the footings. They also isolate wood construction from water, rot, and insects. The minimum height of a foundation wall should be 6 inches above grade. There is no maximum. The thickness of the wall depends on the weight of the structure, but should be at least 4 inches if it is not specified in your building code. The use of reinforcing steel in the foundation walls will increase strength and make less massive walls possible. You will need to look at your building code and observe building practices in your neighborhood to determine the size and spacing of steel to be used in your foundation walls.

Floors of Concrete. Concrete floors that are on the ground or below grade can be cast as part of the foundation wall or footing or they can be poured separately. The usual thickness is 3 1/2 inches and it is wise to indicate welded wire fabric in your concrete floors to prevent cracking and to add strength.

Waterproofing. Any part of the foundation, footing, or floors that comes in contact with the ground must be protected from water. You will always draw your waterproof membrane on the earth side of any structure or part of a structure that goes below grade. This membrane can be a chemical solution painted on or can be a plastic film. Specify whatever is common and found to be an effective barrier in your locality. Draw in a layer of gravel or crushed rock below any concrete floors you plan (to keep water from being drawn up to it by capillary attraction in the soil).

As you observe construction in your area, see if builders are fighting excessive groundwater or poor soil drainage. If drainage is a problem, be sure to design drain tiles into your footings to conduct water away from the load bearing bottom surface of the footing. Figure 8-1 shows a section with forces acting on the footings and drain tiles to keep the ground from getting too soft.

Floors. Floors are designed not only to carry their own weight, but to transmit live loads (people walking around) and dead loads (the weight of the building) to the footings. The simplest kind of floor to design (and build) is the common wood-joist floor. Table 8-1 gives the size and spacing of joists for any span and load. The size and spacing of joists is a factor of the load and the span between supports and the type of wood being used. No fancy floor framing plan need be drawn for this type of floor because carpenters understand this type of construction and could probably build one in the dark! Figure 8-2 shows a section of a wood-joist floor.

A plank-and-beam floor is not uncommon. It differs from the joist floor in that the supporting

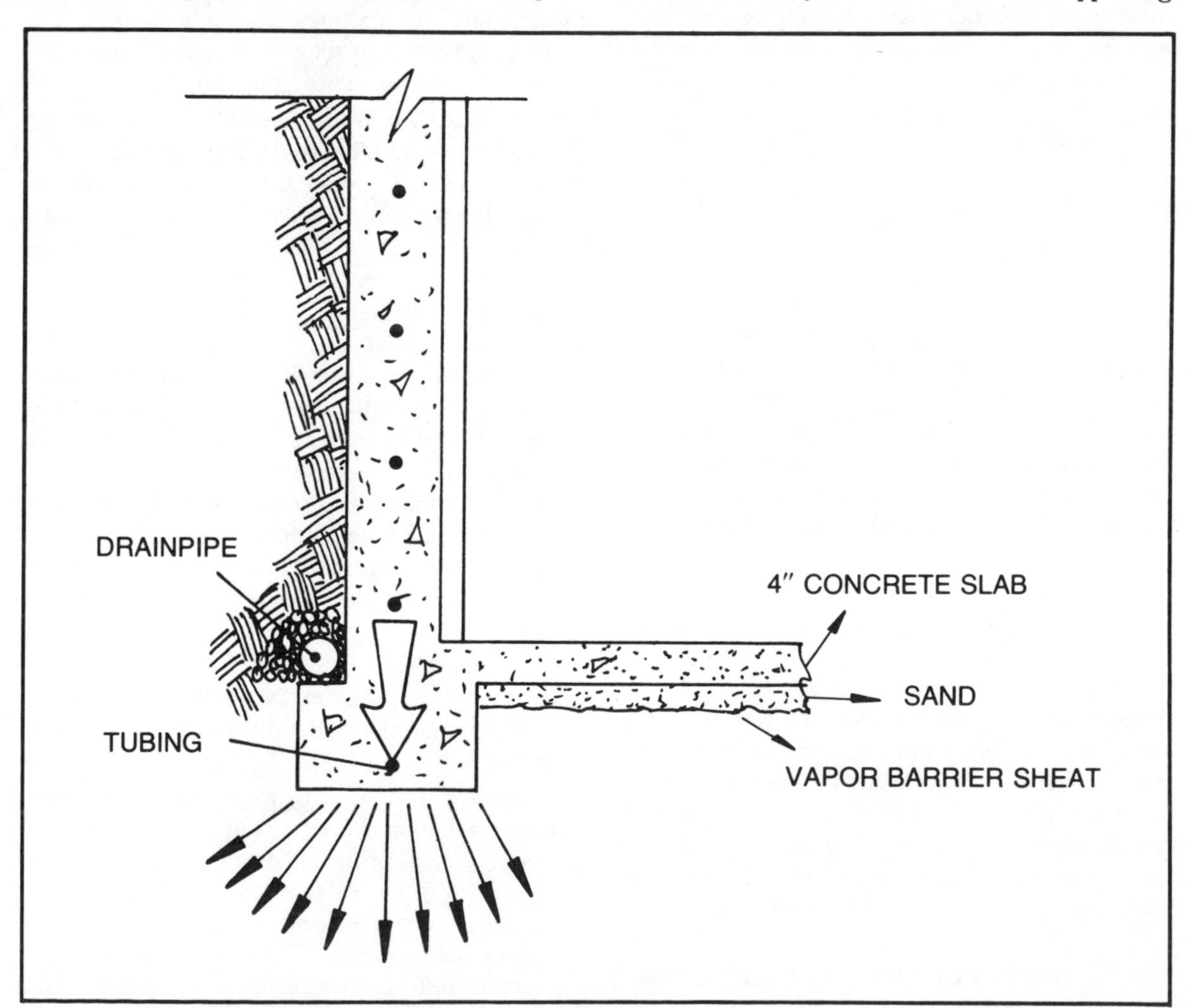

Fig. 8-1. A section with lines showing forces acting on the footing.

Table 8-1. Sizes and Spacing for Floor Joists.

	Grade	2×6		2×8		2×10		2×12		Size
		12	16	12	16	12	16	12	16	Span
Ponderosa Pine	#1	10-9	9-0	14-2	12-10	18-0	16-5	21-11	19-11	
	#2	10-5	9-6	13-9	12-6	17-6	15-11	21-4	19-4	
	#3	8-6	7-4	11-3	9-9	14-4	12-5	17-5	15-11	
Red Pine	#1	11-0	10-0	14-6	13-2	18-6	16-10	22-6	20-6	
	#2	10-9	9-6	14-2	12-6	18-0	15-11	21-11	19-5	
	#3	8-4	7-3	11-0	9-6	14-0	12-2	17-0	14-9	
Western Pine	#1	11-3	10-3	14-11	13-6	19-0	17-3	23-1	21-0	
	#2	10-10	9-4	14-3	12-4	18-2	15-8	22-1	19-2	
	#3	8-4	7-3	11-0	9-6	14-0	12-2	17-0	14-9	
Spruce/pine/fir	#1	11-7	10-6	15-3	13-10	19-5	17-8	23-7	21-6	
	#2	11-0	9-9	14-6	12-10	18-6	16-4	22-6	19-11	
	#3	8-6	7-4	11-3	9-9	14-4	12-5	17-5	15-1	
Douglas Fir	#1	12-3	11-2	16-2	14-8	20-8	18-9	25-1	22-10	
	#2	12-0	10-11	15-10	14-5	20-3	18-5	24-8	22-5	
	#3	10-4	9-6	13-8	11-10	17-5	15-1	21-2	18-4	
Western Cedar	#1	10-5	9-6	13-9	12-6	17-6	15-11	21-4	19-4	
	#2	10-1	9-2	13-4	12-1	17-0	15-5	20-8	18-9	
	#3	8-8	7-6	11-6	9-11	14-8	12-8	17-9	15-5	
Hem-Fir	#1	11-7	10-6	15-3	13-10	19-5	17-8	23-7	21-6	
	#2	11-3	10-3	14-11	13-6	19-0	17-3	23-1	21-0	
	#3	9-3	8-0	12-2	10-6	15-6	13-5	18-10	16-4	
Eastern Hemlock	#1	11-0	10-0	14-6	13-2	18-6	16-10	22-6	20-6	
	#2	10-5	9-6	13-9	12-6	17-6	15-11	21-4	19-4	
	#3	9-7	8-3	12-7	10-11	16-1	13-11	19-7	16-11	
Coast Silka Spruce	#1	12-0	10-10	15-10	14-4	20-3	18-3	24-8	22-3	
	#2	11-6	10-0	15-2	13-2	19-4	16-9	23-6	20-5	
	#3	8-8	7-6	11-6	9-11	14-8	12-8	17-9	15-5	

Floor joists 30 PSE Live Load

beams are widely spaced and heavy flooring material such as tongue-and-groove flooring is used to span the wide spacing. Figure 8-3 shows a section of plank-and-beam floor that would explain what you have in mind to any carpenter.

Walls. Walls support roof loads as well as serve as room dividers and separators in conventionally built houses. Not all interior walls are roof supporters (so called load bearing walls), but all exterior walls are. Figures 8-4 and 8-5 show a section of a conventional 2-×-4 inch wall. Its components are a bottom plate, studs, and a double top plate. Fire blocking 2 × 4s are placed between studs as intermediate bracing. The double plates should overlap at all joints and corners. This gives good support to those roof rafters that fall between studs and in general makes a strong ring around the top of the wall. Standard spacing of studs is 16 inches on center. Walls are finished by nailing or gluing finish materials, such as paneling, to the studs.

In post-and-beam wall construction, the bottom plate is the same, but 4 × 4s are (in place of studs) spaced as much as 48 inches apart. The top plate is one beam, 4 × 4, 4 × 6, or 4 × 8; it depends on the dead weight above and the span of large openings.

Some home designers prefer post-and-beam

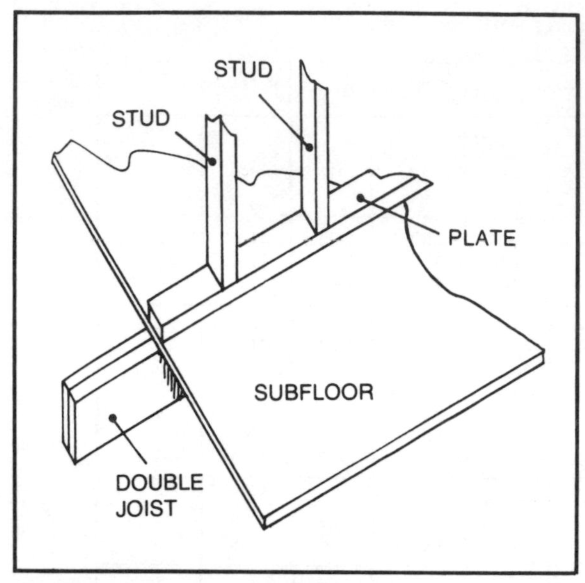

Fig. 8-2. A section of a wood-joist floor.

construction because it saves on materials and labor and results in a building that has a neat modular appearance. Nonbearing walls need not be constructed to carry more than their own weight so it is permissible to use a single 2 × 4 top plate and place studs 24, 36, or even 48 inches on center. Do not design in light nonbearing walls where noise between rooms is an important factor.

Ceiling/Roof Structures and Openings. All openings such as windows and doors must be designed with lintels above them to carry the load of the weight above them. You will need to consult your local building code for the allowed material and sizes.

The ceiling and roof structure above the walls ties together walls as well as providing shelter and support. A simple span system with rafters and 1-inch sheathing to form a gable or hip roof is the simplest to design and it is understood by all carpenters. For open beam ceilings, rafters and 2-inch tongue-and-groove decking are quite satisfactory. This type of roof will span up to 24 feet with no engineering problems. Tables 8-2 and 8-3 will help you figure out the sizes of roof parts.

If you go out to observe the construction of large tracts of residences, you might find wholesale use of trussed systems for roofs. Trusses are made and put in place by special equipment in order to speed up construction. We do not recommend that you design trusses into your house plans unless you plan to build homes by the dozens.

Single pitch (shed or flat) roofs are easy to design and construct, but they tend to leak badly after a short time. The same goes for butterfly roofs. We do not recommend that you consider these types in your designs unless it almost never rains where you will be building. Your roof needs a good slope to it and it should be covered with metal or one of the many overlapping materials such as shingles and shakes, slate, or tiles. Consult your local building code for degree of slope required for these various surface materials.

DRAWING THE STRUCTURAL SECTION

Accuracy and scale are the most important considerations when it comes to drawing structural sections. The sizes, clearances, and spacings of

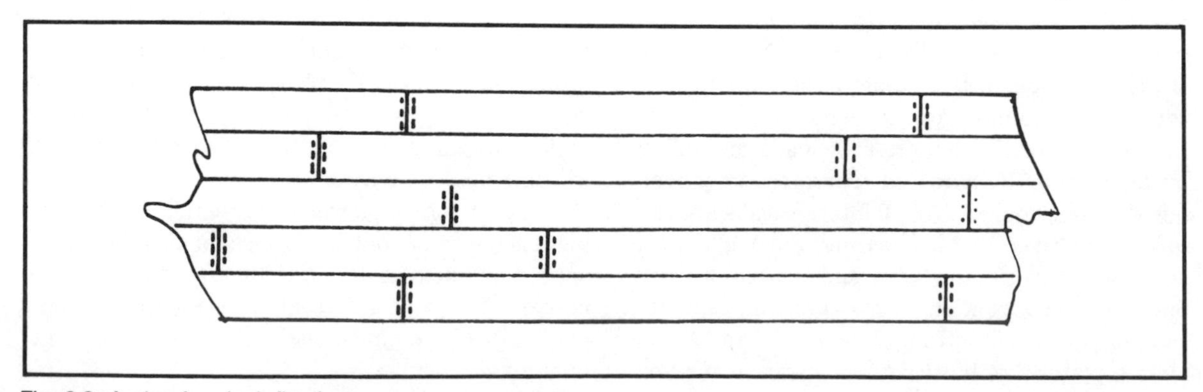
Fig. 8-3. A plan for plank flooring.

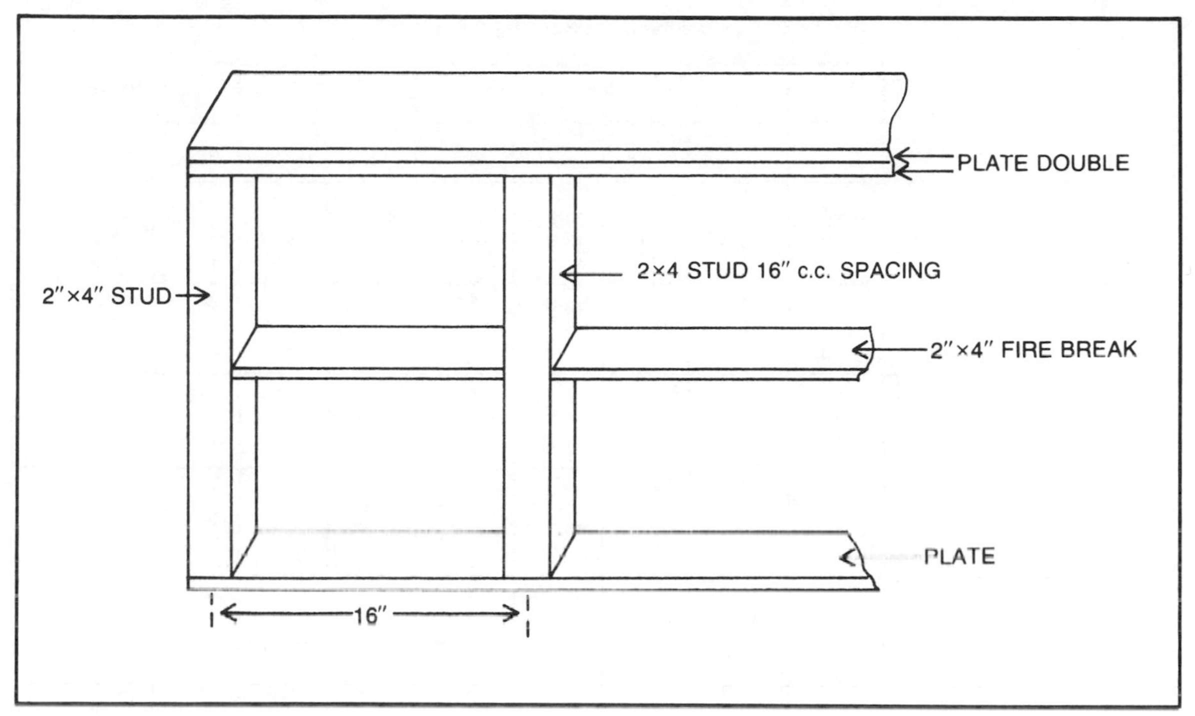

Fig. 8-4. A section of a wooden stud wall.

materials and appliances are furnished by manufacturers. They are available from catalogues and standard architectural reference books such as Sweet's Catalogues.

Rough lumber can be drawn to exactly the sizes given. *Finished lumber* must be drawn smaller than given. The actual sizes vary from one area to another, but they usually run 1/2 to 3/8 inch less than sizes given or quoted. A 2 × 4, for example, is only a 2 × 4 when it is rough lumber. Once it is milled it comes down to 1 1/2 by 3 1/2 inches. *Concrete blocks* and *bricks* all vary in size and nominal size according to locality. *Plywood* will always be the same size as its nominal size. Check around your own locality and make yourself some tables of the actual sizes of construction materials so that you can draw them to scale accurately.

The actual drawing of a section is not at all difficult if you follow these steps.

☐ Decide on the type of foundation, wall, floor, ceiling, and roof you will want to use in the final construction.

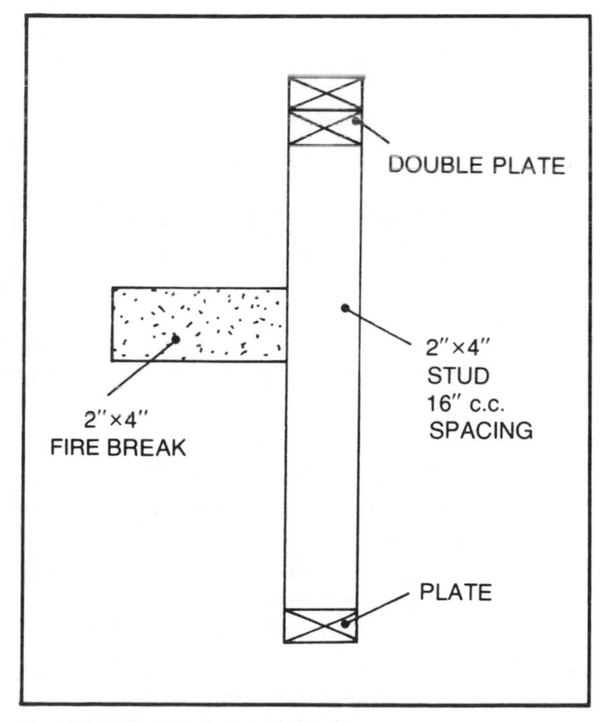

Fig. 8-5. Wooden stud wall detail.

Table 8-2. Ceiling Joist Spacing and Sizes.

	Grade	2×6		2×8		2×10		2×12	
		12" c.c.	16 c.c.	12 c.c.	16 c.c.	12 c.c.	16 c.c,	12 c.c.	16 c.c.
Ponderosa Pine	#1 #2 #3	10-9 10-5 8-6	9-9 9-6 7-4	14-2 13-9 11-3	12-10 12-6 9-9	18-0 17-6 14-4	16-5 15-11 12-5	21-11 21-4 17-5	19-11 19-4 15-1
Red Pine	#1 #2 #3	11-0 10-9 8-4	10-0 9-6 7-3	14-6 14-2 11-0	13-2 12-6 9-6	18-6 18-0 14-0	16-10 15-11 12-2	22-6 21-11 17-0	20-6 19-5 14-9
Western pine	#1 #2 #3	11-3 10-10 8-4	10-3 9-4 7-3	14-11 14-3 11-0	13-6 12-4 9-6	19-0 18-2 14-0	17-3 15-9 12-2	23-1 22-1 17-0	21-0 19-2 14-9
Spruce-pine-fir	#1 #2 #3	11-7 11-0 8-6	10-6 9-9 7-4	15-3 14-6 11-3	13-10 12-10 9-9	19-5 18-6 14-4	17-8 16-4 12-5	23-7 22-6 17-5	21-6 19-11 15-1
Douglas Fir	#1 #2 #3	12-3 12-0 10-4	11-2 10-11 9-0	16-2 15-10 13-8	14-8 14-5 11-10	20-8 20-3 17-5	18-9 18-5 15-1	25-1 24-8 21-2	22-10 22-5 18-4
Hem-Fir	#1 #2 #3	11-7 11-3 9-3	10-6 10-3 8-0	15-3 14-11 12-2	13-10 13-6 10-6	19-5 19-0 15-6	17-8 17-3 13-5	23-7 23-1 18-10	21-6 21-0 16-4
Western Cedar	#1 #2 #3	10-5 10-1 8-8	9-6 9-2 7-6	13-9 13-4 11-6	12-6 12-1 9-11	17-6 17-0 14-8	15-11 15-5 12-8	21-4 20-8 17-9	19-4 18-9 15-5
Eastern Hemlock	#1 #2 #3	11-0 10-5 9-7	10-0 9-6 8-3	14-6 13-9 12-7	13-2 12-6 10-11	18-6 17-6 15-1	16-10 15-11 13-11	22-6 21-4 19-7	20-6 19-4 16-11
Coast Silka Spruce	#1 #2 #3	12-0 11-6 8-8	10-10 10-0 7-6	15-10 15-2 11-6	14-4 13-2 9-11	20-3 19-4 14-8	18-3 16-9 12-8	24-8 23-6 17-9	22-3 20-5 15-5

30 PSE Live Load

☐ Be sure you have or can compute the actual sizes of all structural members to be used in construction.

☐ Establish the grade line. Measuring to scale, draw in a line for the bottom of the footing, top of the foundation wall, finish-floor level, and top plate (see Fig. 8-6).

☐ Establish widths and thicknesses. The total width of your section comes from your floor plan, of course, and the width of the footing and foundation walls, floor joists, wall, roof (and its pitch) all come from Tables or the building code.

☐ Darken all lines representing outlines of structural members. Then draw in all structural connections and all sheathing and finish materials.

☐ Print in all dimensions of structural parts.

☐ Write or print in any necessary notes.

☐ Draw any needed large scale details of structural connections.

Figure 8-6 illustrates all of the above instructions.

PROJECTING SECTIONS FOR DRAWING ELEVATIONS

Although it is entirely possible to make your elevation drawings by a process of transferring measurements from your floor plan and your vertical sections, you can make a more accurate and satisfactory elevation drawing (as a beginner, anyway) by what is called the *direct projection method*. This method avoids the many errors that tend to creep into your drawings when you transfer measurements. It also gives you a better appreciation of the close relationship of the three basic sets of drawings you need to translate your ideas into construction plans: floor plan, section, and elevation drawings.

Your floor plan is really a horizontal section. The structural section you have just learned how to do is a vertical section. By projecting these two sections onto a third drawing, an elevation emerges that will show exactly what the appearance of our section designs will be. Here is how to do your elevation drawings, step by step.

☐ Make sure all drawings (floor plan and sections) are to the same scale.

☐ Place your floor plan directly above the space on your drawing paper where your elevation drawing will appear. One outside wall (the one you want an elevation of) should be facing down. Do not project upward or you will get a mirror image of the wall.

Table 8-3. Rafter Spacing and Sizes.

Size / Spacing		2×4		2×6		2×8		2×10	
	Grade	12″	16″	12″	16″	12″	16″	12″	16″
Ponderosa Pine	#1	9-10	8-11	15-6	13-11	20-5	18-4	26-0	23-6
	#2	9-7	8-8	14-6	12-6	9-1	16-6	24-4	21-1
	#3	7-5	6-5	11-1	9-7	14-8	12-8	18-8	16-2
Red Pine	#1	10-1	9-2	15-8	13-7	20-8	17-11	26-5	22-10
	#2	9-10	8-6	14-3	12-4	18-10	16-3	24-0	20-9
	#3	7-5	6-5	10-10	9-5	14-4	12-5	18-3	15-10
Western Pine	#1	10-4	9-3	15-8	13-7	20-8	17-11	26-5	22-10
	#2	9-7	8-3	14-1	12-2	18-7	16-1	23-8	20-6
	#3	7-3	6-3	10-10	9-5	14-4	12-5	18-3	15-10
Spruce/pine/fir	#1	10-7	9-7	16-1	13-11	21-2	18-4	27-0	23-5
	#2	10-0	8-8	14-8	12-8	19-4	16-9	24-8	21-4
	#3	7-6	6-6	11-1	9-7	14-8	12-8	18-8	16-2
Douglas Fir	#1	11-3	10-3	17-8	16-1	23-4	21-2	29-9	27-1
	#2	11-1	10-10	17-4	15-4	22-11	20-2	29-2	25-9
	#3	8-11	7-9	13-6	11-9	17-10	15-5	22-9	19-8
Hem-Fir	#1	10-7	9-8	16-8	15-0	22-0	19-10	28-0	25-3
	#2	10-4	9-3	15-8	13-7	20-8	17-11	26-5	22-10
	#3	7-11	6-10	12-1	10-5	15-11	13-9	20-4	17-7
Western Cedar	#1	9-7	8-8	15-0	13-8	19-10	18-0	25-3	22-11
	#2	9-3	8-5	14-7	12-8	19-2	16-9	24-6	21-4
	#3	7-6	6-6	11-4	9-10	15-0	12-11	19-1	16-6
Eastern Hemlock	#1	10-1	9-2	15-11	14-5	20-11	19-0	26-9	24-3
	#2	9-7	8-8	15-0	13-8	19-10	18-0	25-3	22-11
	#3	8-4	7-3	12-5	10-0	16-5	14-3	20-11	18-2
Coast Silka Spruce	#1	11-1	9-9	16-5	14-2	21-7	18-9	27-7	23-11
	#2	10-3	8-10	15-0	13-0	10-10	17-2	25-3	21-11
	#3	7-8	6-8	11-4	9-10	15-10	12-11	19-1	16-6

Rafters: 3-12″ slope; 20 PSE Live Load, PSE Dead Load

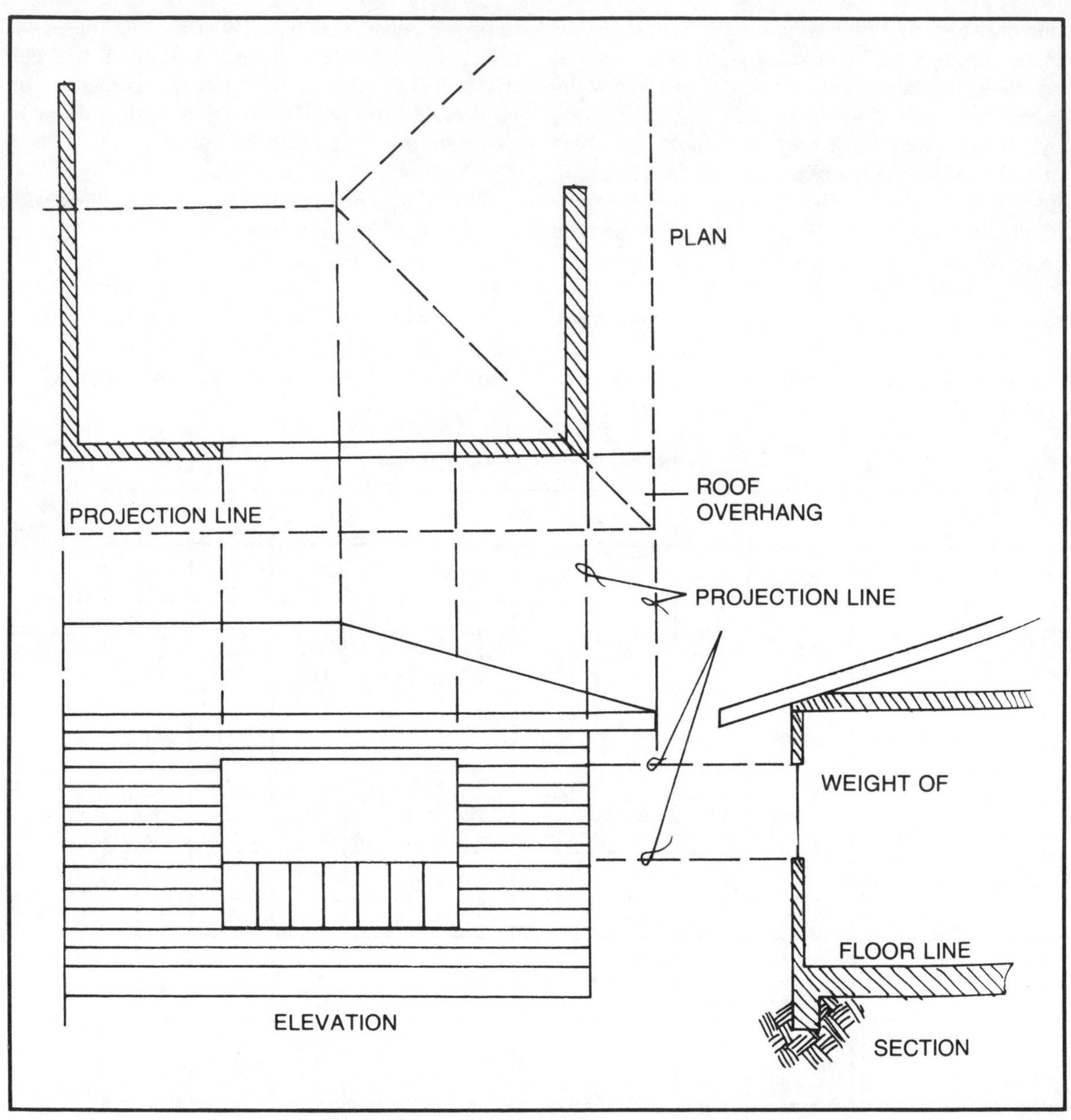

Fig. 8-6. The relationship of a floor plan to an elevation drawing.

□ Over to one side, and directly in line with the space where your elevation drawing will appear, place your structural section drawing.

□ Now project, as a series of light lines, from the floor plan, lines representing the length and all horizontal features of your building such as ends of walls, windows, doors, eaves, etc.

□ Project lines across from the section drawing that represent height and all vertical features of your building. Don't overlook grade line, floor line, window and door heights, plate line, and ridge line.

□ Darken all the lines that represent the features of the elevation that are important for you to show. Erase all other lines from the drawing.

Repeat the above steps for all other elevation drawings you need (usually four if you are planning a four-sided house and only two, perhaps, for a dome-type structure) rotating the floor plan for each side and putting the proper section drawing in place. After you have projected these drawings, you will have a set of elevations that lack only the details of porches, chimneys, trim and other details that you might now want to complete.

There are many sources from which you can draw information for the completion of these details. Besides your local building code, that will give you such things as your chimney height, you can consult the sources most architectural draftsmen use: *Sweet's File*, *Time-Saver Standards*, *Architectural Graphic Standards*, and good old manufacturers' catalogs. We find large mail-order company catalogs very helpful because they show such details as window and door units, the arrangement of window muntins, etc.

You will want to print in all your notes, reference symbols, title and scale, and perhaps even texture materials on the elevation drawing.

INTERIOR ELEVATIONS

Elevation drawings of interior walls can be very helpful in preventing errors during construction. All interior walls do not need to be drawn. Just draw the important ones such as kitchen and fireplace walls and walls on which there is special cabinet work to be done.

The procedure for projecting elevations can be used on all interior walls, step by step, just as you did with the outside walls. The same sources can be consulted for details of drawing appliances, cabinets, and other interior fixtures.

ANALYZING ELEVATIONS

Elevations should be analyzed to see how the house you have designed is going to fit into its natural environment. The elevation interfaces the environment and shows such important features as roof overhang and angle from which can be determined how much or how little sunlight will strike the house and enter the doors or windows. All of the elements of the elevation will show the relationship of the house, not only to the sun, but to the view, the topography of the site, and the climate of the area. By analyzing the house in relation to these considerations, ways to improve its beauty, economy and comfort will come to you.

CIRCULATION AND LIGHT

Ask yourself some questions and then see how these questions are answered by the elevation and section drawings. How much sun do you want in the interior spaces of your house? How is it allowed to enter and at what time of day? These are important questions to ask of your design. If your elevations show unsatisfactory answers to these questions, change the drawings (working back from elevations to sections to floor plans) until you like the answers you get.

The sun's path in the sky depends on latitude and the season of the year. In most of the United States and Canada, the winter sun stays low in the southern sky, rising in the southeast and setting in the southwest. In the summer, the sun appears to shift to the northeast for rising, arch across the northern part of the sky, and set in the northwest.

Rooms within the house should take into consideration the path of the sun in the sky. You will want your breakfast dining area brightly lighted by the morning sun, for example, but you would not want the sun in your eyes in your study during the afternoon.

Analyze your elevations in relation to the views you get from all directions on your building site. Your window arrangement should be such that it fully exploits the views from your building site.

Building sites with steep slopes call for designs of one kind and flat land requires another. Your elevation drawing will tell the story of how well your design fits the building site.

What do your elevations tell you about your house and its relationship to your climate? Exposures that are very desirable in cold climates can be terribly uncomfortable in hot ones. How are heating and ventilating influenced by your design?

Hot and humid areas require houses that are sheltered from the sun, but that remain open to the slightest cool breeze that might happen along. Elevations should show long, wide overhangs to act as barriers against the sun on the western and eastern facades. Air will need to circulate underneath the house.

In climates with extremely cold winters, windows need to be few and small to minimize heat loss, all rooms might wrap around a central chimney, and any large openings would be restricted to the elevation facing south. A row house with only two narrow walls exposed to the outside is a good design concept for energy conservation.

Work with your elevation drawings until you are satisfied that your house blends with its natural environment. Change your sections and floor plans accordingly.

The good feeling that you get when you open your front door and walk inside your home is one of the most important single things your house can offer you. That feeling is part of the mystique of owning a home. Part of that mystique is to have your guests feel welcomed when they cross your threshold.

ANALYZING YOUR SITE

That good feeling that comes with the opening of your front door must really begin before your front door is reached. You have to feel good about the environment in which you choose to live. You will not long maintain a homey atmosphere inside your house if you despise the neighborhood in which you live.

We have some friends who found a beautiful, big, wooded lot the whole family fell in love with. The price was right and they bought it without giving a thought to the neighborhood except that it seemed quiet and nice.

After much excitement and planning, they built their dream house and moved in. Slowly they discovered that their neighbors were almost all retired folks living on pensions of one sort or another. Our friends had four boisterous children, a big dog, and two cats. The older residents didn't really hate urchins, but they did appear to resent their intrusion. Living on fixed incomes, they predictably voted down every bond issue that came up for school improvement for the very good reason that they could not afford the higher taxes. Our friends just simply could never feel "at home" in that neighborhood even though their site and house were otherwise ideal.

The moral is clear. Pick your community first. That is of vital importance. Where do you and your family feel most at home? On a desert island away from it all? Smack in the middle of the city with a pulsing social life? How do you feel about high taxes to support good schools and long commutes? Once you have searched your soul and found a suitable community then, and only then, are you ready to look for a building site around which you can draw your house plans. Make up a check list and then check out each and every item on the list. It should include answers to the following questions.

☐ Are there children nearby you will be happy to have your children play with?

☐ Are your future neighbors, those with children, satisfied with the schools?

☐ If you are active in community affairs or politics, do your prospective neighbors share your interests? If not, will you be happy proselytizing or miserable without a support group?

☐ If you are connected with a religious group, is there a place of worship nearby?

☐ Does the community provide for garbage disposal or is this on their do-it-yourself list?

☐ Are there parks and playgrounds and free library services nearby?

☐ How does the tax rate stack up against the services your taxes pay for?

☐ Is the police protection good and efficient?

☐ Is the fire department well organized and can they get to your site in less than 10 minutes?

☐ Are the roads and streets well manintained? Who pays for this? Is the snow plowed in winter? Is the area fogged for mosquitoes in summer? Who pays?

☐ What about mail? Is it delivered or will you

have to pick it up every day from the post office?

☐ Are there non-noxious industrial plants nearby to provide jobs for residents, customers for stores, and to carry a big share of the community's taxes?

☐ Study the houses in the neighborhood and notice how they are maintained. Is this a community once deteriorated, but now being revitalized? If so, you can very often find a real land bargain. Check to see if there is any reason why the community might be disrupted in the future. An example would be the building of an expressway or highway.

HOW MUCH LAND?

The amount of land you purchase is usually determined by how much money you have to spend or how much you can afford to pay down. But it should also include enough space for any special activities such as gardening, basketball or tennis practice areas, or a swimming pool. It should also afford privacy and, if possible, a view. All of these things need to be considered when it comes to making a site plan. The amount of land, however, is not nearly as important as some other site considerations such as available utilities, access, sunlight, wind, and zoning. A small lot with good, workable physical characteristics is a more livable and satisfying homestead property than a lot of acreage you can get to only on foot, or one that would require a thousand-foot well to bring up that life sustaining necessity, water!

WATER AND DRAINAGE

Most home sites in or near a city are supplied with water and sewage disposal as utilities. Out in the country, neither are usually available as utilities; you are on your own. You will have to dig a well for your drinking, bathing, and washing water. Before buying any rural property, always inquire of neighboring land owners how deep they had to go to find safe drinking water and what the cost of such wells might run to. If you must go to a depth of a hundred feet or more, the cost of a well can run to thousands of dollars. And that is just for the hole and casing. You will still need a pump and plumbing lines to get the water to where you need it.

You will want a home site that gives you a slope to build on. All water should drain away from your house so you do not end up with a swimming pool in your living room. Building at the crown of a gentle slope is also the most economical way to go for several reasons.

☐ You need no expensive drainage system because rain water will run off in response to gravity.

☐ No fancy machinery for excavation is needed when building on a nearly flat site.

☐ No elaborate underpinnings are needed to support your structure such as might be necessary when building on a steep site.

☐ There is more leeway in making use of the sun and wind.

SUNLIGHT AND WIND

Building on the crown of a slope will keep you warmer than building at the bottom of a valley. Warm air rises because it is lighter than cool air so the tops of hills and slopes are always several degrees warmer than where the cool air "pools" in the valleys.

You might want to build just below the crown of the slope in order to protect your home from the prevailing wind in winter. If your piece of land is on the windward side of a hill or slope, you will get the full force of the prevailing winds. If you locate your building site on the leeward slope, you will have some protection from the wind.

In addition to finding out the direction of the prevailing winds, you need to find out how the sun travels across your site. Are there tall buildings, a ridge, or a mountain that will put your home in shadows for most of the day? If so, is this desirable? How will the sunlight and shadow patterns change with the seasons? Can you build so as to have the shade in summer and the sun in winter? Architects have sworn to us on a stack of house plans that if you build so as to take full advantage of the sun and wind (both use of and protection from seasonally) you can cut your energy bills by as much as 75 percent!

ACCESS AND SOIL CONSIDERATIONS

Roads cost a lot more to build than houses. Even the crudest of roads will cost you an arm and a leg. This dictates that those great building sites off the beaten track with such a splendid view will not be worth the price even if they are selling for a dime an acre. Not only is the access road expensive to construct, it will cost you lots more money to get your materials and machinery onto your building site if it is 10 miles from nowhere.

Keep in mind that land is not nearly as stable as we tend to think. There are lots of desirable land and beautiful building sites now being offered for sale that are disappearing by inches per year. This erosion can whittle your property down considerably in the time span of a single generation. If the area is subject to flooding and severe storms, this erosion can come in quantum leaps of acres per year in the form of mud slides.

Find out about any flooding and erosion problems that neighboring property owners have experienced. Try to determine the rate of erosion. It may be that the erosion is so slow that you will be long dead and buried before there is any problem. If so, you might decide to partake of the "temporary" charm and advantages of a magnificent waterfront location.

Other soil considerations may not be so easy to deal with. If you are really serious about a piece of property, by all means have the soil tested by a soil-analysis lab. Get a surveyor to supervise percolation tests, borings, installation of a well point, and to determine if there are large boulders under the topsoil. If there are large boulders present, excavation for a foundation can be expensive and excavation for a basement might be extremely difficult.

If you do plan a basement, then your well point will give you the fluctuations in the water table that will tell you if you are going to have flooding problems in your basement. If the level of water is only a foot or two below ground, you might decide you don't want to deal with the problems a basement under water will entail.

Percolation tests will tell you how porous your soil is and thus how well it can absorb water and sewage effluent. There are almost always city or county—sometimes even state regulations—about the soil conditions in relation to sewage disposal systems such as septic tanks or evaporative transpiration installations.

The strength of the soil is another important factor that you will need to determine. If the soil is weak, you will face a choice of enlarging your foundation, at considerable added expense, or risking serious settlement and cracking problems in your future home. Very poor bearing soils require putting your house on stilts or piles of tapered, rot-resistant telephone poles.

The best kind of soil to look for is compacted sand and gravel or a sand and clay combination. Clay alone and silt and noncompacted sand are less desirable. Clay without sand is very slow to absorb water and tends to puddling and flooding. It does absorb the water after awhile, however, then it swells to a much greater volume than when dry. This can give your house quite a ride.

Don't forget to have your drinking water tested if you will be needing a well. Pollution is showing up everywhere and you will want to assure yourself that the level of pollution you must deal with is within safe limits.

Don't overlook the value of trees for their shade and cooling effect. Then again, if you are a sod busting gardener or a lawn fancier you may prefer few trees. Keep in mind that they do give you the best insurance against erosion you can buy.

ZONING AND COVENANTS

Three terms you need to become familiar with, before you sign on the dotted line and write a check for your building site, are *zoning laws*, *covenants*, and *easements*.

A trip to the planning department of your city, town, or community will get you all of the information you need. If anything seems unclear or confusing, it never hurts to pay an attorney a small legal fee to counsel you before making your commitment. You will want to check especially any recent amendments to zoning laws to be sure that an airport or anvil factory has not been just approved for your prospective neighborhood.

Zoning laws are, in general, a very good tool for protecting a community from mixing up homes, factories, offices, massage parlors, convenience stores, etc., without regard to the effect that one might have on the other. But zoning is not forever and the laws are often bent to accommodate large land developers. Neighborhood associations have been formed in cities and towns all across the country to fight for maintaining the zoning they had when they first bought their property. If there is a neighborhood association operating around your proposed building site, by all means consult with the members to get as much information as you can.

Once you are satisfied with the zoning situation around the property you are thinking of buying, take a good look at how the zoning relates to your building plans. Zoning regulations restrict the height of what you build in order to keep the neighborhood in scale. They will also tell you the minimum distance from the street or from your neighbor's house that your structure must be built. These *set-back* lines will determine the shape and dimensions of your yards—front, sides, and back. Always be sure your building site is large enough to accommodate the size house you plan to build once these set-back lines have been established. Is there enough room left to add on a new room, a car port, a tool or a storage shed?

After the planning department, the next place to visit is the building department of your local community government. These folks will issue you your building permit and they will be concerned about your plans to use approved materials and techniques of installation. You will be required to conform to their minimum standards for electric wiring, heating, plumbing, etc.

Keep in mind that you might be able to get a variance of the zoning laws and building restrictions to suit your special needs if you can show good cause and no harm to the community. You will need to check out the procedure with the town clerk and go before the town council or planning commission. You would be well advised to seek legal council.

In some small communities where you find no planning departments or commissions, you might find that the zoning and building restrictions are established by covenant. Restrictions are established by covenant and apply to all the property within the boundaries of the town or village. A covenant is simply an agreement between buyers and sellers of property. Sometimes they require some off-beat things such as that you may only paint your house you build on this property some shade of pink—no other color—pink!

Covenants are legally binding no matter how serious or silly they might seem. Be sure to check out any covenants that are in effect on the property you are buying.

Beware of easements. Almost all deeds have clauses written into them giving someone else the right to make use of your property. This might be the government, a utility company, another landowner needing access to his property through yours, etc. Even though you own the property, you can do nothing that will interfere with the access necessary to the other party having the easement right.

Other deed restrictions might be written into your deed.

MAKING A SITE PLAN

Once you have done all of the preceding, you are ready to draw a site plan. The surveyor, who measured your parcel of land and set stakes marking the exact corners of your property, will give you a site map (or plot plan) from which to work. Figure 8-7 illustrates a site map we received on a piece of property we once bought. It was while drawing plans for this building site that we discovered some simple ways to save money.

If you order your survey of the lot for a line drawing before purchasing the property, you not only can use the drawing for buying the property, but also as a *topographic map* with which to build and as a mortgage survey. Your mortgage company will require a final survey before giving you your money. Be sure to give them a copy of your topo-

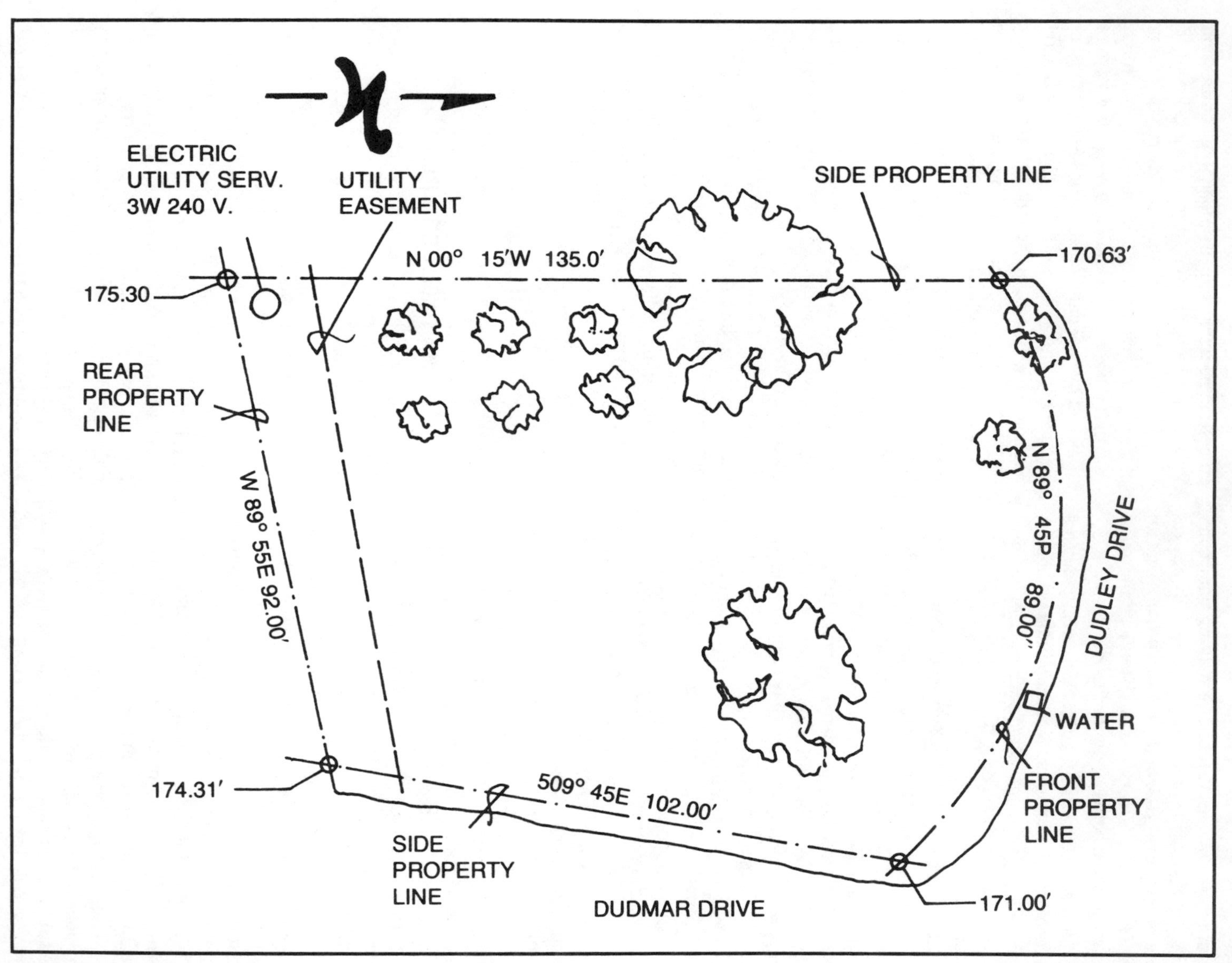

Fig. 8-7. A site map.

graphic map early on. Otherwise they will order a *mortgage survey* and you will pay double for your drawing. Maybe it will cost you more than double if they use a surveyor unfamiliar with the area where your property is located. If at all possible, employ a surveyor who has worked on the boundary properties. He can save you money by using his previous notes and check points.

The mortgage company will be satisfied with a simple line drawing of the lot with the house located on it (Fig. 8-8). Your construction people and local building department, however, will want something more elaborate such as Fig. 8-9. To arrive at such a finished site plan, first lay a piece of tracing paper over the surveyor's drawing and trace it off with a soft pencil. Trace off the contour lines that show the elevations of the property. Practice visualizing the shape of the land from these contour lines. A good exercise for this is to imagine pulling the plug in the middle of a lake and then imagine drawing lines at different levels as the water recedes. See all this happening from directly above the lake and the lines you draw will be contour lines.

As you trace off the lot lines, recognize that your surveyor laid them out using polar coordinates. Each line is described by *distance* and *bearing*; these are its length and its angle in relation to true north and south. The distance is measured in hundredths of a foot (instead of inches) so that, for example, 16.25 feet would equal 16 feet 3 inches. Bearings are measured with true north and south reading 0° (Fig. 8-10). Although your surveyor should have noted bearings down to minutes and seconds (there are 60 seconds in a minute and 60 minutes in a degree) you can round them off to the nearest half degree. That is all the accuracy required on a site plan and it is about as small as you can measure with an ordinary protractor anyway.

Now trace in your setbacks and easement lines so that you will know where you cannot build. Use a symbol or notation for any prominent landscape features such as trees or wild flower beds you want to preserve. Take your tracing to a gathering of the whole family and get them involved.

LISTING DESIGN GOALS

Encourage your family to visualize the property from memory and your tracing. Try not to get hung up on details, but do keep them in mind. Get everyone thinking about the things that matter the most to them. As you think and discuss, doodle out a flow diagram on your tracings (Fig. 8-11).

What kind of view would you like and from where? How will you and your family likely move around? How best to get onto and off of your property? Where will the quiet areas likely be? Where is the action? Who has preferences for what? Will there be a vegetable garden? Rock garden? Pool? Badminton court? Tennis anyone?

You will begin to generate some ideas for a floor plan with your flow diagram, but for right now concentrate on the family flow in and out of your future home and around the property. Make a list of these design goals. Take your time and make a long list so you will have a lot of choices when it comes time to pick and choose what you feel you must have in relation to what you can afford. Keep in mind the feeling of the entire space you and your family will occupy—inside and out—and the good feelings you want to create.

At this time, you should settle on the dimensions and shape of your house in order to be able to draw your final site plan.

DRAWING THE SITE PLAN

When drafting your site plan, it is a good idea to draw the project in the order of its construction.

☐ Pick a scale large enough to show detail; 1/16 = 1 foot is a scale most architects use.

☐ Lay out the lot showing everything your surveyor showed on the site map he gave you. Include contour lines (unless it's a flat lot). Be sure to place the north arrow accurately.

☐ Draw in all streets and alleys.

☐ Show locations of all utilities: electric, water, gas, sewer lines, etc.

☐ Draw in the shape of any existing structures that are not to be torn down during construction.

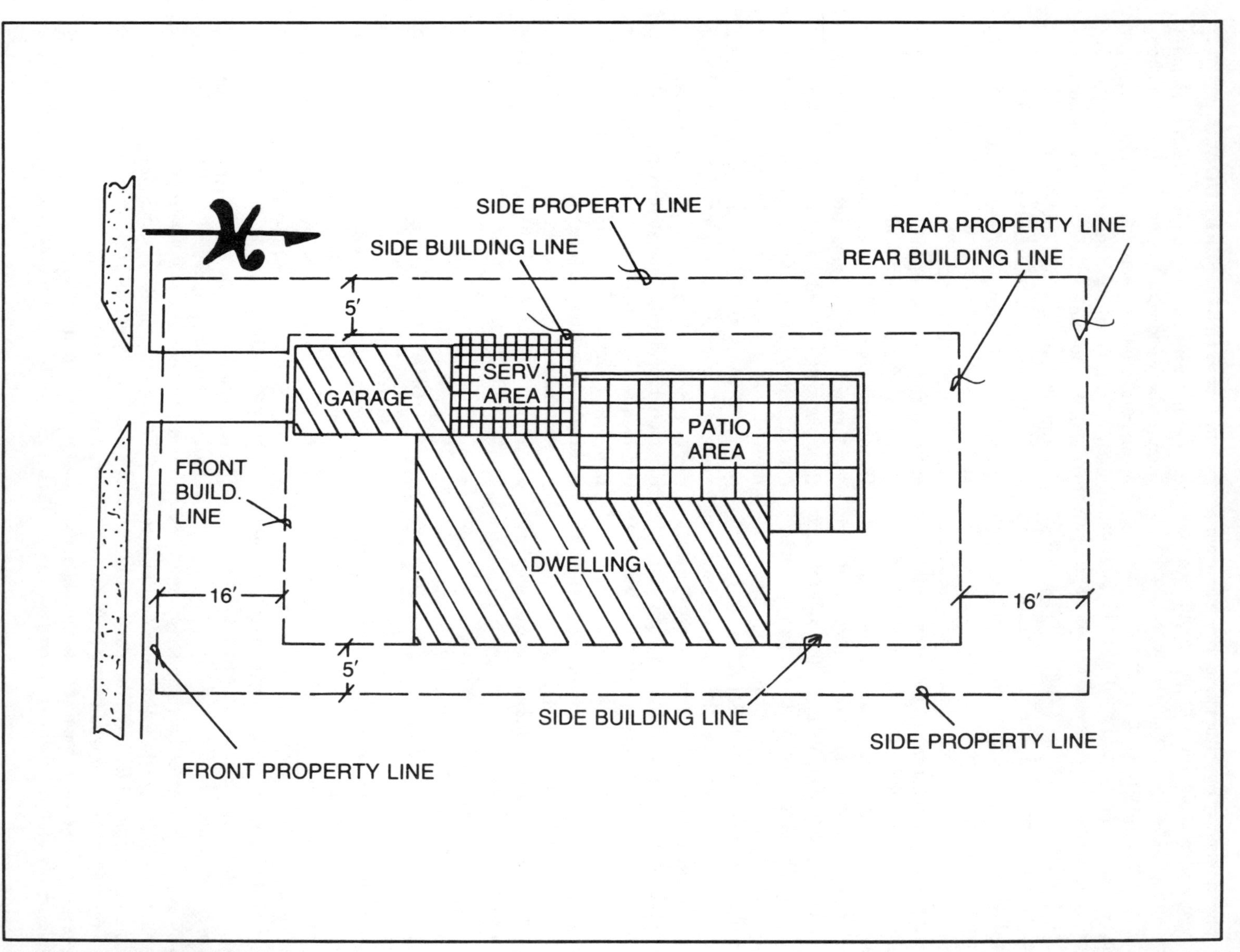

Fig. 8-8. A line drawing of a lot.

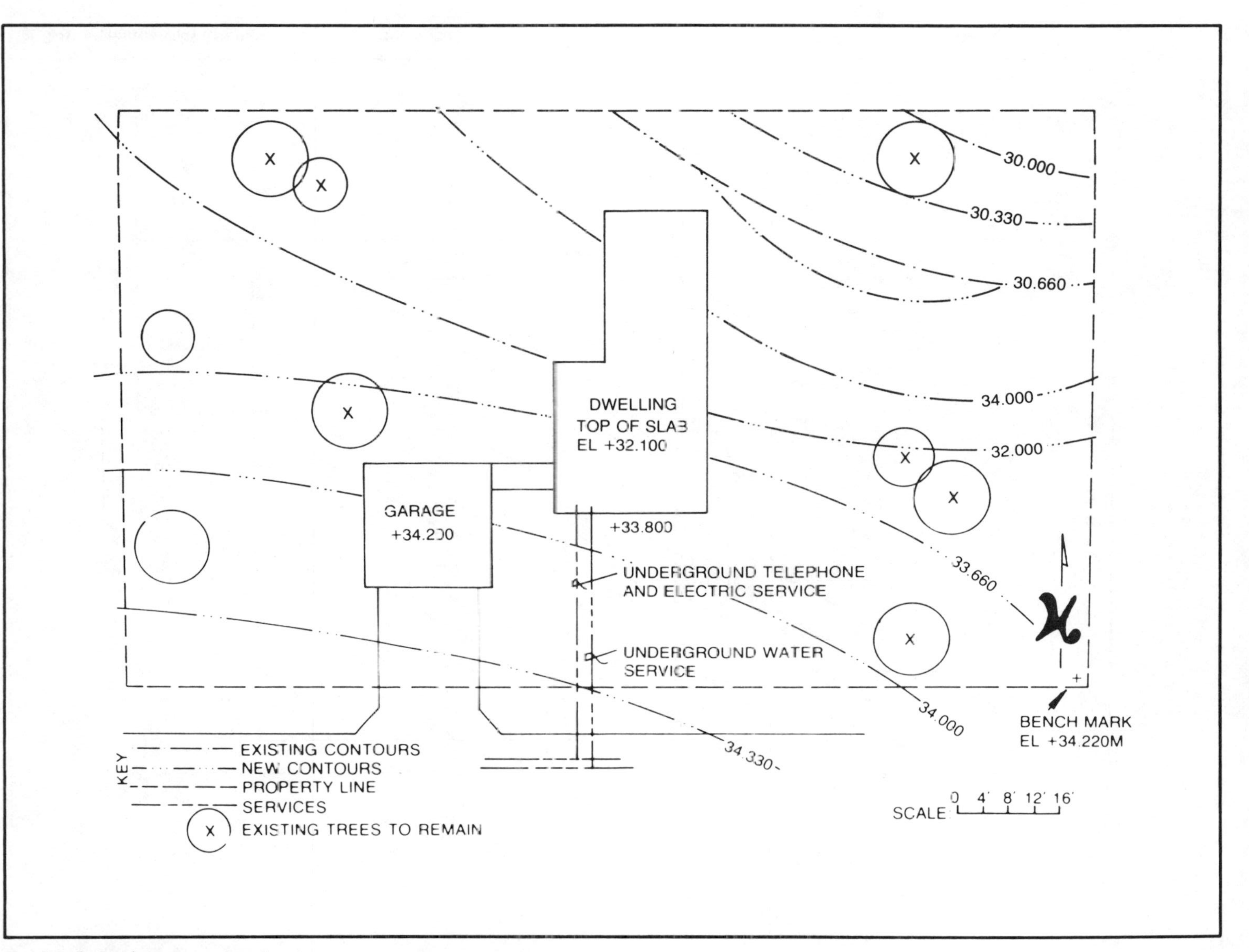

Fig. 8-9. A drawing appropriate for showing to a mortgage company.

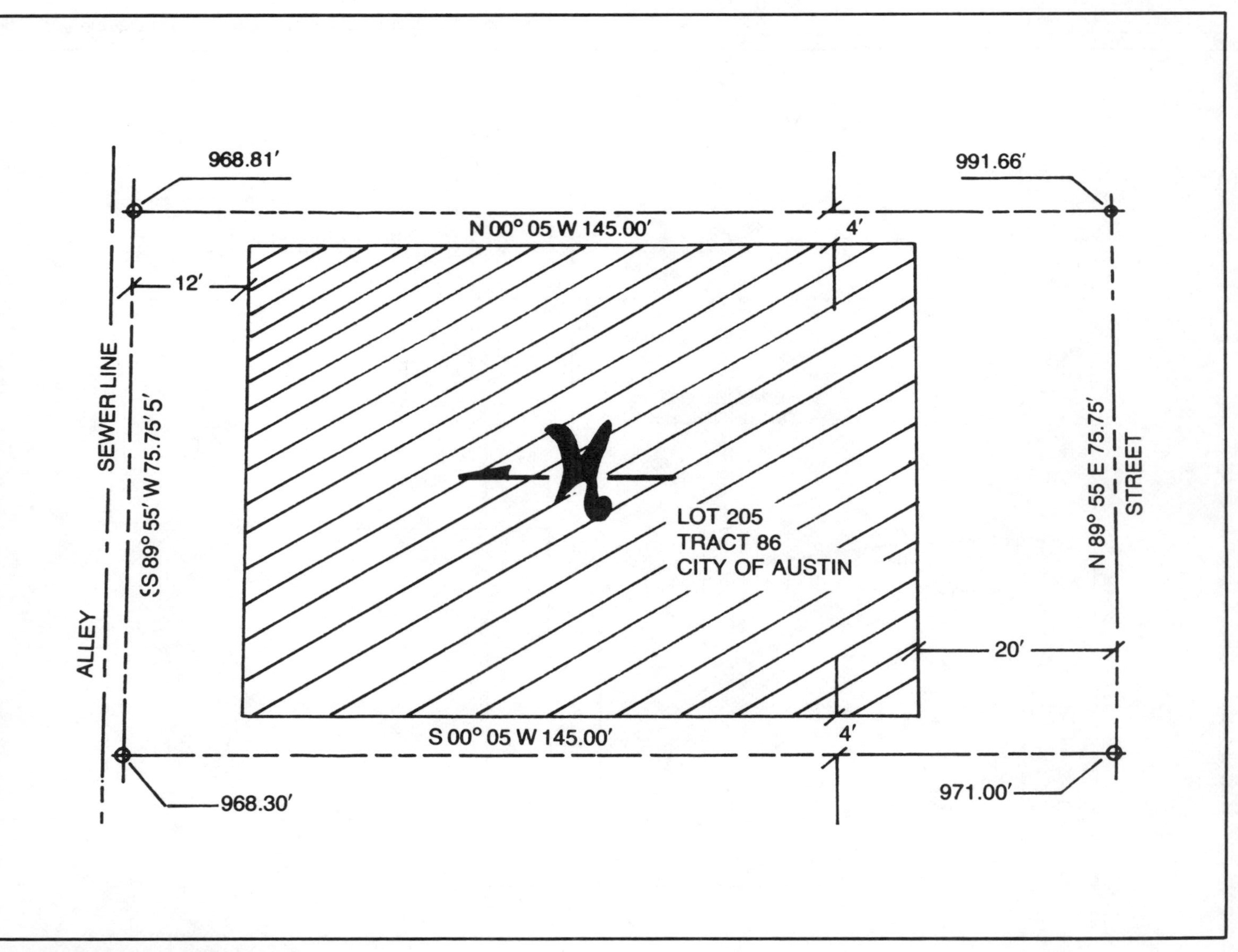

Fig. 8-10. A sketch with distance and bearing notations.

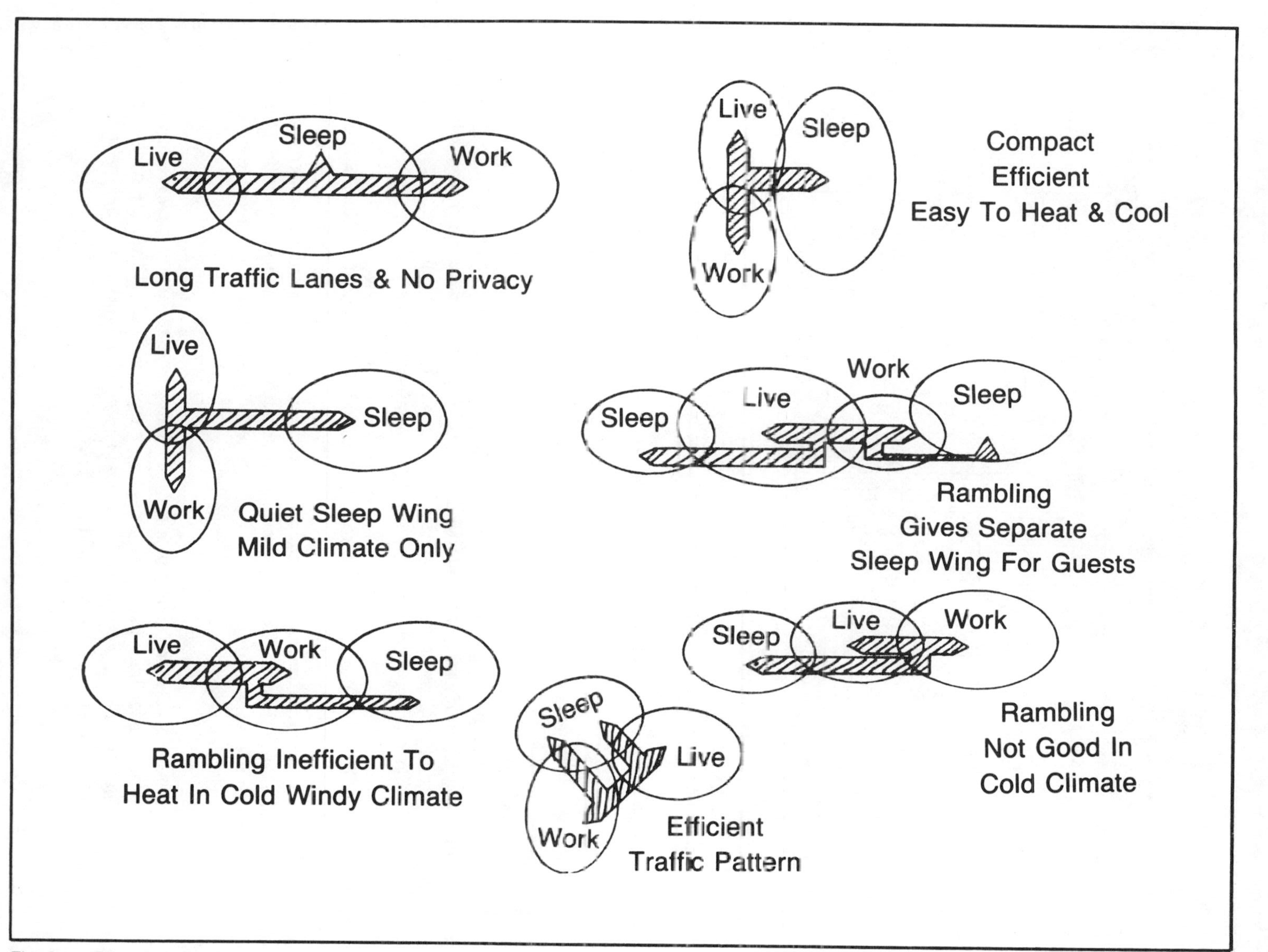

Fig. 8-11. Flow diagrams.

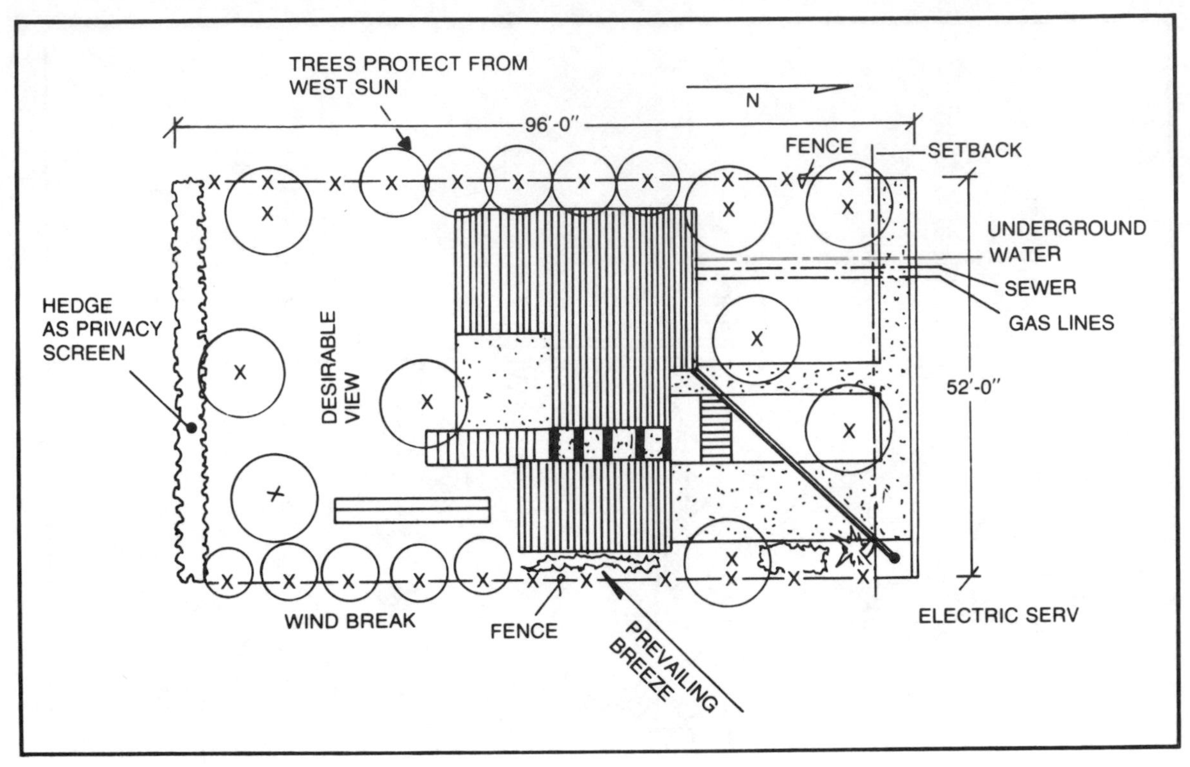

Fig. 8-12. A site plan.

Use dotted lines to draw in any existing structures that are to be demolished. Make a margin note stating by whom this structure is to be removed.

☐ Using a hard-lead drawing pencil, rule off light lines representing front, rear, and side setbacks required by ordinance. Lay off the boundaries of any easements that cross the property.

☐ Using a soft-lead drawing pencil draw in heavier lines the outlines of all buildings and roof overhangs.

☐ Track down and locate on your drawing all the little details like fences, gates, plumbing vents, air-conditioning compressors, and the like.

☐ Draw in all areas of gravel, rock, or concrete work.

☐ Draw in dimension lines as shown in Fig. 8-12. Always draw dimension lines parallel or perpendicular to lot lines. Two dimensions are needed if the lot lines are 90 degrees to each other and the building is parallel to one of them. If the building is not parallel to a lot line, then you will need to locate one corner of the building in relation to two dimensions at a corner of the lot (Fig. 8-12).

☐ Draw your trees and shrubs to the size they will be when they are mature. Use circles to mark the locations of tree trunks.

☐ Identify all plants or planting areas in your notes right on the site plan. Note also the thickness of concrete and paving. Identify all utilities. Show elevations of ground at both property and building corners. If your planned structure is multilevel, show the elevation of each floor. Identify all areas.

Landscape templates are available from drafting supply stores and they are neat to use. They will speed your work along and make it look more professional.

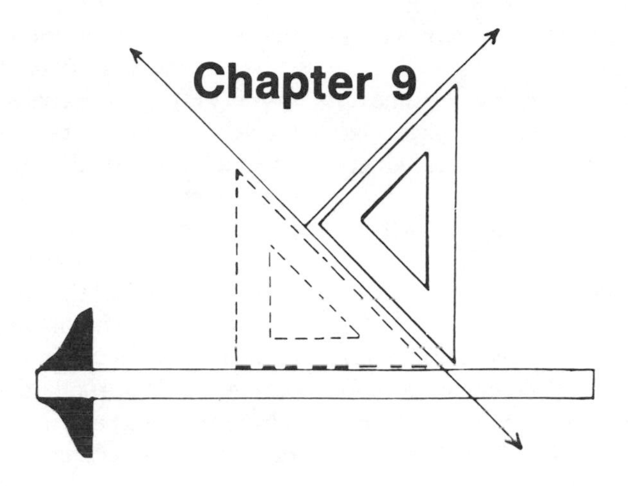

Presentation Drawings

The do-it yourself planner, drawing strictly for his own use, may need no more than the working drawings necessary to obtain a building permit. Even so, it can be worthwhile to make what a professional planner calls a *presentation drawing*, a neat, attractive drawing (either a floor plan, an elevation, a sketch of the completed house, or any combination of these) used chiefly to display important features of a proposed house. A presentation drawing may be just a simple elevation and floor plan or an elaborate perspective drawing done in pencil, ink, or watercolor.

A professional planner may work directly for a prospective homeowner, or he may be employed by a builder whom the client has contacted. In either case, the planner is given instructions about the sort of house that is wanted. He then draws preliminary plans. These preliminary drawings usually are pretty messy—lines have been erased and redrawn, different ideas have been tried. Thus the planner must prepare a presentation drawing for his client by making a clean tracing from the work sheets or by drawing a neat sketch of the proposed house (complete with landscape) or by doing both. If the prospect is satisfied that this is the house he wants, he will order completion of the working drawings.

For the building contractor, such a drawing is an effective sales help. These drawings are often used at reduced size in newspaper advertisements. They may have enough information for an experienced builder to quote a tentative price contingent on completion of the final working drawings.

PERSPECTIVES

Drawing in perspective is the technique of representing objects on paper as they might appear to the eye. Sketches of the completed house are drawn with this technique and are thus called *perspectives*. Architects often employ specialists to do such drawings and their fees can be in the hundreds of dollars.

To draw a perspective, start with a large sheet of paper, such as 24 × 32 inches if you're using the 1/4 inch scale. Projection lines become complicated when using a smaller scale. Refer to Fig. 9-1 and proceed as follows:

1. Draw a portion of the floor and roof plans including the features that will be shown on the perspective (1 in Fig. 9-1).

2. Draw a horizontal *picture plane (PP)* line (2) touching the left front corner of the floor plan. This line is the reference point from which true heights will be measured.

3. Draw a *center of vision (CV)* line (3) halfway between the two *outermost* corners of the house walls. Project this line down to the bottom of the sheet.

4. Determine the *point of view (PV)*, where lines 15° from the CV (drawn from the outermost corners of the house plan) intersect on the CV line (4). This triangle consisting of lines 9 and 2 is called the 30° *cone of vision.*

5. Draw a horizontal *ground line (GL)* a sufficient distance below the PP so there will be room for the perspective view (5).

6. Draw a second horizontal line (6) at a suitable scaled distance above the GL to represent the *horizontal line (HL).* If you were standing at the PV in front of the house, your eyes would be about 5 feet above the ground and this would be the height of the HL. If you wanted a bird's eye view showing mostly roof, the HL might be 20 feet above the GL.

7. From the PV (4) on the CV line (3), draw lines 30° and 60° from the CV (7). One line should be parallel to the left end wall of the floor plan; the other should be parallel to the front wall of the floor plan. These lines should intersect the HL (6). The intersections are the left and right vanishing points (VP-L and VP-R).

8. Draw an elevation on the GS (5), like the one shown at 8. It should be drawn to the same scale as the floor plan.

9. From the PV, draw lines to the corners of the various features on the floor plan and mark their intersection points on the PP. From these points, drop vertical lines down into the perspective area (the area where the house is to be drawn). The vertical line running along the forward corner of the house (11) will be a vertical reference line (the *true height line or TH*).

10. Draw horizontal lines from the same features on the elevation (8) over the TH (11). Mark the points of intersection.

11. Draw lines from these intersection points to the VP-L. The intersection of these lines with the vertical lines marks the correct position of the various features on the perspective. To avoid confusion, determine each point separately until the entire picture is developed. For example, let's say you want to determine the position of the far left end of the roof ridge of the house in Fig. 9-1. Draw line a from the left end of the ridge on the floor plan to the PV. Line a will intersect PP. From this point of intersection, drop line b down. Next, draw line c from the roof peak on the elevation to TH (11). From the intersection of c and 11, draw line d to VP-L. The intersection of d and b is the left end of the ridge line in the perspective. After you determine the position of the left end of the ridge, finding the right end is easy. Draw a line from VP-R to the left end of the ridge. This line runs along the roof ridge. Drop line e down to this line. The intersection of these lines is the position of the right end of the roof ridge.

All features of the house (which are visible on the front and left ends) are determined using the above methods.

Because the resulting picture is rather small, less prominent features can be drawn in freehand. Shadow lines are worked out and trees and landscaping are sketched in.

SHADED ELEVATIONS

Shaded, or rendered, elevations are often used as presentation drawings. The shaded elevation is fairly simple to do.

Start with a preliminary work sheet of tracing paper large enough for both front and end elevations, as well as the floor plan.

Trace an outline of the front elevation. If done with a black-ink felt-tip pen, the outline will stand erasures of later pencil work. Next, in exact horizontal alignment, draw the left end elevation which has features that will appear in the front elevation (Fig. 9-2).

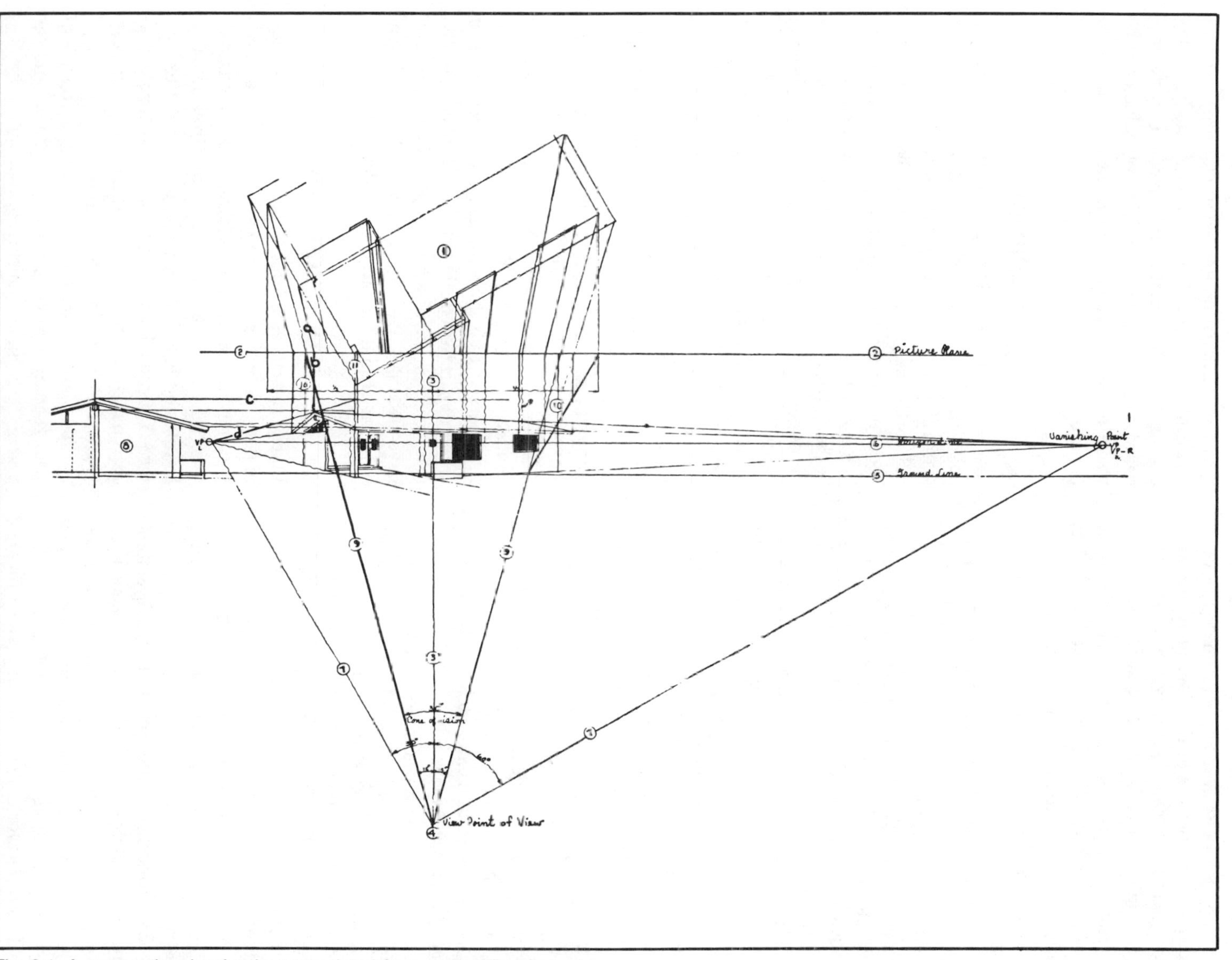

Fig. 9-1. A perspective drawing is extrapolated from a simplified floor plan.

Now in exact vertical alignment, draw in enough of the outline of the floor plan so that its doors, windows, roof overhangs, etc. can be projected up onto the front elevation.

Assume that the sun is shining from the left and from the front at a 45° angle so that your 45° triangle can be used to determine the lines where the overhanging roof will cast shadows on the front wall.

To shade elevations like those shown in Fig. 9-2, first use a sharp pencil and lightly draw lines 1 and 2, which indicate the edges of shadows that strike the front wall. Project these lines across the front elevation by drawing lines 3 and 4 with the T square. Using the 45° triangle, draw lines 5, 6 and 7, which define the limits of the high and low roof lines. Because these lines intersect at a window area they will not be visible on the finished drawing.

At this point it will be helpful if you indicate window drapes and crossbars, and give a sense of depth by blacking in the glass areas. Indicate the shaded wall areas by vertical lines or some other way of indicating shadows. Note also that the little wall around the entrance casts a shadow too.

Remember that every indentation (as at doors and windows) and projections (such as window sills) all throw shadows on vertical surfaces. Doors and windows are set in from 4 inches to 6 inches and sills project out 3 inches.

Next, consider how the textured wood wall *inside* the carport will be shaded. To do this it will be helpful to draw the shadows thrown on the carport floor, though they will not be seen in the finished drawing. From the floor shadow it can be seen that only a small area of this carport wall will be lighted. All the rest will be in the shadows.

Further understanding of shades and shadows will be gained by casting the shadow of the gable end roof overhang.

The drawing is completed by sketching suitable shrubbery and trees into both foreground and background to make a contrast between the light-toned stucco walls and the darker shadows. A few clouds in the sky make all of it more authentic.

From the work sheet you can trace a finished drawing, the house outlined in black, shrubbery and trees and other parts drawn in soft pencil. From this tracing, black line prints can be taken. Using spirit-based colors, such as Magic Markers which do not wrinkle the paper, you can make very pleasing pictures.

DIMINUTION

Go out on a street where people are walking and hold your hand upright with your arm extended forward. Notice that persons close by will be about as tall as the height of your whole hand, but what about people more distant? Some in the middle distance will be the height of your thumb and those in the far distance less than the height of your little fingernail! Now take a large piece of drawing paper and draw the outline of your hand (Fig. 9-3). To one side, draw a series of stick figures: one the same height as your hand, another the height of your little finger, and still another the size as your fingernail.

With your T square, draw a straight line across the paper running through the middle of the smallest figure. Put a dot on this line (you have just located a *vanishing point*) and with your triangle draw several lines from the bottom and sides of your paper to this dot. You have just created your first perspective drawing using the principle of diminution. Objects always appear smaller as the distance from the observation point increases. You have also made use of *convergence* which is a combination of diminution and *foreshortening*.

FORESHORTENING

Lines or surfaces parallel to your face show their maximum size, but as they are rotated away from you they appear to grow shorter. To demonstrate this, take a cardboard core from a roll of paper towels and hold it at arms length parallel to your face. You would draw this view as a rectangle, perhaps 6 inches long (Fig. 9-4). Rotate the core away from your face about 30 degrees.

To represent what you see now, the top line of your rectangle spreads out into an ellipse and the rest of the rectangle must be drawn much shorter—

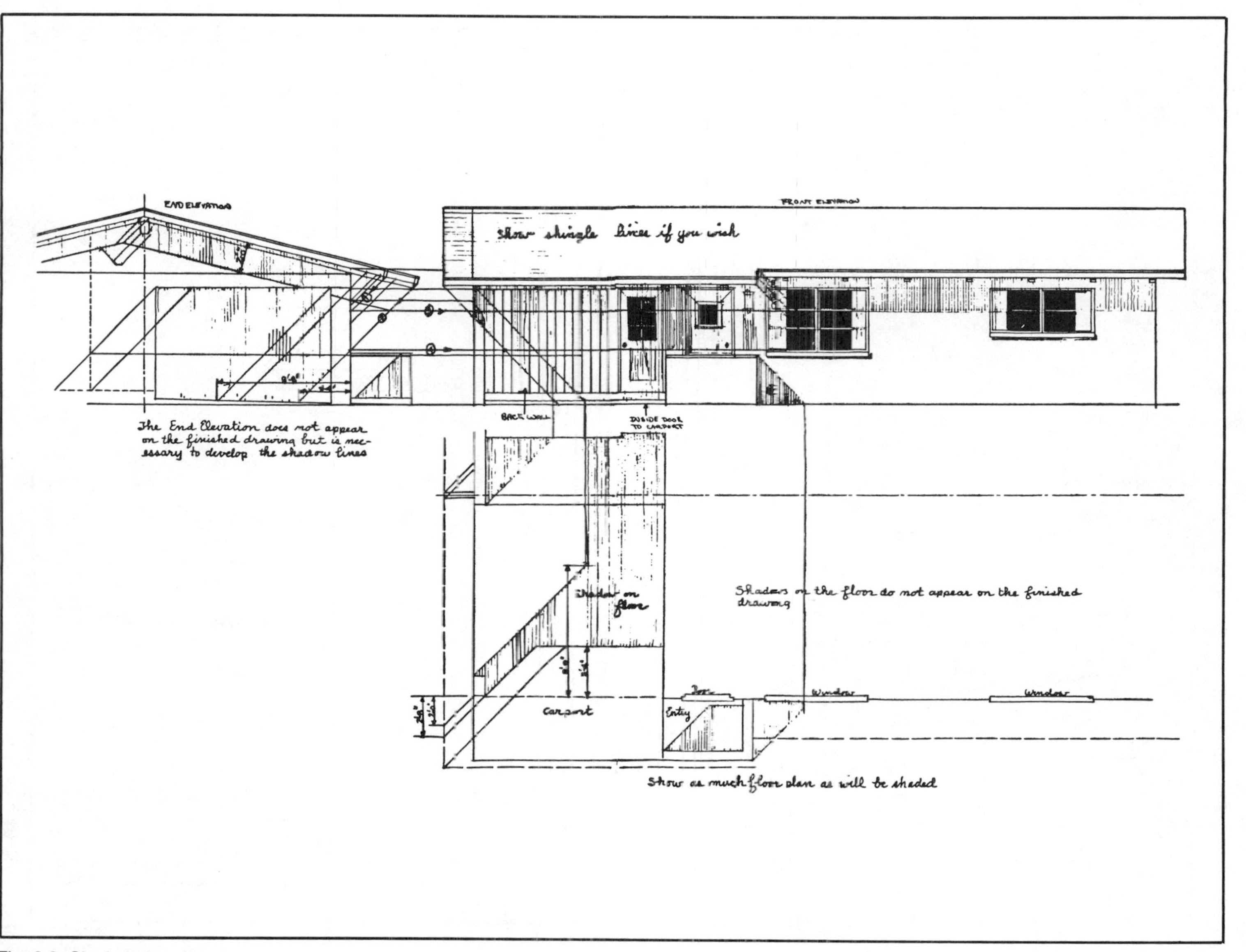

Fig. 9-2. Shaded elevations.

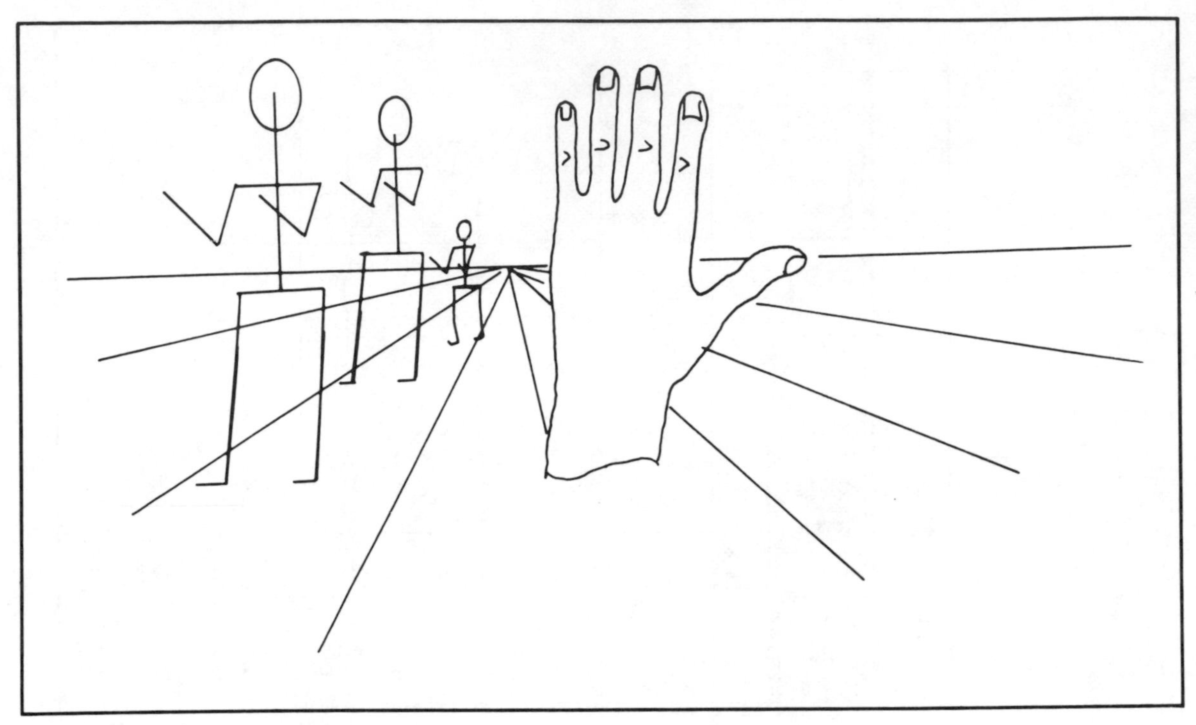

Fig. 9-3. Diminution.

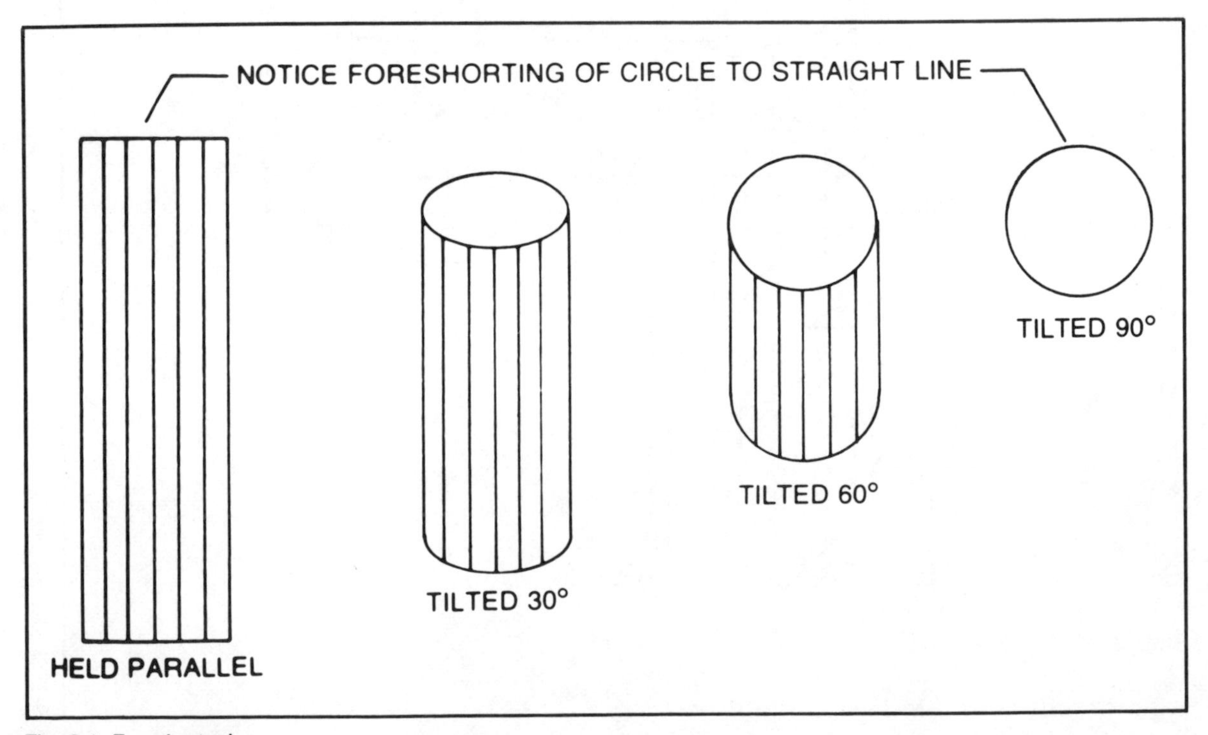

Fig. 9-4. Foreshortening.

only about 4 inches long. Rotate the cardboard tube another 30 degrees and what you see is foreshortened once more. This is represented by a rectangle only about 2 inches long. Rotate it a full 90 degrees and the rectangle disappears entirely. What you see can now be represented by a circle. The foreshortening is 100 percent complete.

CONVERGENCE

As parallel lines or edges of objects recede from an observation point, they appear to come together. Figure 9-5 illustrates that this convergence is equal to diminution plus foreshortening. Viewed "head on," the pickets are evenly spaced and all the same apparent height. Horizontal lines are parallel. But look at this same picket fence at an angle, down the fence line. How does it look if represented in a two-dimensional drawing? Each picket appears shorter as distance from the observer increases and the width and spacing of the pickets appears narrower. The horizontal lines now angle in such a way that they will eventually converge.

DEPTH AND SPACE

Figure 9-6 shows how the sense of depth and space can be achieved in a drawing by the simple technique of overlapping objects instead of drawing them spaced out and separate. In Fig. 9-7, you can see how it is that shadows give a sense of the third dimension to drawings. Keep in mind that all objects and scenes must be viewed in some kind of light. It is really the shades and shadows created by three dimensional objects in a field of light and gives them their appearance of shape and structure.

Pretend your object or scene is being illuminated by a single source of light and sketch in your shadows accordingly. In real life, pay close attention to how the sun "paints" scenery with its light. Notice how the shade and shadow fall in relation to the light. Use color value, or grey values if you are drawing in black and white, to enhance depth in your drawings.

Figure 9-8 illustrates how values (the range of colors, or the range of the grey scale from black to white) are bright and crisp and clear up close and then grow weaker and more neutral in the distance.

Details and patterns are more discernible close up than at a distance (Fig. 9-9) and "focus" is fuzzier or less sharp at a distance away from the observer. Actually, when your eyes focus on objects close to your eyes, background objects become blurred and unclear (Fig. 9-10). This effect can be used to spotlight objects in your drawings.

Figure 9-11 illustrates how *convergence, overlapping, shadow, grey-scale value,* and *pattern* all contribute to achieve the illusion of depth and space in a two-dimensional drawing. We suggest you copy this drawing and be aware of each of the above fundamental principles as you use them.

HORIZON LINES AND VANISHING POINTS

Almost all architectural artwork represents views of something in real life or something that is due to appear in real life when constructed. In this type of artwork, you rarely see horizon lines and you never see vanishing points. Yet these unseen things must be understood and kept in mind constantly. Often you will need to sketch them in temporarily to arrive at a realistic looking drawing.

The *vanishing point* is that point where any two or more parallel lines appear to meet. Figure 9-12 shows the classic example of a vanishing point by picturing a railroad track running back to the *horizon line*. The rails are parallel, but—as do all parallel lines extended indefinitely—they appear to converge at a single point. Note that this is the point also where a line connecting the bottoms of the telegraph poles and the telegraph lines also meet.

There is one exception to this rule and that is when the parallel lines are also parallel to the observer's face and to the plane of the picture. Such a case is the brick wall shown in Fig. 9-13. Except for that situation, the basic rule of drawing what you see is that all parallel lines meet at a *single* vanishing point.

Most pictures are not as simple as the railroad tracks, shown in Fig. 9-12, with only one set of parallel lines. Most pictures you draw will have many sets of parallel lines going in different directions. But the rule still applies. Each set of parallel lines will meet at a single vanishing point. You will sim-

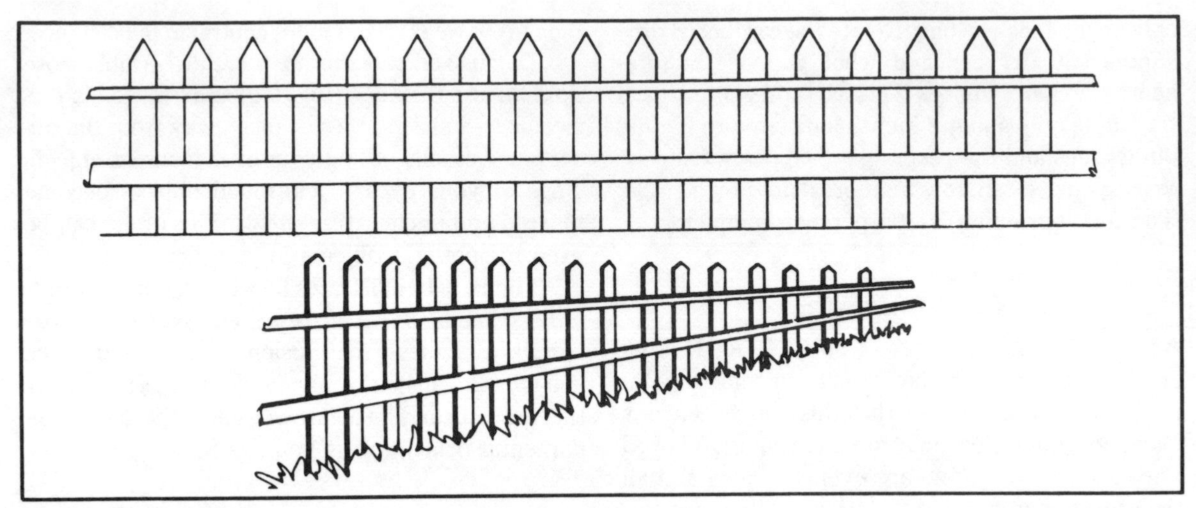

Fig. 9-5. Convergence.

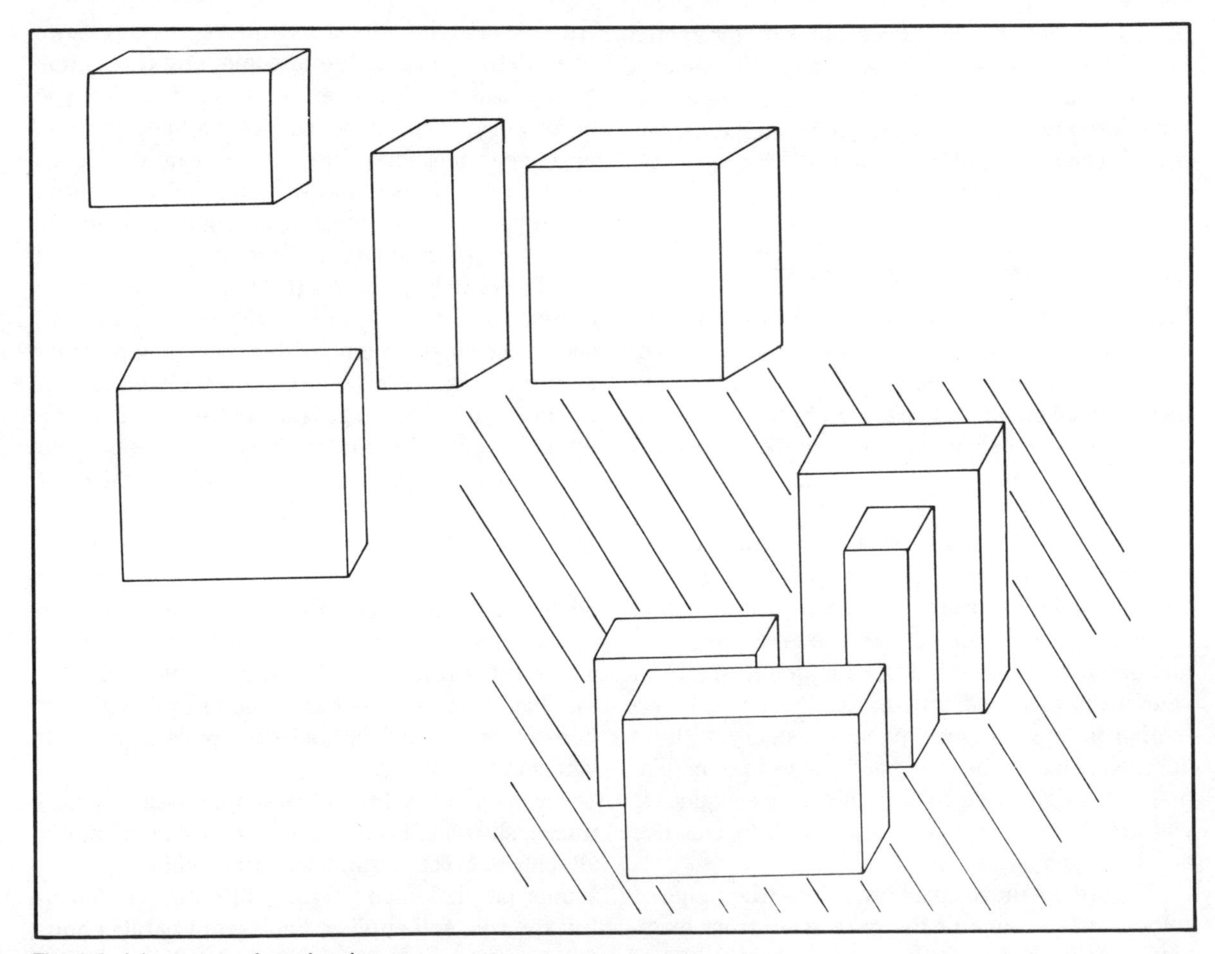

Fig. 9-6. Advantages of overlapping.

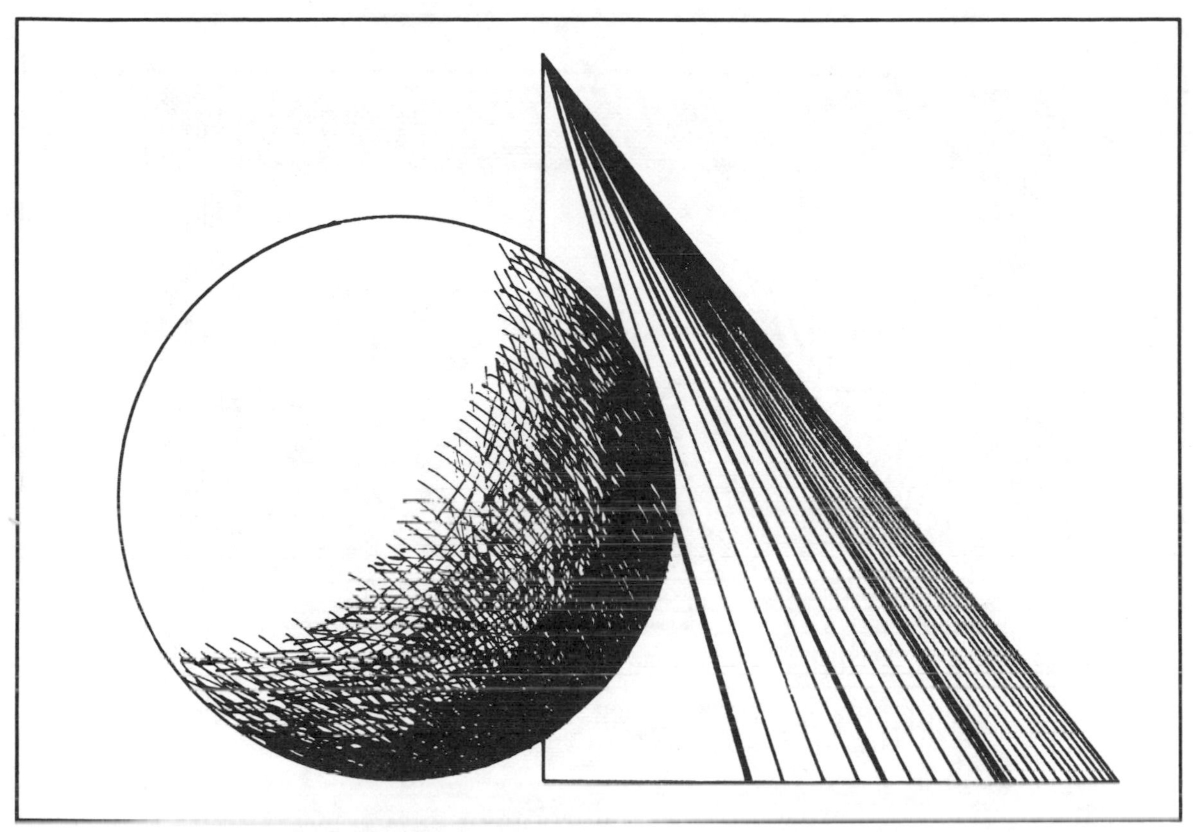

Fig. 9-7. Shadows for three-dimensional effect.

Fig. 9-8. The use of value.

Fig. 9-9. Spotlight effect.

ply have a separate vanishing point for each set of parallel lines. Some of the vanishing points will exist far away from the picture, but they do exist. Each set of parallel lines will meet at its own vanishing point.

Take any photograph of a building or object and, with a straightedge and pencil, you can extend all the converging lines until they meet at their vanishing point. Try this exercise on the building shown in Fig. 9-14. Locate vanishing points on a large piece of drawing paper. Using your straightedge, angle back from the vanishing points and draw in all the parallel lines in the building.

With a triangle and a T square draw in the vertical lines. From the perspective of the drawing, they converge so slightly that you can draw them parallel. They converge somewhere in the stratosphere. You have now made a perspective drawing of something architectural by mechanically drawing nothing but straight lines. And you probably thought you couldn't draw! You can learn to draw as well as a professional architect if you give it a lot of practice and come to know a few more simple things about horizon lines and vanishing points.

If you will now connect all the vanishing points of all the converging lines that are parallel to the ground (horizontal), you will discover that they line up along a single horizontal line. This is called the vanishing line for all converging horizontals and it will always be at eye level. This eye-level horizontal vanishing line is the secret of good perspective drawing. Where does it come from? How do we use it? How does it change the appearance of a drawing?

What locates the vanishing line in all pictures is the eye of the observer. The vanishing line for a worm's-eye view and the vanishing line for a hawk's-eye view of the same scene will be in two quite different places. It is always the eye level of the observer that determines the location of this important straight and horizontal line. It is an imaginary plane at eye level and parallel to the ground, as shown in Fig. 9-15.

Because the eye level is the vanishing line for all horizontal lines, specific vanishing points can be located by "pointing" the eyes to a given set of lines. You simply point in the same direction as the lines and discover the vanishing point right there on the vanishing line.

When working with T square and triangle, the technique of *parallel pointing* to locate vanishing points is very useful. Try the exercise suggested in Fig. 9-16 and you will see what we mean. First construct your plan (top view of an object). Show the object, the picture plane (drawn as a line) and the position of an observer. Now draw your sight lines from the observer "pointing" parallel to the object's lines to locate the vanishing points on the picture plane. The picture plane line will show the relationship of the objects apparent size to the vanishing points. Notice that when you transfer this "measurement line" to the picture it is superimposed on the horizontal vanishing line at eye level. Notice also that whether you draw the object above or below this line, the relationship of apparent size to vanishing points remains the same.

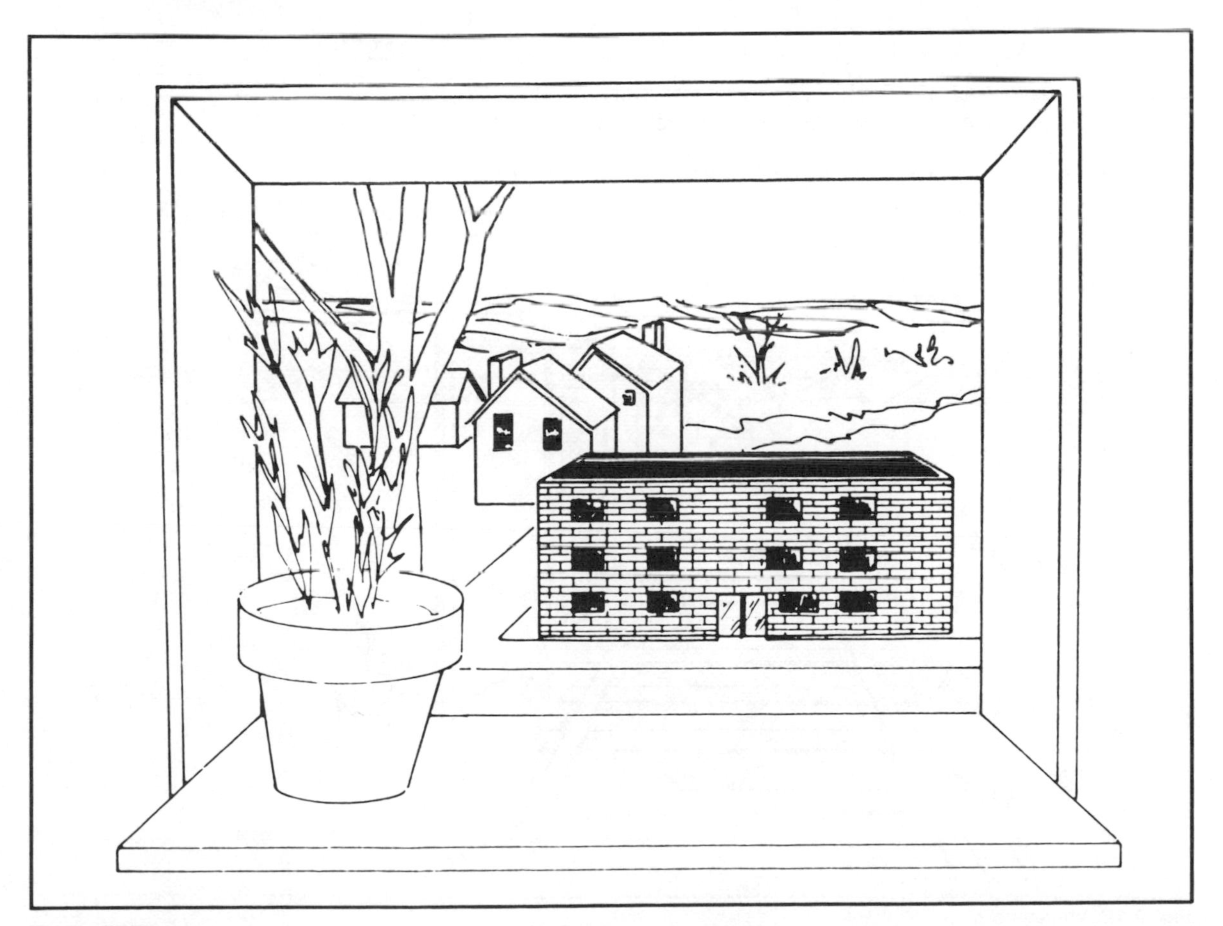

Fig. 9-10. Focus.

Fig. 9-11. Convergence, overlapping shadow, grey-scale value, and pattern.

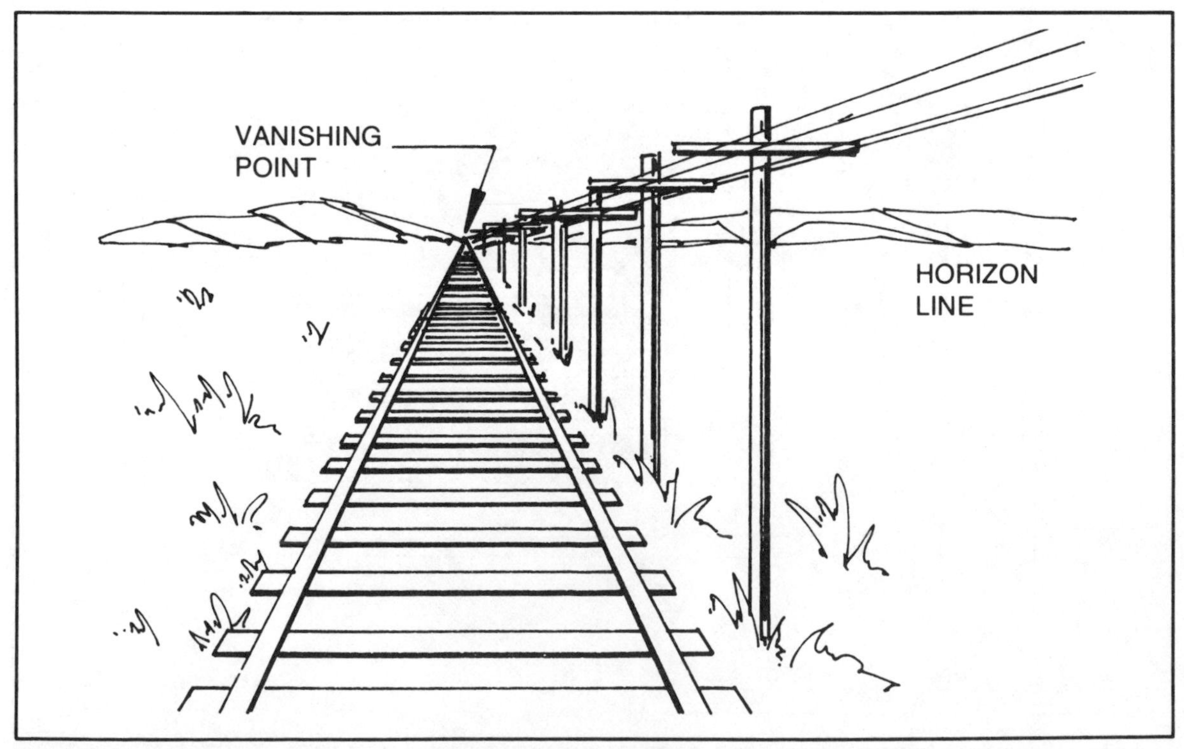

Fig. 9-12. Vanishing point.

Fig. 9-13. The exception to the rule.

When you begin to observe nature through the eyes of one who draws, you will soon observe that nature always supplies a horizon line right at the eye level of the observer. When you are descending in a roller coaster, the horizon line descends with you. Lie flat on the ground and you observe that the horizon line is still exactly at the level of your eyes. Notice that the big difference in the view from the top of a roller coaster and the view from flat on the ground is the amount of ground seen before you. The amount of foreground in a picture diminishes with a decrease in elevation.

Fig. 9-14. Locating the vanishing point.

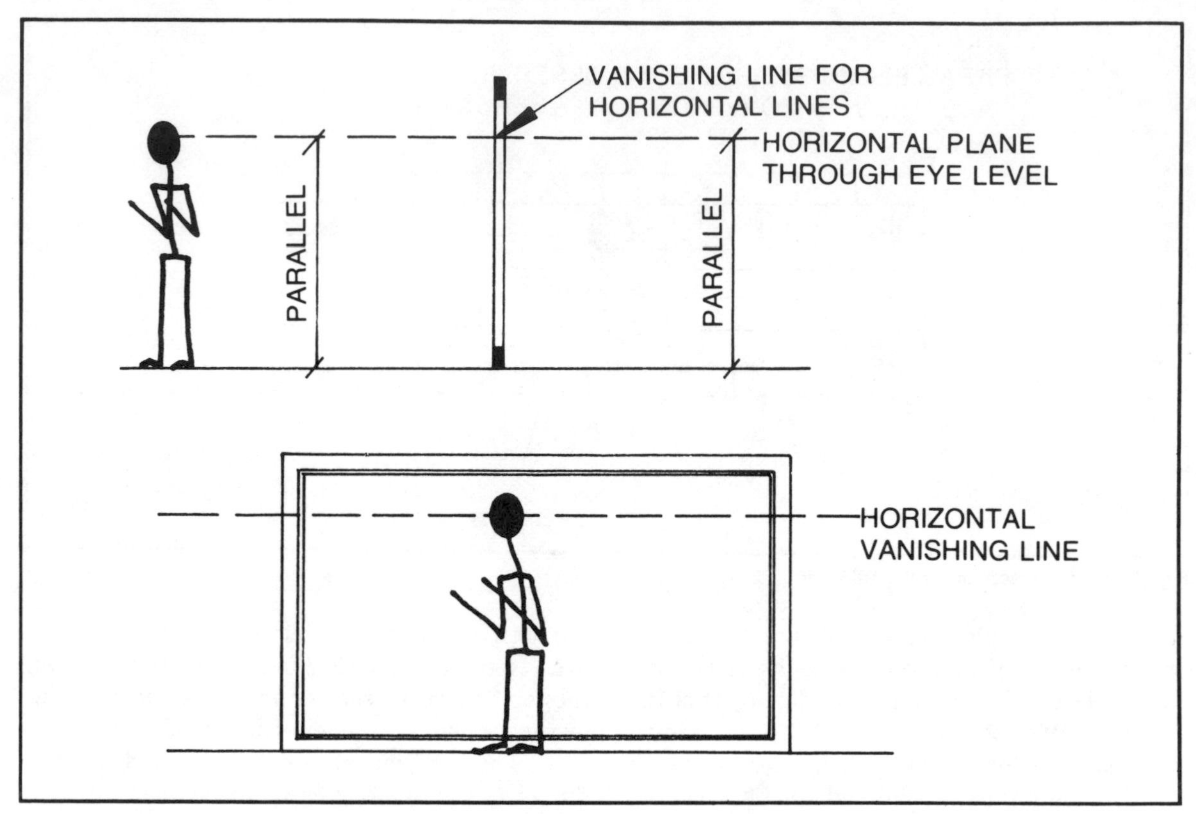

Fig. 9-15. Horizontal and vanishing lines.

Nature's horizon line, whether you can see it or not, (it is often obscured by people, hills, and buildings) always appears at the eye level of the observer and is the vanishing line for all horizontal lines. On it are located all vanishing points for sets of horizontal lines.

If you continue to observe nature, you will notice that the amount of foreground also changes when you look up, down, or straight out. When you are looking up at something in the sky or at a roof top, you barely see the eye level plane and so the horizon line is close to the bottom of your cone of vision. When looking down, on the other hand, the cone of vision just does include the eye level plane near its top and there is a lot of foreground visible.

From these observations you have discovered the inverse rule for view point. When you place the horizon line high near the top of your drawing paper, you will have to draw a view looking down on your subject. Place the horizon line low, near the bottom of your paper, and you must draw a view looking up. Place the horizon line in the center of your drawing paper and your view will be as if you are looking straight out at your subject matter.

Most of your drawings will probably be as seen by looking straight out. When it is necessary to show the underside or topside of something, you will want to shift your horizon line and thus your vanishing points accordingly.

TRACE-OF-THE-VERTICAL-PLANE DRAWINGS

One of the easiest systems for converting a floor plan to perspective drawings with triangle and T square is the one illustrated in Fig. 9-17 which we call *trace-of-the-vertical-plane* drawings. It is a system often used by TV and stage set designers to make quick perspective sketches from scale floor plans of their sets. What you draw is what you will get from your scale floor plan. You can even measure to scale back "inside" your drawing. Or you can draw in your vertical scale just the way it will look in reality.

Use the top half of your drawing paper to make your scale floor plan of the room you want to draw in perspective. Orient the scale plan so that the "invisible wall"—the one you wish to "look through"—is at the bottom of the drawing. The line representing this wall becomes your trace-of-the-vertical-plane (TVP) as seen from above. We call this line CD. From points C_1 and D_1, drop verticals down to the scale equivalent of the height of the walls in the room you are picturing. Connect the bottom ends of these lines and you have a rectangle ABDC (as shown in Fig. 9-17).

You can now measure to scale on the perimeter of this rectangle as well as on your floor plan. You will need to do this now to establish a vanishing line (horizon) for your drawing. Because most people view a room from an eye level between 5 and 6 feet from the floor, we draw in this line temporarily. Draw lightly because you will want to erase this line later. In Fig. 9-17, line GH is drawn through points G_1 and H_1. These points are measured to scale (6 feet) up from points A and B on the floor line.

We know that the vanishing point (VP) will fall along this line. Its location will depend on the placement of our position of view (PV) along the TVP. In our Fig. 9-17, we have placed it in the exact center of the room. This will make your drawing symmetrical, just as the room would appear viewed from the center of the "invisible wall" by a person about 6 feet tall. Your vanishing point will be anywhere along the VL farthest away from the wall you want to emphasize. Figure 9-18 shows what the

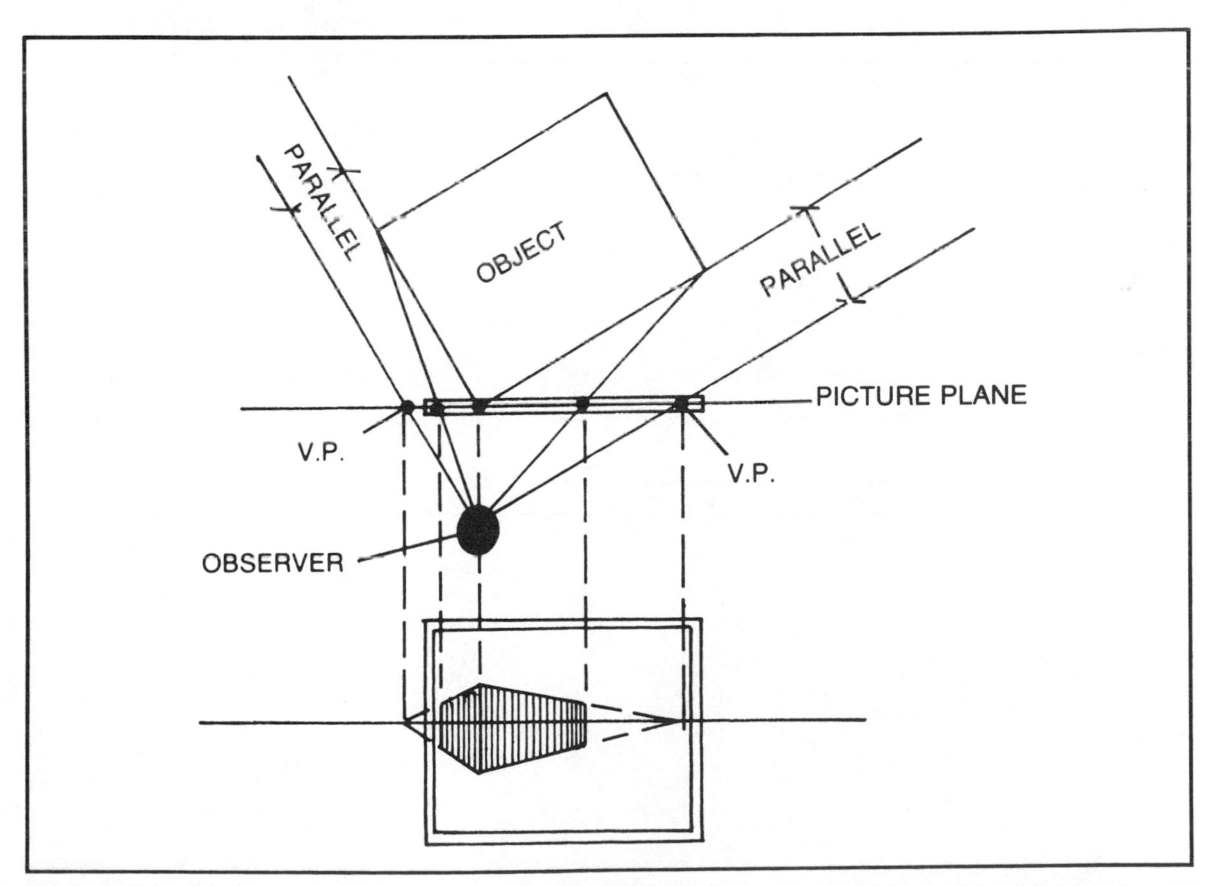

Fig. 9-16. A parallel pointing technique to locate vanishing points.

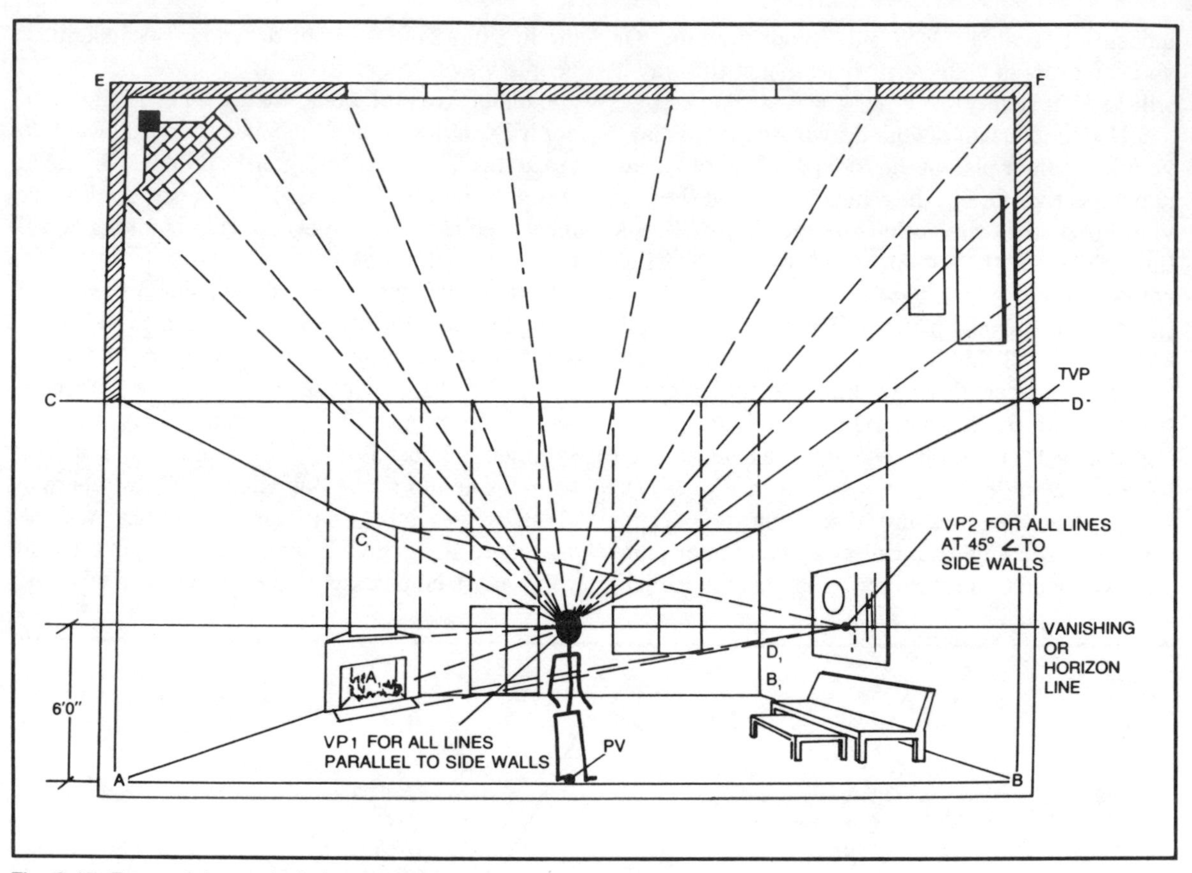

Fig. 9-17. Trace-of-the-vertical-plane drawing.

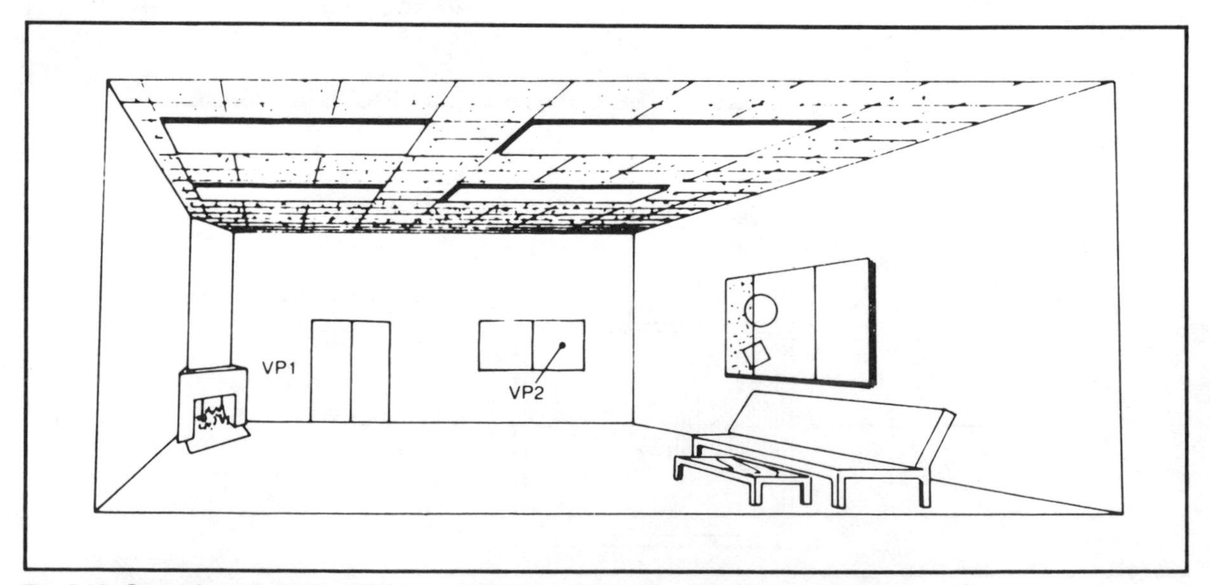

Fig. 9-18. One point perspective: VP2 is a vanishing point for lines that run at a 45-degree angle.

drawing will look like if the VP is located below a PV far to the left.

Now you are ready to locate all the vertical lines in your drawing. Place a straightedge connecting a point on the floor plan that locates a vertical line (such as point E which is the back corner of the room). Where the straightedge crosses the TVP, make a light line through the TVP and drop a perpendicular line straight down from the point of intersection of the two lines. Do the same from point F to your VP and drop a vertical line perpendicular to the TVP to establish the other corner of the back wall. This is the way you establish the location of all the vertical lines for your drawing.

To locate the horizontal lines, such as ceiling and floor lines, place your straightedge between the appropriate point on the periphery of the rectangle representing the TVP and the vanishing point. In Fig. 9-17, the ceiling lines are established on a line from C_1 to VP and from D_1 to VP. The floor lines are drawn from points A and B to the vanishing point. The back ceiling and floor lines connect the points of intersection with the perpendicular line you dropped from the TVP EE_1 and FF_1.

Follow this same procedure to establish the location and size of the windows on the left-hand wall and the large painting on the right-hand wall. Then establish the outlines of the furnishings and accessories in the room the same way. To establish the height of anything in the room, measure the actual height of the object on the outside lines of your picture frame (the rectangle ABD_1C_1 representing the TVP) and locate it along a horizontal line between that point and your vanishing point. Its position "in depth" in the room will be determined by the vertical lines you drop from your TVP line. Study the lines in Fig. 9-17. You can study this further by using your straightedge on the drawing in Fig. 9-18. Most of the trace lines have been erased.

Figures 9-17 and 9-18 show what architects call *one-point perspectives*. All horizontal lines converge on one vanishing point. Sometime, in architectural drawing it is desirable to draw a room or a building in *two-point perspective*. Two-point perspective drawings often look more natural and tend to emphasize the focal point in a room.

In our one-point perspective drawing (Fig. 9-17) with a vanishing point in the center of the drawing, the focal point becomes the couch. Yet the focal

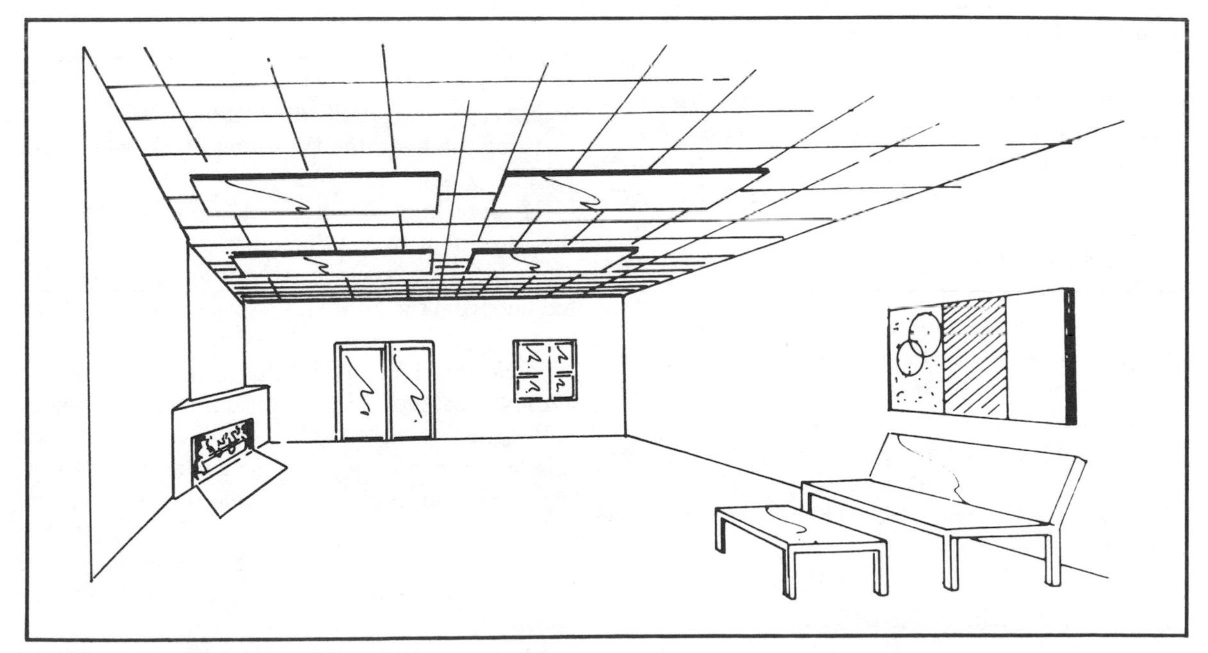

Fig. 9-19. Two-point perspective.

point of the room is not the couch at all, but the painting above the table on the right-hand wall. The other one-point drawing (Fig. 9-18) with the vanishing point moved to the far left does tend to emphasize that wall more and it is more suggestive of the focal point of the room. But the VP is so far to the left that the horizontal lines that run parallel to the TVP tend to appear distorted. They should all converge on another vanishing point to the far right of the picture frame.

Figure 9-19 shows the same room in two-point perspective with one point several inches off the page to the right and the other falling correctly within the picture. Notice especially the change in perspective of the table top and the ceiling tiles and fixtures. Some would say this is a more "honest" representation of the finished room.

To make these two-point perspective drawings, simply extend your vanishing line and locate another vanishing point to the far right or left off the picture along this line. Draw your back ceiling and floor lines and all other lines parallel to your TVP so that they converge on this distant vanishing point.

Other vanishing points along the vanishing line can be used for accurate rendering of objects sitting at oblique angles in the room (such as the corner fireplace). Each *set* of parallel lines will have its own VP; remember? For all practical purposes, however, you can sketch freehand. Keep the other vanishing points in mind. Locate them mentally once the position of the object is established by the TVP method.

Technically, all of your vertical lines should be converging on a common vanishing point. In drawings of rooms and small structures, this vertical convergence is so slight that it can be ignored and all lines can be drawn parallel just as you do the set of horizontal lines parallel to the TVP in one-point drawings. If you were making a drawing of city skyscrapers from a low angle, however, their vertical lines would converge noticeably on a vanishing point far above them.

OVERCOMING DISTORTION

Distortion in perspective drawings relates to the spacing of vanishing points. Placement of vanishing points also relates to the distance effect in the drawing. Placing vanishing points close together tends to distort the edges of a drawing and pull the viewer in close to the subject matter. Spreading out the vanishing points will eliminate the distortion and give the effect of the observer having backed off for a more distant view.

Distortion due to excessively close vanishing points is the most common because it is easier to work with vanishing points that are close together. If you work with your vanishing points spread too far apart, however, your drawing will go "flat" and appear to have little dimension due to minimal convergence of lines.

Generally, convergence is less near the center of the drawing and increases toward the edges. If you try to draw too wide an angle of view, distortion at the edges of your drawing becomes unacceptable.

DETERMINING HEIGHTS AND WIDTHS

When drawing in furnishings and wall hangings, doors and windows, etc., most heights can be found by relating to your wall height. The standard wall height is 8 feet. Measure off eight equal lengths, or 8 feet to scale, on the outside vertical edge of your wall and use this as a guide to find the proper heights. As in Fig. 9-20, a 7-foot high door is drawn by using the line from the 7-foot mark back to the vanishing point.

A door knob 3 feet off the floor is located on the three foot to VP line. If you want to draw a 6-foot grandfather clock in the foreground, the dotted line from the 6-foot mark provides you with its height.

The vanishing line for this drawing was picked at the 4-foot level. This makes our clock four-sixths below the two-sixths above eye level. Notice that this ratio remains the same in all parts of the drawing. A six-foot high cabinet against the back wall will appear much smaller than the grandfather clock in the foreground, but four-sixths of the cabinet will still be below eye level and two-sixths above.

Drawing a three-foot chair at the same "depth"

as the door knob requires carrying the door knob height across the room to the place where you want the chair. A 1-foot diameter lamp hanging from a 1-foot long chain directly over the chair is measured by chasing the 6- and 7-foot lines in to the "depth" of the door knob then coming out with parallel lines to a position over the chair. (See Fig. 9-20).

The width of objects is similarly gauged by measuring to scale across the bottom of the picture frame and "chasing" lines back to the vanishing point to get the attenuation of the width of objects. The grandfather clock is 3 feet wide and so is the door on the back wall. Notice (Fig. 9-20) how the width attenuates as objects are placed farther back.

FINISHING TOUCHES

Don't forget that round objects and circles will foreshorten and appear to be ellipses when drawn in perspective. The only exception is when the circle is exactly parallel with the face of the observer (when seen "front face"). Cylinders tend to become cone shaped in perspective drawings. This will not be noticeable in vertical cylinders unless you are drawing huge storage tanks from a worm's eye view. But horizontal cylinders, even in room-size drawings, will take on the shape of a section of a cone with the apex of the cone being the vanishing point.

Shade and *shadow* make up the final touch to a good perspective drawing of any kind. Shade is defined as darkening that exists when a surface is turned away from the source of light. Shadows exist when a surface is facing the light source, but receives no light due to the presence of an intervening object.

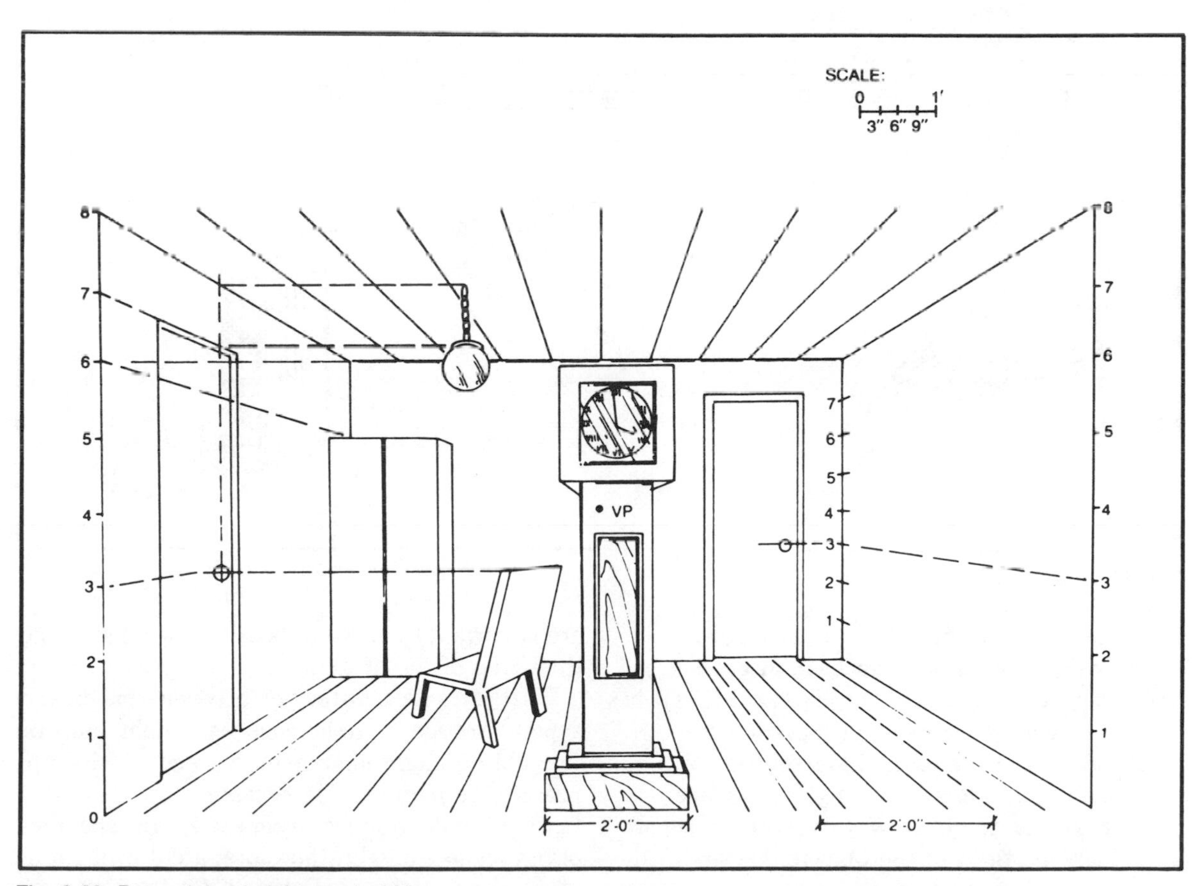

Fig. 9-20. Determining heights and widths.

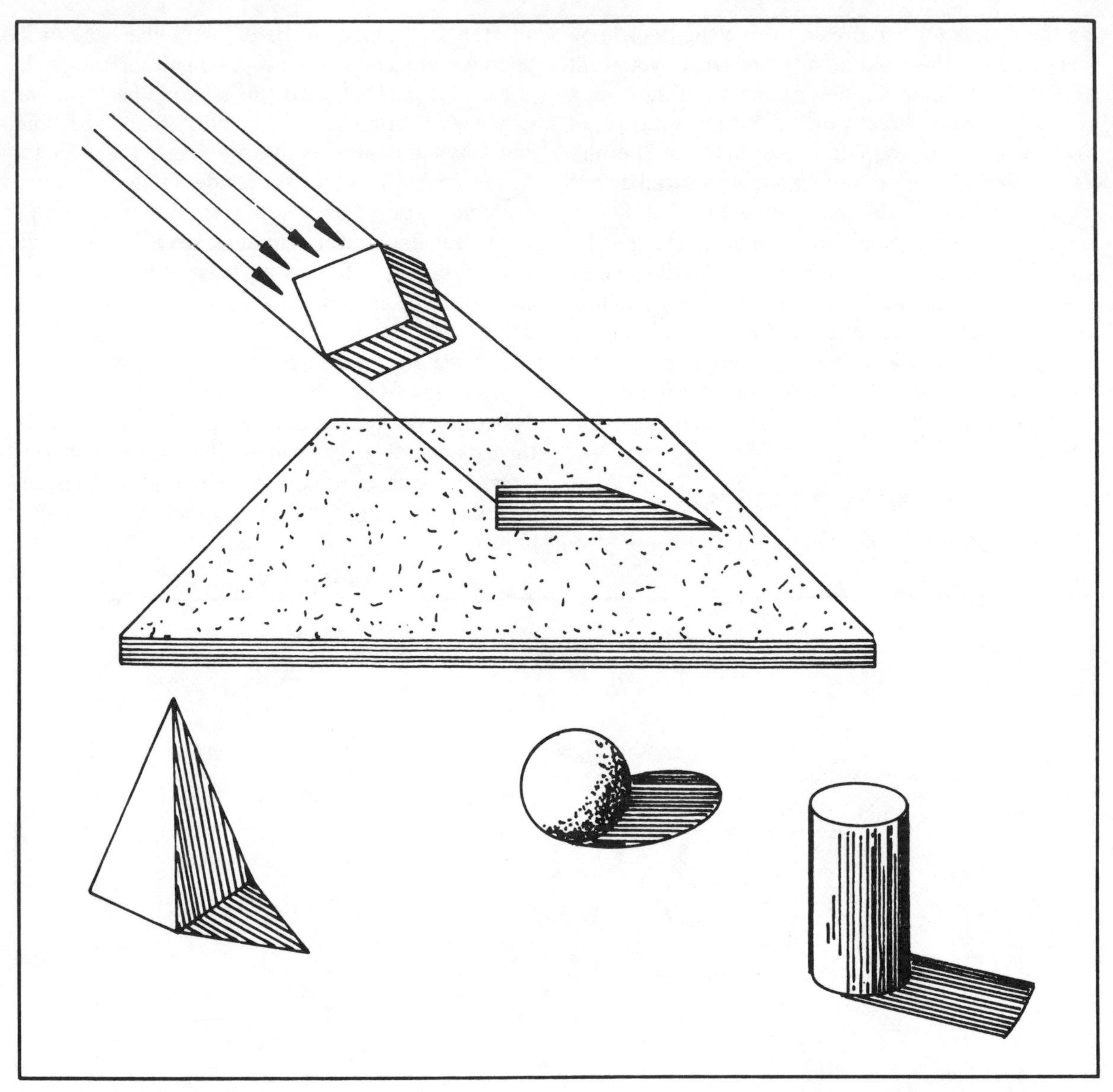

Fig. 9-21. Shade and shadow.

The shapes in Fig. 9-21 all have several surfaces in light and some surfaces in shade. The surface they are sitting on is all turned toward the light, but is in shadow where the shapes intervene between the surface and the light source. Each object casts its shadow on the lighted surface.

Notice that there is always a *shade line* separating those portions of the objects that are in the shade from those that are in the light. Shade lines are important to an artist because they determine the shape of the shadow.

Artists make a distinction between sunlight and a point source of light such as a light bulb or firelight. Sunlight illuminates a scene by what appears to be parallel rays because of its great distance (93 million miles) from the Earth. Shadows of figures in sunlight, therefore, are drawn parallel with one another (as in Fig. 9-22). The closeness

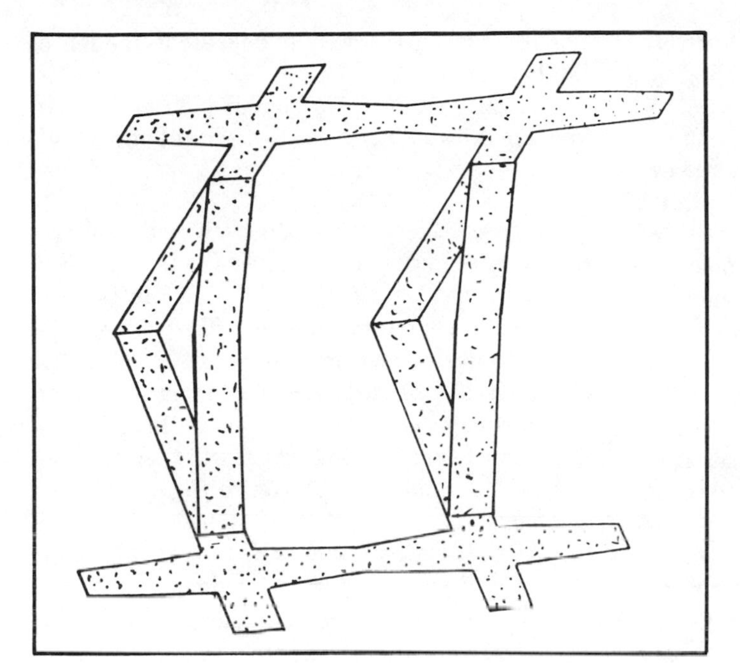

Fig. 9-22. Drawing sunlight shadows.

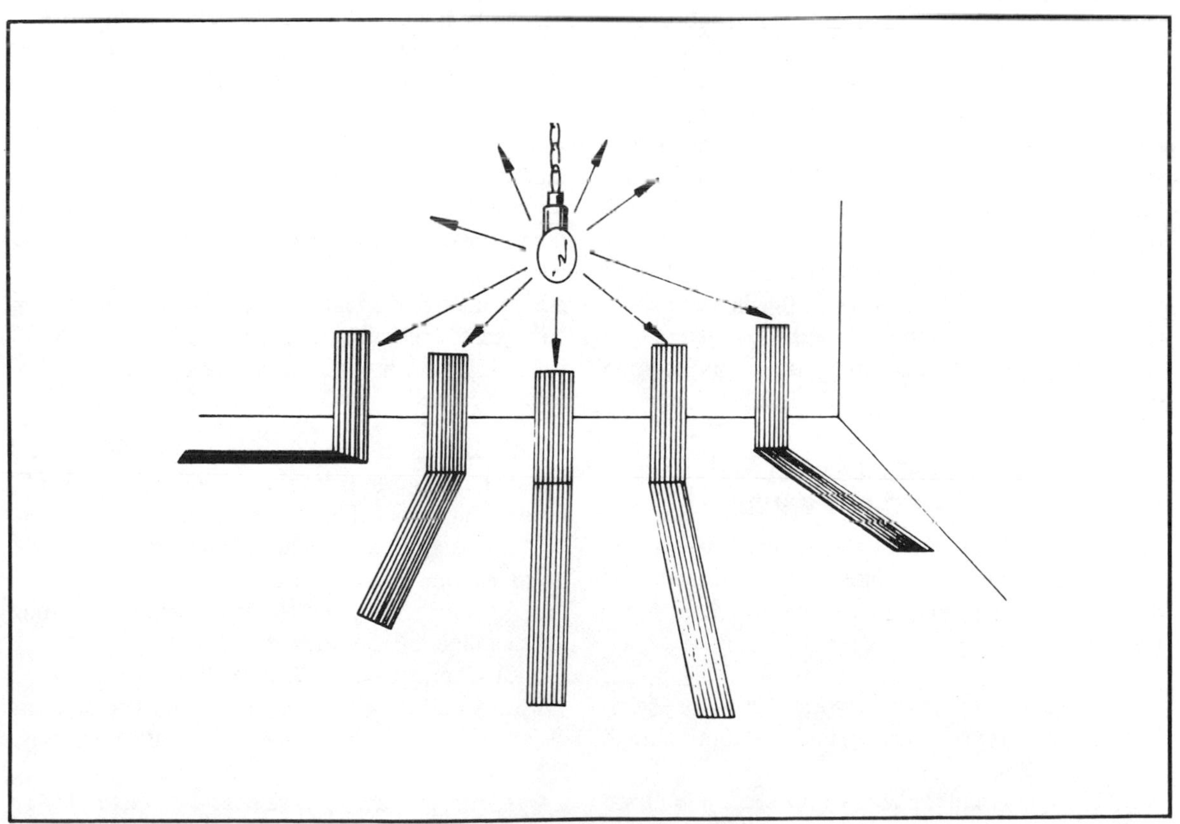

Fig. 9-23. Shadows in a radial pattern resulting from a close source of light.

of point sources of light causes shadows to diverge into radial patterns around the source of light (as shown in Fig. 9-23).

The vanishing point for the light rays illuminating your drawing will always be located at the light source. Therefore, the vanishing point for sunlight is 93 million miles away. The shadow, however, will have a vanishing point of its own. It will follow the rules for any other object: all sets of parallel lines converging on a common vanishing point lying along a vanishing line that is always at the eye level of the observer.

Study the shade and shadow created by the point source of hanging light above the chair shown in Fig. 9-24. Single-arrowed lines are used to indicate diverging light rays used to locate shadows. Dotted lines represent guidelines that project the light source onto floor, wall, and ceiling surfaces, etc. Double-arrowed lines (coming from large dots) are guidelines to determine the directions of shadows cast by shade lines (perpendicular to the surfaces).

From studying Fig. 9-24, you can set down this rule about shade and shadow: a point source of light can be thought to be radiating diverging rays which locate all shadow points. Shadow direction is determined by lines diverging from a point that is directly below the source of light.

Observing shade and shadow in sunlight and under artificial light and sketching what you see is the best way to develop expertise at finishing off your drawings with neat and professional-looking shading and shadowing.

SIMPLE RECTANGULAR ISOMETRIC DRAWINGS

Another way to represent the appearance of what you are drawing is through the use of isometric drawings. The principles can be stated in three rules.

☐ All vertical lines in your section or elevation drawings remain vertical in the isometric drawing (Fig. 9-25).

☐ All horizontal lines in your section or elevation drawings are drawn 30 degrees to the left of the horizontal.

☐ All vertical lines in your plan view drawings are drawn 30 degrees to the right of the horizontal.

Figure 9-25 shows a plan view and an elevation represented in an isometric "perspective" in which you can identify lines AB, CD, EF, BC, etc. A 30/60/90-degree triangle and T square are essential for making isometric drawings. You can use dividers to step off distances on your plan and elevation drawings. You can then transfer them directly to the isometric drawing which will then turn out to have the same scale as your orthographic plan and elevation drawings. Proportional dividers can be used to transfer distances if you want to change the scale of your isometric drawing.

To transfer lines that are neither vertical nor horizontal from orthographic to isometric drawings, you must first find the end points of the line in terms of a vertical and horizontal line and then connect the end points with a straightedge.

To make a practical application of drawing non-isometric lines in isometric drawings, make an isometric drawing of a rectangular house by projecting the roof plan and front and side elevations (as suggested in Fig. 9-26). The hip roof construction will give you four lines that are neither horizontal or vertical in the plan and elevation drawings. To transfer them to the isometric drawing, do the following step by step.

☐ Using your dividers, step off the vertical distance measured from the ground line to point C, level with the roof ridge shown in the side elevation.

☐ Using this vertical distance, locate point C in your isometric drawing.

☐ Find point A on the isometric by stepping off distance CA on your plan view.

☐ Connect points A and B.

☐ Use the same procedure and transfer the other three roof ridges to your isometric drawing.

Isometric drawings are not as "true to life" as

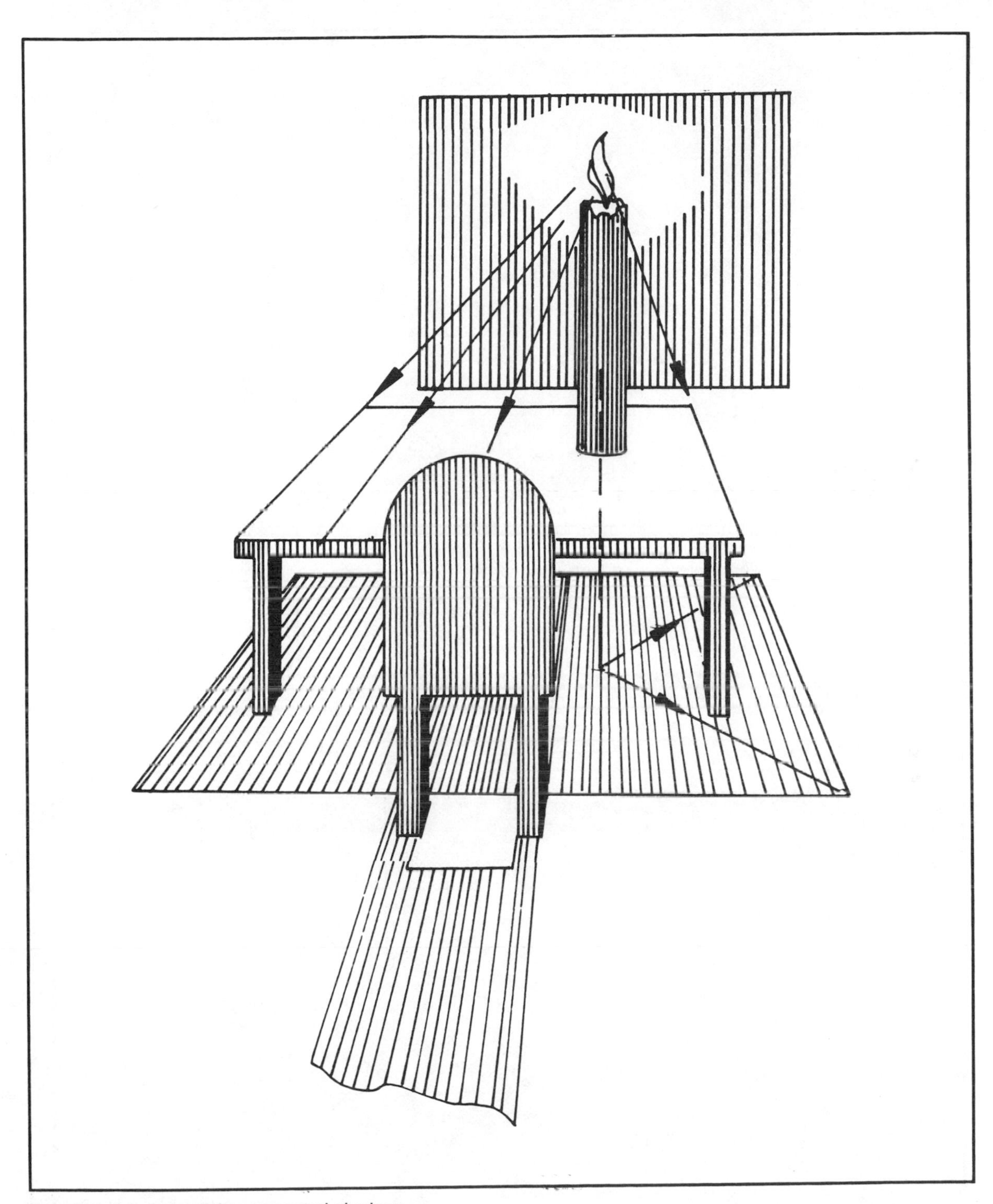

Fig. 9-24. A hanging light source and shadows.

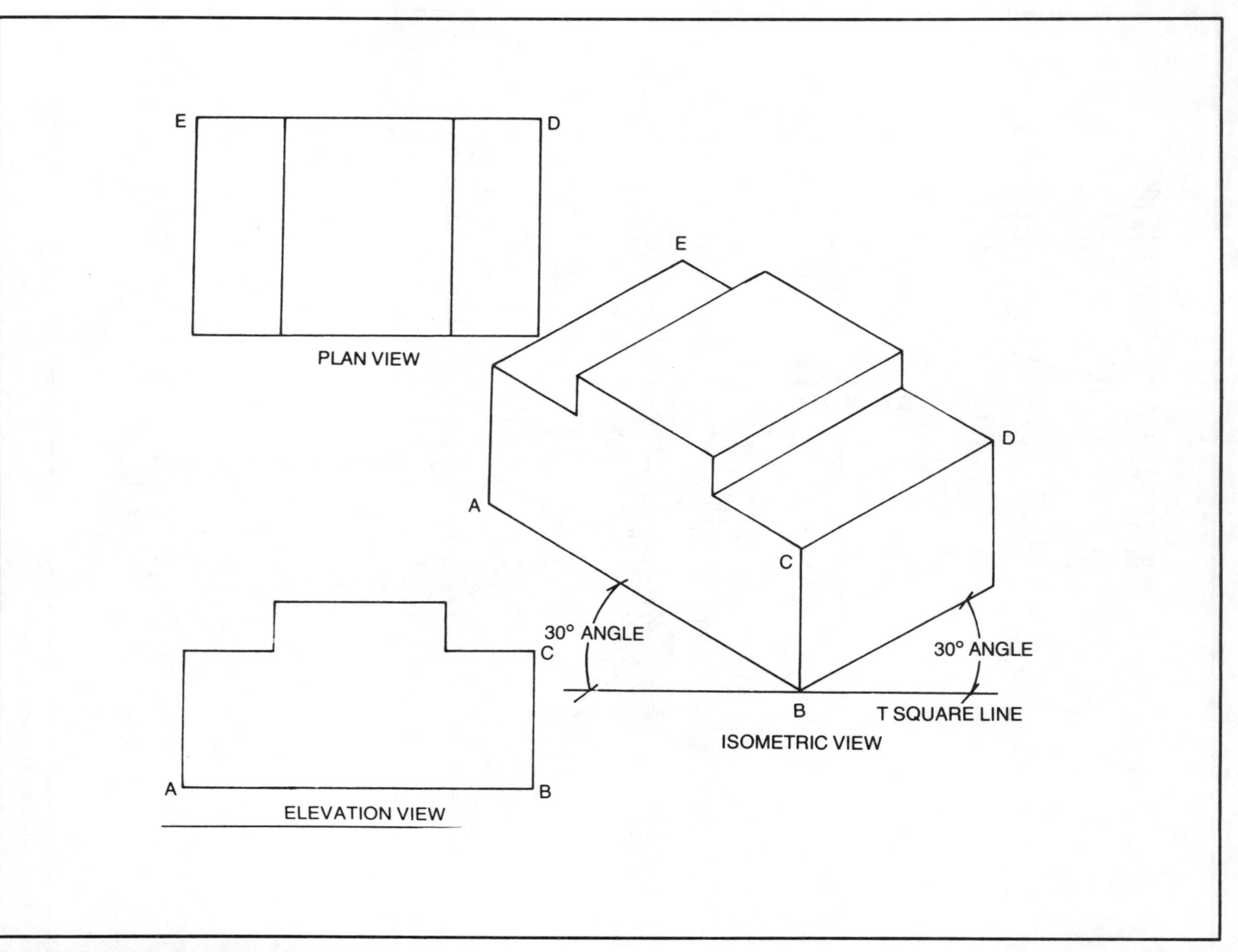

Fig. 9-25. A simple rectangular isometric drawing.

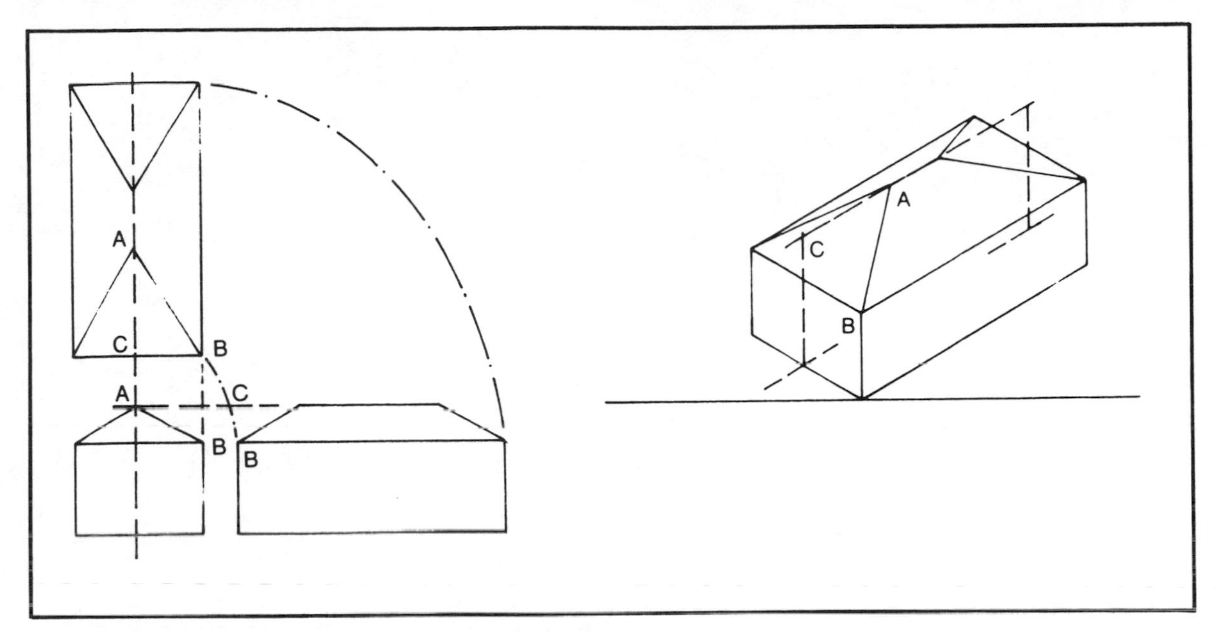

Fig. 9-26. Drawing non-isometric lines on isometric drawings.

perspective drawings and they have a "mechanical" feel, but they are true to scale and they do a beautiful job of communicating information to the builder. As a rule, you use isometric drawings to communicate details about the house plans and perspective drawings to communicate the feeling you want to design into your plans.

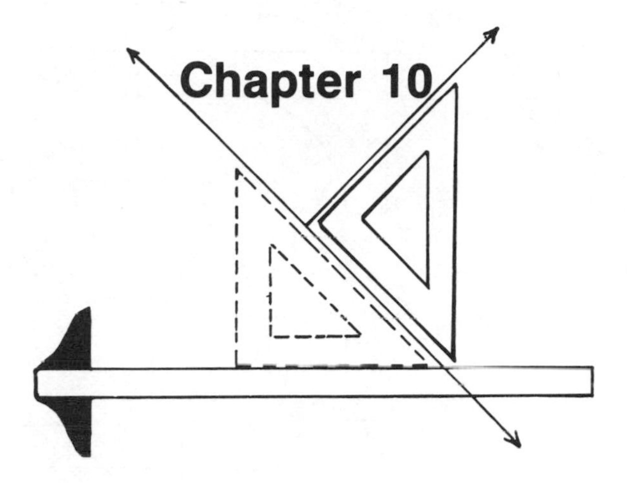

Millwork

Millwork is the term applied to building materials made of finished wood. It includes doors, windows and their frames, shutters, entrances, porchwork, stairs and their railings, mantels, moldings, interior trim, cabinet work, paneling, and all built-ins. It does not include finished wood flooring, ceilings, and siding.

Lumberyards and building supply firms usually sell both rough lumber and millwork, and there are those that sell millwork and finished building products only. Some of the larger lumberyards may have their own millwork shops that manufacture specially designed architectural millwork.

Most of the millwork you will need in house construction is standardized material manufactured elsewhere but retailed in the lumberyard.

For the do-it-yourself planner, a catalog of millwork is invaluable.

Figure 10-1 shows cross sections of several stock moldings sold at many lumberyards. Using these shapes in various combinations it is possible to build almost any type of finished wood structure.

There are millwork firms that manufacture beautifully designed building components such as entrances, stairways, fireplace mantels, doors, and windows. Most lumberyards handling millwork will have their catalogs and details available and may loan you a catalog.

Some people prefer to design their own built-ins rather than buy standard, ready-made ones. Usually, kitchen cabinets can be planned to fit your family needs better, in many ways, than ready-made units. There are other units that can also be planned: beds with storage drawers below, built-in dressers, built-in headboards, etc.

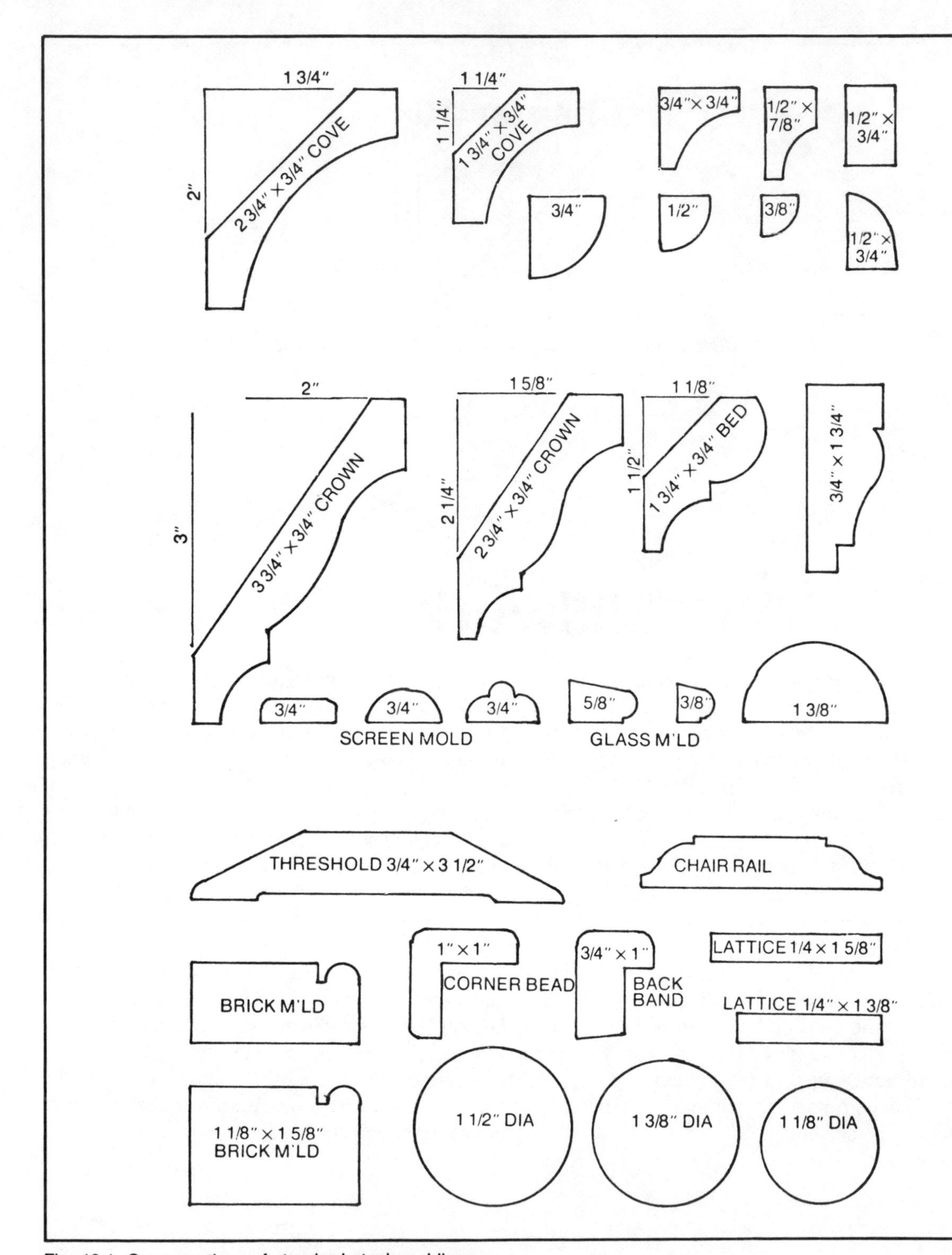

Fig. 10-1. Cross sections of standard stock moldings.

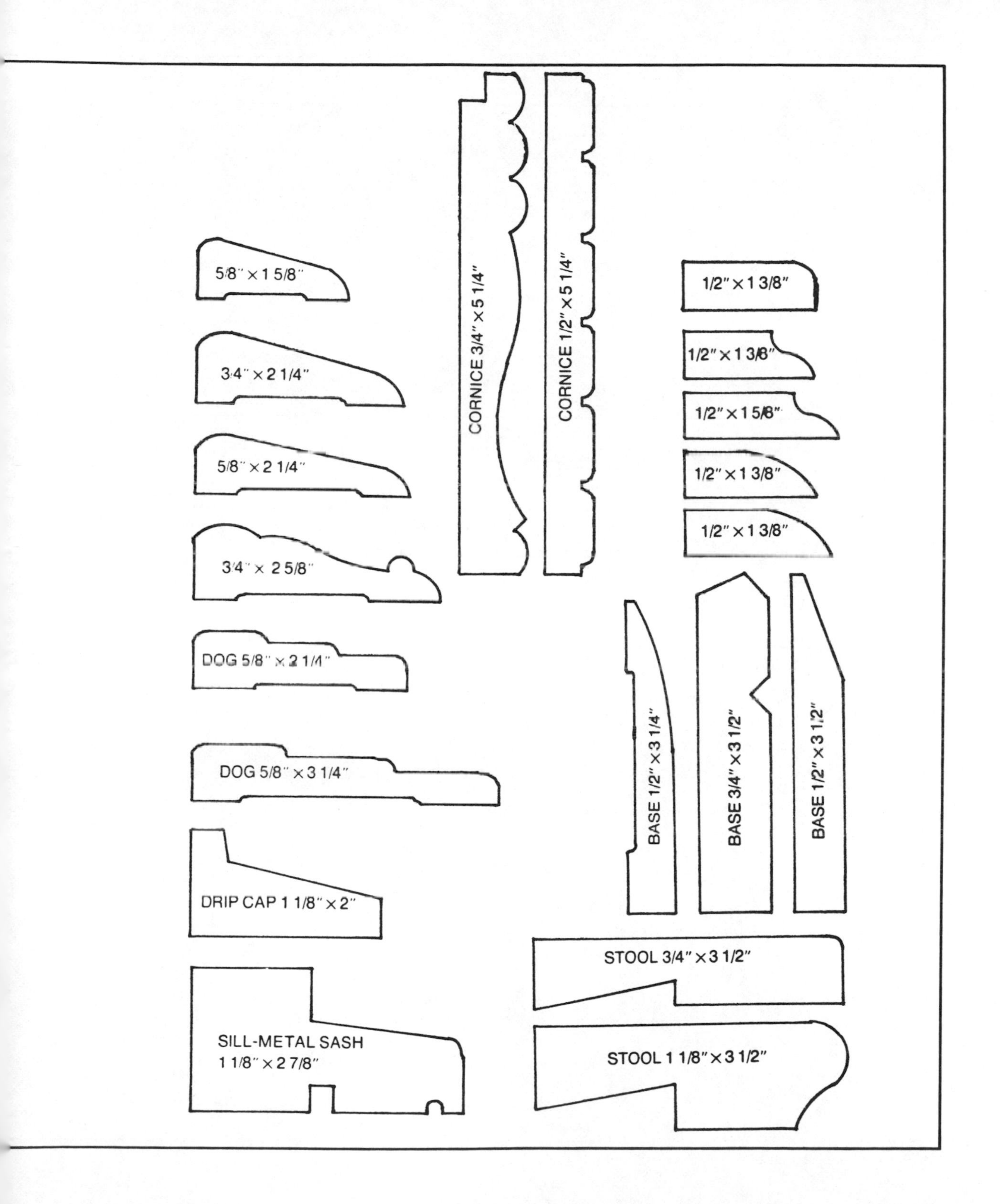

5/8" × 1 5/8"
3/4" × 2 1/4"
5/8" × 2 1/4"
3/4" × 2 5/8"
DOG 5/8" × 2 1/4"
DOG 5/8" × 3 1/4"
DRIP CAP 1 1/8" × 2"
SILL-METAL SASH
1 1/8" × 2 7/8"
CORNICE 3/4" × 5 1/4"
CORNICE 1/2" × 5 1/4"
1/2" × 1 3/8"
1/2" × 1 3/8"
1/2" × 1 5/8"
1/2" × 1 3/8"
1/2" × 1 3/8"
BASE 1/2" × 3 1/4"
BASE 3/4" × 3 1/2"
BASE 1/2" × 3 1/2"
STOOL 3/4" × 3 1/2"
STOOL 1 1/8" × 3 1/2"

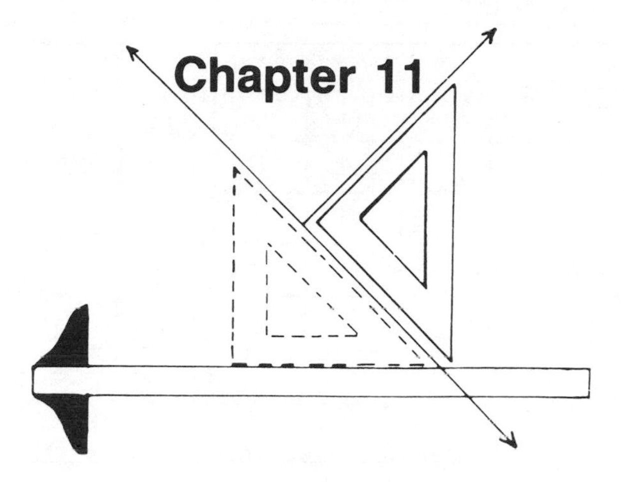

Room Sizes

If, like most prospective home planners, you have been studying plans found in the Sunday newspaper and home betterment magazines, you have gained many ideas about suitable room sizes.

Governmental and private studies have been done on such matters, and these have resulted in the establishment of minimum standards, such as those shown in Table 11-1, which were taken from the U.S. Department of Housing and Urban Development book on minimum standards. These standards are mandatory on houses which are financed or guaranteed by the federal government.

Large families were once raised in log cabins with one room no larger than 15 × 20 feet, and most were smaller. Such living was not healthy; thus more space and better living conditions are required by law.

When planning your own home it is unlikely that you will have rooms as small as those listed in Table 11-1. In planning your own home it is probable that you will have special conditions that require more space. Often it is helpful to make cutouts from paper of your principal pieces of furniture (scaled to the 1/4 inch of a floor plan) and then shift them about on a preliminary floor plan. You are then better able to draw a final floor plan that accommodates your possessions.

A home in which you and your family will spend so many hours of your lives should be as well planned and efficient for living as a factory where the proper location of machinery can mean profit or loss.

Unless building money is not a factor, it's a good idea to consider lumber lengths when planning room size. Lumber costs increase dramatically after 16 feet.

Particularly at the present time when building costs have risen faster than average incomes, planners must give careful consideration to the most economical uses of materials. Planning becomes something of a juggling act: somehow you must balance needs and costs and dreams.

Table 11-1. Minimum Sizes for Separate Rooms.

NAME OR SPACE	MINIMUM AREA (IN SQ. FT.) (6)					
	LU WITH 0 BR	LU WITH 1 BR	LU WITH 2 BR	LU WITH 3 BR	LU WITH 4 BR	LEAST DIM.
LR	—	160	160	170	180	11′-0″
DR	—	100	100	110	120	8′-4″
BR(PRIMARY) (1)	—	120	120	120	120	9′-4″
BR (SECONDARY)	—	—	80	80	80	8′-0″
TOTAL BR AREA	—	120	200	280	380	—

Minimum Room Sizes for Combined Spaces (2)

NAME OF SPACE (3)	MINIMUM AREA (IN S. FT.) (6)				
	LU + WITH 0 BR	LU + WITH 1 BR	LU + WITH 2 BR	LU + WITH 3 BR	LU + WITH 4 BR
LR + DA	—	210	210	230	250
LR + DA + K	250	—	—	—	—
LR + DA + K (4)	—	270	270	300	330
LR + SL	210	—	—	—	—
K + DA (5)	100	120	120	140	160

Abbreviations:

LU = Living Unit
DR = Dining Room
BR = Bedroom
B = Bath
LR = Living Room
DA = Dining Area
SL = Sleeping Area
K = Kitchen

(1) Primary BR to have at least one uninterrupted wall space of 10 ft.

(2) The minimum space of a combined room shall be the sum of the dimensions of the single rooms involved except for the overlap or combined use of space.

(3) For two adjacent spaces to be considered a combined room, the horizontal opening between spaces shall be at least 8 ft. wide except that between kitchen and dining room functions, the opening may be reduced to 6 ft.

(4) A combined LR-DA-K shall comply with the following: (a) The food preparation-cooking area shall be screened from the living room-sitting area. (b) The clear opening between the kitchen and dining area shall be at least 4 ft. wide.

(5) These required minimums apply only when eating space is in the kitchen.

(6) The floor area of an alcove, or recess off a room, having a least dimension less than that required for the room, shall be included only if it is useful for the placement of furniture.

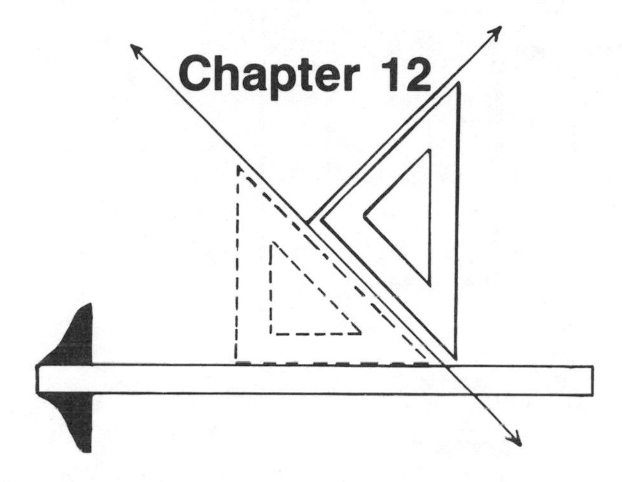

Kitchens

Less than 150 years ago, the kitchen with its fireplace was the only regularly heated room in a home. It was, necessarily, the focus of the family's activity. For the most part, the kitchen is still an important place, and careful consideration should be given to its planning and equipment. Unlike other parts of the house, kitchen cabinets and equipment are permanently installed and cannot be easily rearranged.

Kitchens may be located at the front of the house where you can view the street, or at the back so you can watch over the children's play or have access to a porch for outdoor dining and informal entertainment.

Study how your kitchen will be used and make it as adequate, efficient, and good looking as possible; someone will be spending a lot of time there.

KITCHEN USES AND TYPES

Basically the uses of a kitchen are:

☐ The storage of foods.

☐ Food preparation. This requires a cooking range, hot and cold water supply, counter work space, and storage for pots and pans.

☐ Food service. This requires serving counters and storage space for dishes and tableware.

☐ Cleanup. This requires a sink, towel bar, often a dishwasher, disposal, possibly a compactor, or a pantry.

The above uses will differ from family to family. A family with young children will have different needs from an elderly couple. A family in which all adults are working will require a unique kind of kitchen. Certainly the family that loves to entertain requires special facilities.

The person planning his own home must take all these factors into account. He must visualize how the kitchen work will be accomplished and arrange the facilities to the best advantage.

The usual types of kitchen arrangements are:

☐ The U-shaped type (Fig. 12-1).

☐ The L-shaped type (Fig. 12-2).

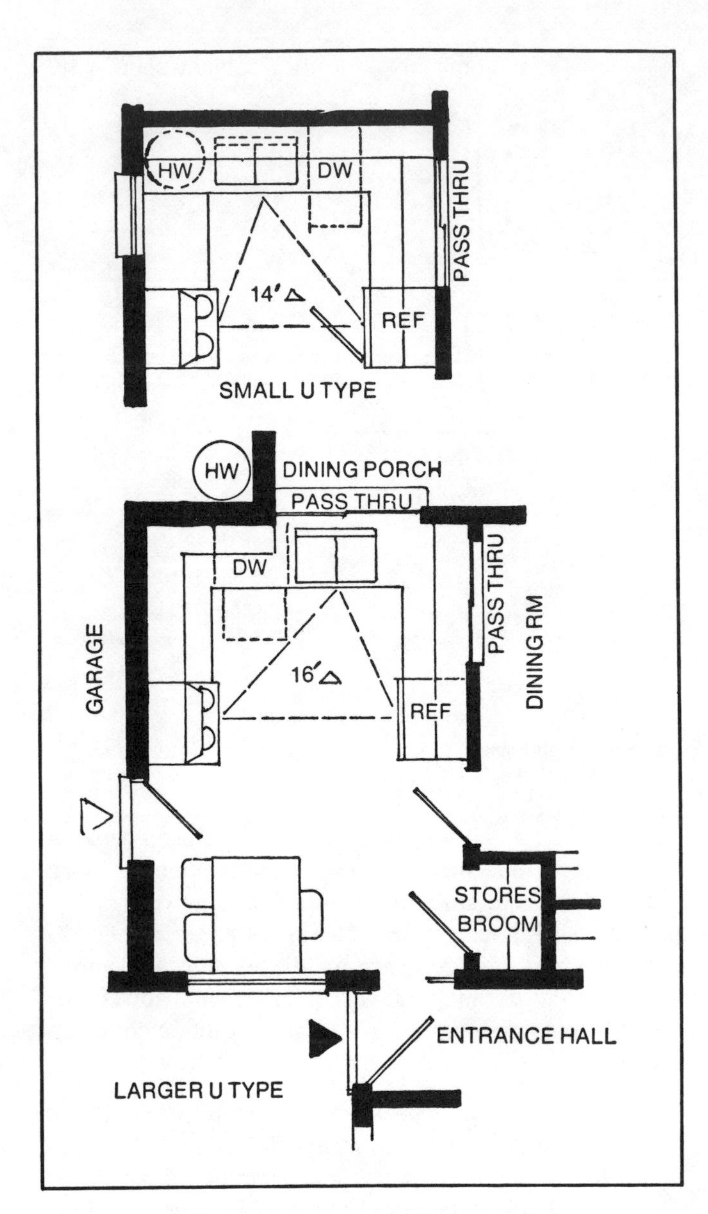

Fig. 12-1. Two U-shaped kitchens.

□ The two-wall type (Fig. 12-3).

□ The one-wall type and variations of all the above.

BASIC CONSIDERATIONS

The principal activities in a kitchen are largely controlled by the placement of the range, sink, and refrigerator. If these appliances can be kept within a triangle (Figs.12-1, 12-2, 12-3), it will restrict travel to within the triangle and thereby lessen the work. Home economists call this the *work triangle*. They have found that if the perimeter of this triangle can be kept within 15 feet or 20 feet, the kitchen will be a step-saver. If this perimeter is greater than 20 feet in a kitchen you are planning, try to rearrange these three appliances.

The range and refrigerator can be located without much restriction, but it is usually wise to place the sink so as to make use of the plumbing pipes

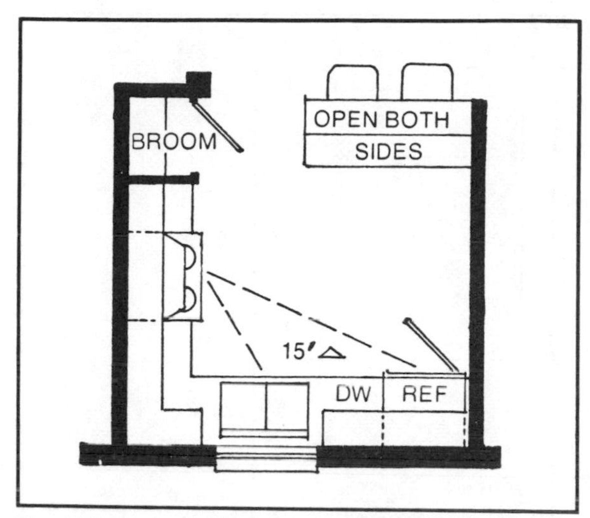

Fig. 12-2. An L type kitchen with peninsula.

of an adjacent or upstairs bath. Do not, however, let this consideration wreck an otherwise very suitable plan. The piping is a one-time cost, while a satisfactory plan may please for a lifetime.

As far as possible, avoid making the kitchen a passageway to other areas of the house, or at least keep the passage out of the work triangle. Ideally, groceries should be brought directly from their point of entry, usually the garage, into the kitchen where they will be stored.

Plan a place where someone can keep the cookbooks, sit to study them, plan a meal, make a phone call, or wait for a pot to boil.

If there are small children, try to plan some play space just outside the work area. Or at least have a window looking out on outdoor play areas.

Ask yourself questions about your proposed kitchen. Will you be able to easily transport food from the kitchen to the dining room? Is there a kitchen pass-through where food or dishes can be passed from the kitchen to the dining room? If you prefer to eat on the porch in good weather, can you serve foods there without going through other rooms? These are some of the things you should consider when planning a kitchen.

KITCHEN FURNISHINGS AND EQUIPMENT

Most kitchen furnishings become standardized as to dimensions. And whether your kitchen furniture will be job-built or made in a cabinet shop, you should use standard sizes in most instances.

Counter tops are usually 24 inches wide and stand 36 inches high (Fig. 12-4). Under-counter (base) units range in length from 12 inches to 48 inches in 3 inch steps. Wall cabinets are usually deep enough to hold 10 inch plates; such cabinets usually are 15 inches, 18 inches, 24 inches or 30 inches tall. Lengths match those of the base units.

The Department of Housing and Urban Development provides minimum standards for the size of kitchen furnishings (Fig. 12-5). These requirements are good guidelines and should help you keep your plans within reasonable bounds.

A glance through a cabinet manufacturer's catalog will show many specialized cabinet units, both base and wall types. There are blind corner units that make corners more accessible and units with doors on two sides for use between the kitchen and dining area. There are door height cabinets to accommodate built-in ovens.

Because most building contractors generally use only factory built cabinets, it is usually wise to

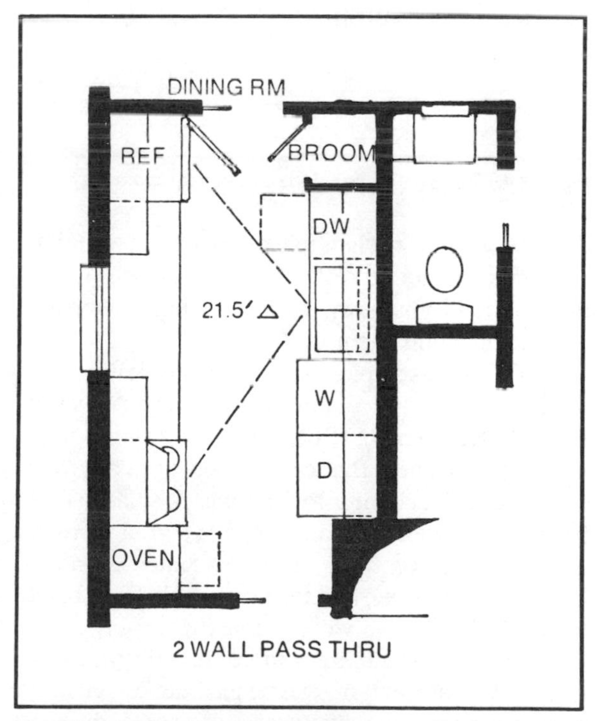

Fig. 12-3. A two-wall type of kitchen.

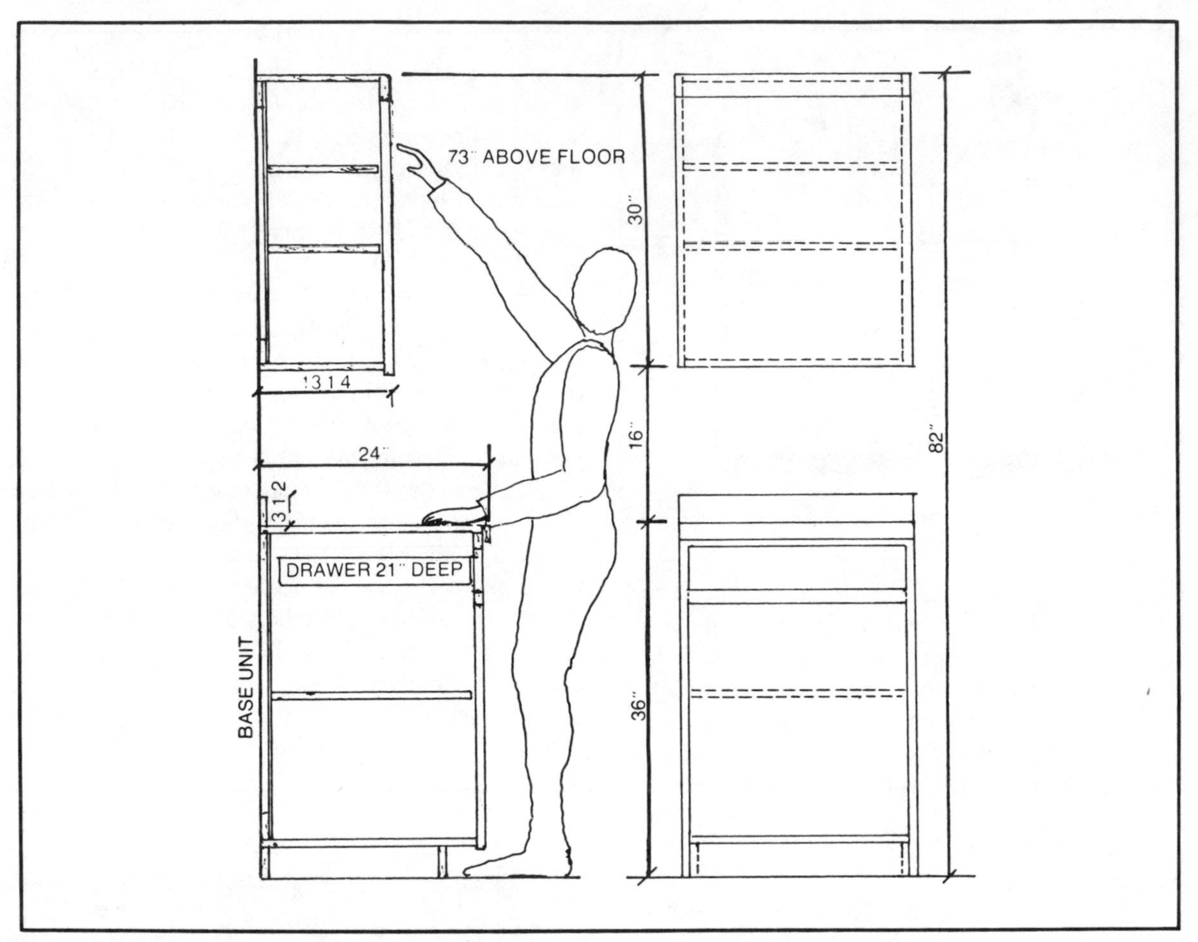

Fig. 12-4. Standard dimensions of kitchen cabinets.

dimension your kitchen to use standard size units. But you need not be restricted to stock sizes if you plan to build your own kitchen or have someone competent build them on the job

Tops of base units are usually specially made to fit available space and are commonly covered with plastic laminate (PL). The PL is a thin, somewhat flexible sheet that can be molded to run over the edge of the counter to form a back splash. This PL is waterproof and can be had in a great variety of colors and patterns. It even comes in simulated-wood finishes, and thus is suitable for cabinet fronts and doors as well as tops.

Drawers, which are a more difficult part of a cabinet to make and fit, can be job-built or you may use factory-made polystyrene one-piece drawer units with seamless rounded easy-to-keep-clean corners. They come with metal slides to be attached to the side walls of cabinets and need only front pieces and pulls matching other cabinetwork. Figure 12-6 shows drawer opening widths and recommended minimum heights.

Adequate removal of heat and cooking odors is desirable and can be accomplished by a small in-wall exhaust fan over the range. A hood is not necessary if the underside of the cabinet over the range is protected by 1/4 inch of asbestos board and sheet metal. A good combination is a hood with exhaust fan and light. Such fans tend to be noisy, but this can be overcome if they are in a weatherhood above the roof.

Another type of range hood requires no con-

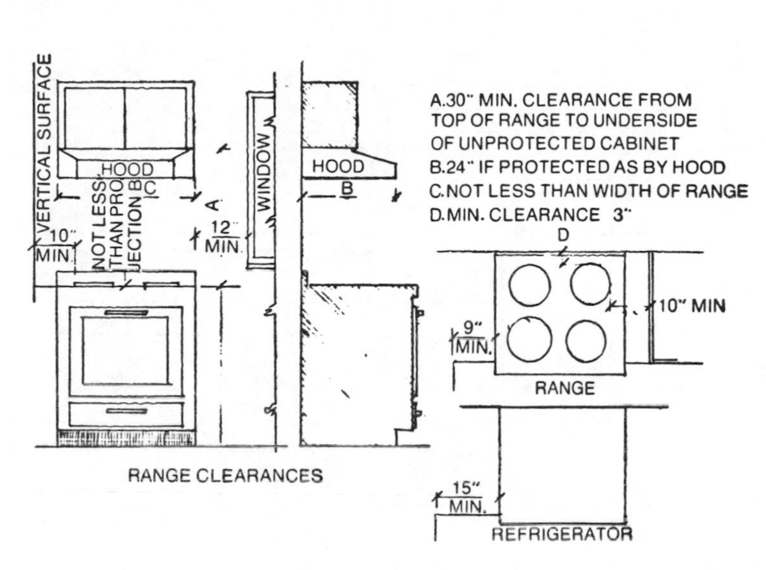

Countertop and Fixtures

WORK CENTER	NUMBER OF BEDROOMS 0	1	2	3	4
	MINIMUM FRONTAGE IN INCHES				
Sink	18	24	24	32	32
Counter Top Each Side	15	18	21	24	30
Range or Cooktop Space	21	21	24	30	30
Counter Top One Side	15	19	21	24	30
Refrigerator Space	30	30	36	36	36
Counter Top One Side	15	15	15	15	18
Mixing Counter Top	21	30	36	36	42

1 When a dishwasher is provided, a 24" sink is acceptable.
2 Where a built-in wall oven is installed, provide an 18" counter top adjacent.

Storage Area

	NUMBER OF BEDROOMS 0	1	2	3	4
Minimum shelf area (sq. ft.)	24	30	38	44	50
Minimum drawer area (sq. ft.)	4	6	8	10	12

1 A dishwasher may count as a 4 sq. ft. of base cabinet storage.
2 Wall cabinets over refrigerators or above 74" do not count.
3. Count only 50% of shelf area of blind corner cabinets. However, count actual area of revolving shelves.
4 For drawer area in excess of minimum amount, count as shelf area if 6" deep or more.

Fig. 12-5. FHA-HUD minimum requirements for kitchens.

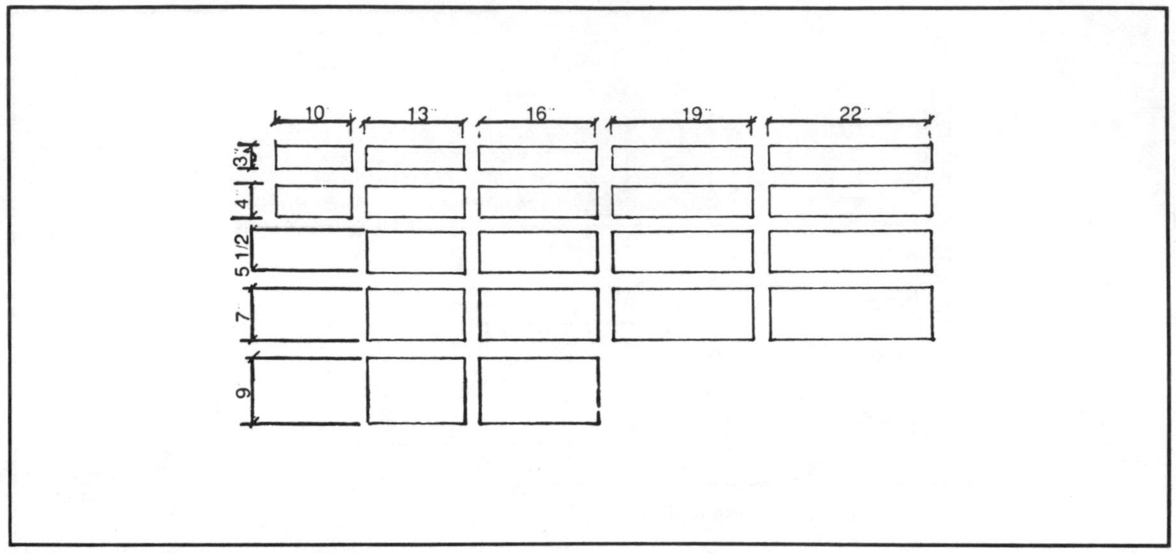

Fig. 12-6. Standard drawer dimensions.

nection to the outside: it simply recirculates the air through a filter that removes odors and grease particles. This type does not take away heat and is less desirable.

A TYPICAL KITCHEN PLAN

Look at the floor plan in Fig. 12-7. Even though the kitchen arrangement is comparatively simple, a lot of decisions had to be made in order to produce such a plan.

For example, by placing an under-counter water heater in the blind corner next to the sink, pipes to both kitchen and bath fixtures are kept short, which saves on the cost of piping as well as heat and water loss. Note, however, that such a concealed heater must be readily accessible in case of leakage, so it will be necessary to make the base cabinet (between the water heater in the corner and the range) movable.

The window (for light and ventilation) is placed in the only available wall space; its right side is placed on an 8 inch modular line, with space allowed for a wall cabinet above the water heater (Fig. 12-8). Because a range must be located at least 12 inches away from a window, the location of the range is established and this allows 27 inches for the width of the movable base unit.

The double sink drops into a 36 inch counter space. Since there are no shelves under the sink, a front with doors is required.

Whether or not a dishwasher is to be installed, it is usually best to plan for a 24 inch wide space, just in case. If no dishwasher, a 24 inch base unit, with or without drawers, will be useful.

Matching the 27 inch wide base unit next to the range, on the opposite wall there is a 27 inch base unit. There is also space for a freestanding refrigerator.

The wall cabinets follow the lengths of the base units. Except where the sink, range and refrigerator are, 30 inches high, 12 inches deep wall cabinets are standard. The top of the cabinets are level with the head of the door trim. This leaves about 16 inches between the counter top and the underside of the wall units.

There is a 24 inch high two-shelf wall cabinet above the sink. It is 36 inches wide, a good place for glassware.

A wall cabinet above the refrigerator must allow space for air circulation (between the cabinet and refrigerator top). So the cabinet is only 15 inches tall.

The cabinet above the range must be at least as long as the range. But it is only 15 inches high with a single shelf because of necessary clearance above the cooking area.

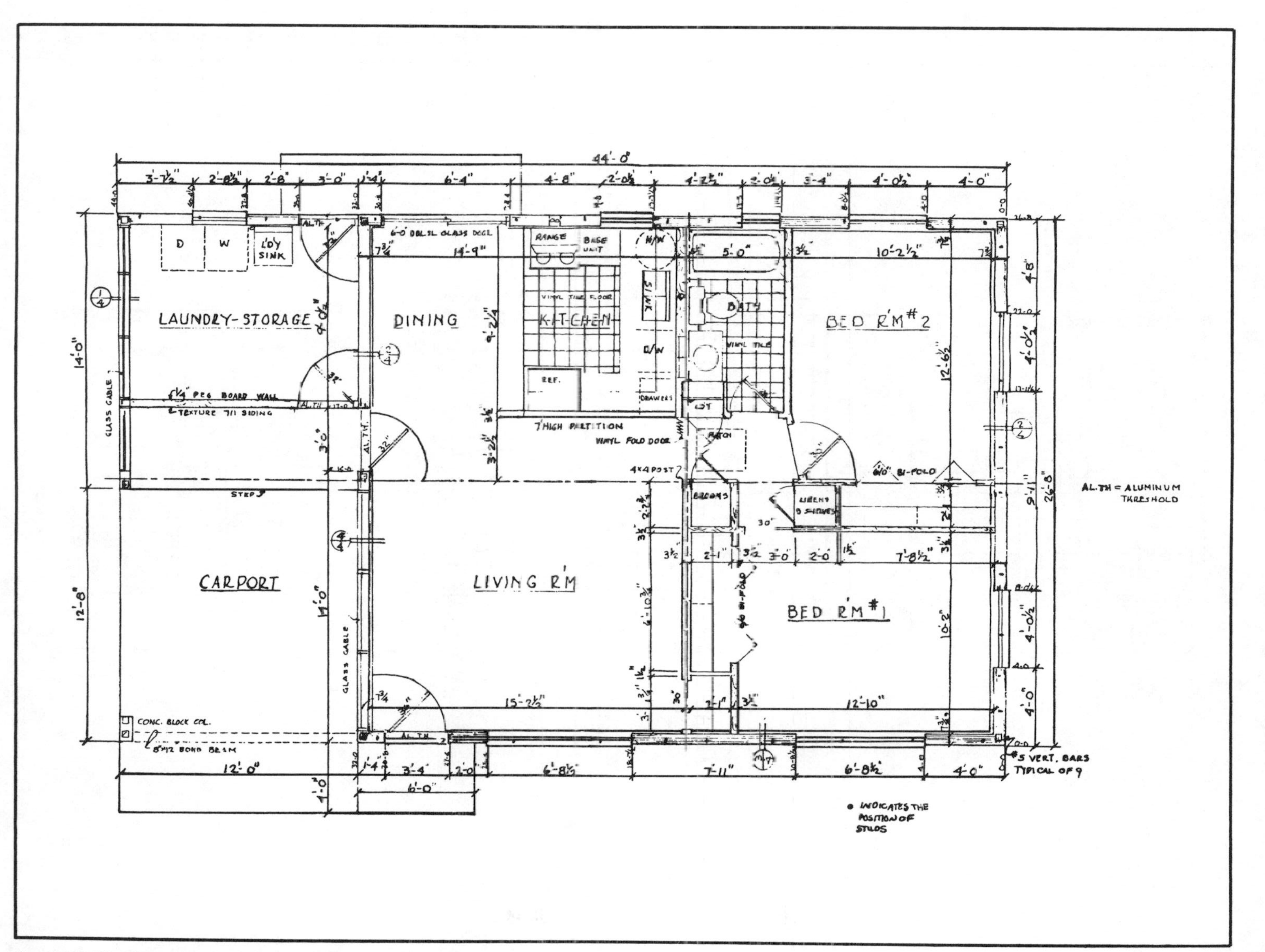

Fig. 12-7. The floor plan for a small, single-story house.

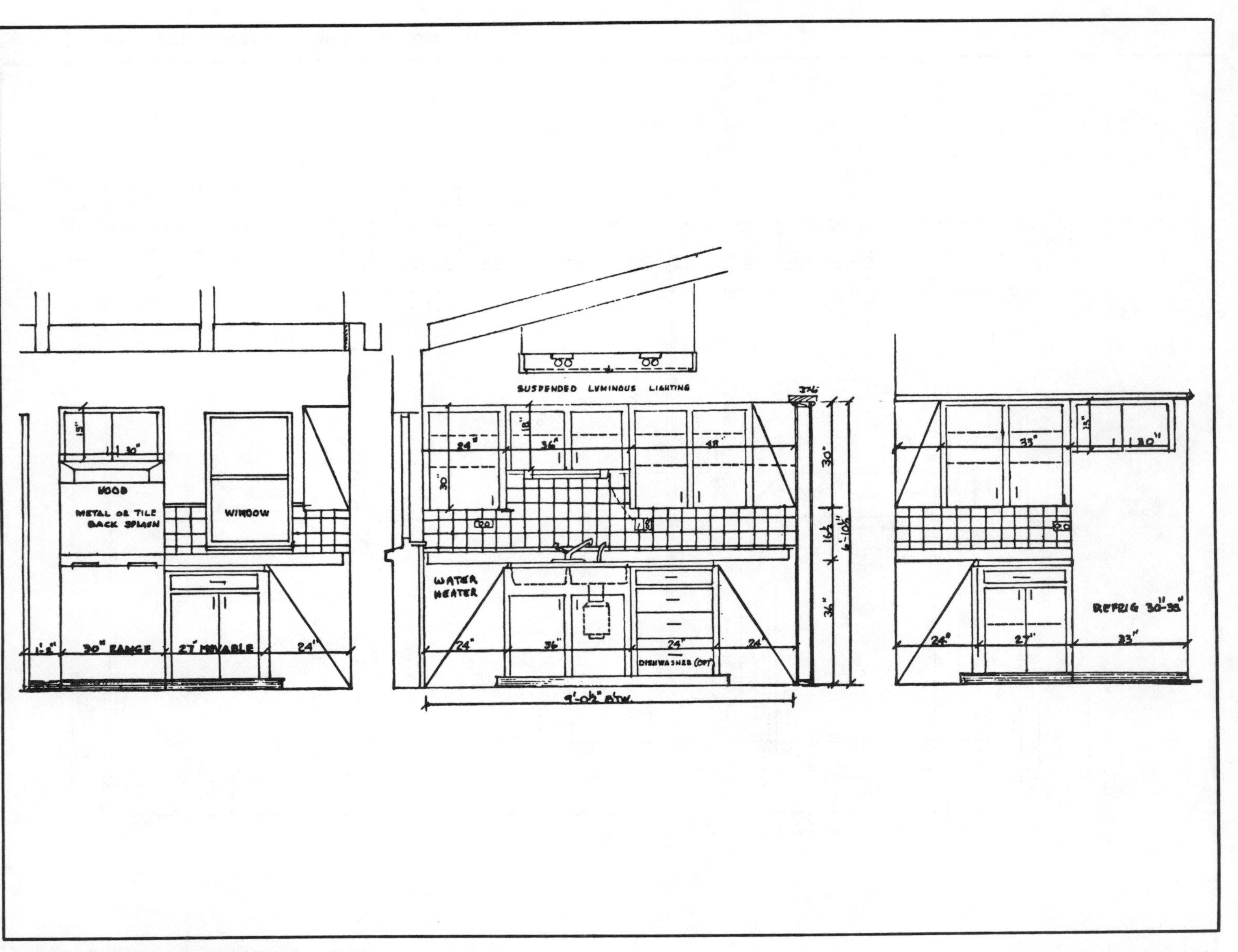

Fig. 12-8. Elevations showing kitchen cabinets.

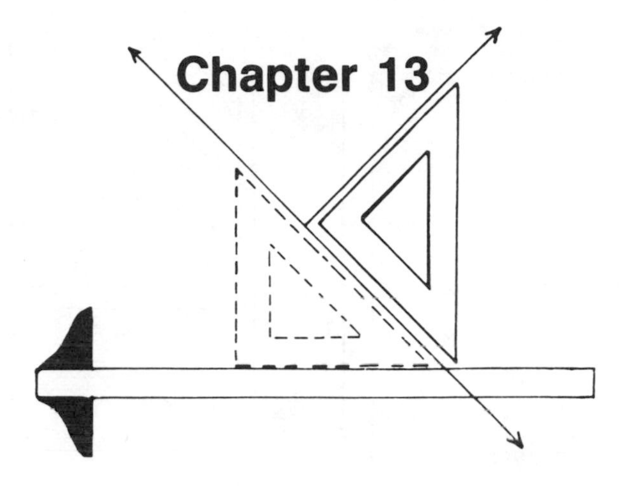

Bathrooms and Plumbing

A few years ago a single bathroom was acceptable; nowadays the ideal is a private bath for every bedroom plus a handy powder room with a toilet and lavatory for the occasional guest. In more frugal times, this objective may be approached by better bathroom planning, such as placing the fixtures in separate compartments, permitting their use by more than one person at a time. Although likely to require more piping, there is common sense in placing a lavatory in each bedroom, as is common in some European countries.

A glance through any home magazine shows how luxurious bathrooms have become in this country. They are the envy of most people from abroad.

For reasons of personal comfort and to make your house more salable, you should provide multiple bath facilities to the best degree you can afford.

Needless to say, two-story homes should have facilities on both floors, as well as in the basement if it will be used to any extent.

A glance through a catalog of plumbing fixtures or a visit to any plumbing supply store will give you an idea of the variety of bath equipment that is available.

BATHROOM FIXTURES

The largest bathroom fixture is the bathtub, which is usually 5 feet long and 30 inches wide, though there is a unit 28 1/2 inches wide for limited space. Figure 13-1 shows some standard 5 foot bathtubs as well as a few fancy alternatives. As a minimum, include a shower head over the tub. The shower head is more likely to be used than the tub filled with water.

It's possible to install two shower heads at different heights. An extra shower head can be installed about 32 inches above the floor of the tub with a diverter between the two heads.

With a shower, there must be some way to confine the spray within the tub. This can be a waterproof curtain hung from a rod or a sliding or folding door.

Of course, there must be a toilet, or water closet. There are three types, each varying in the

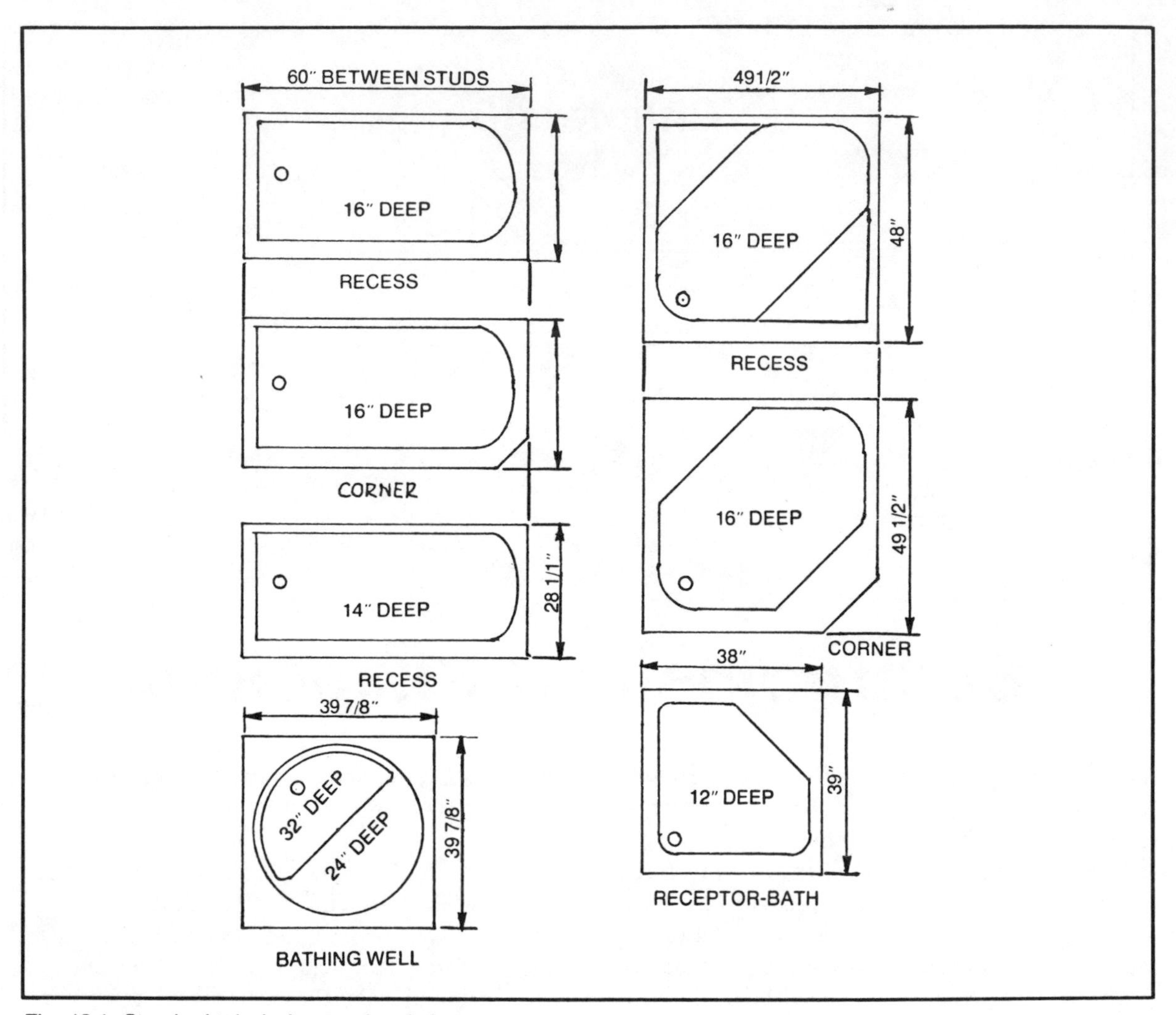

Fig. 13-1. Standard tub designs and variations.

construction of the ceramic bowl. The siphon wash-down type is least costly. In addition to standing toilets, there are wall hung types, easier to clean around but more costly. There are also types that are quieter than the standard sorts, but these are more costly. Progress is being made in the design of toilets that use considerably less water for flushing; these should be available in the near future.

Only a few years ago, wash basins, properly called lavatories, were the wall hung type. Now they are usually set into the counter of a cabinet, often called a vanity, and are much more convenient to use because there is space for laying out toilet articles. Bowls may be china, porcelain-enameled iron, steel, fiberglass, or manufactured marble. Some of the most common types of lavatories are shown in Fig. 13-2.

No mention has yet been made of another bathroom fixture which is considered one of the basics in France and some other European countries. This is the *bidet* (Fig. 13-3). It is coming into increasing use in this country. It takes about the same space as a toilet, beside which it is usually placed. It requires the same waste piping and may have hot water as well as cold.

Figure 13-4 shows the standard dimensions of

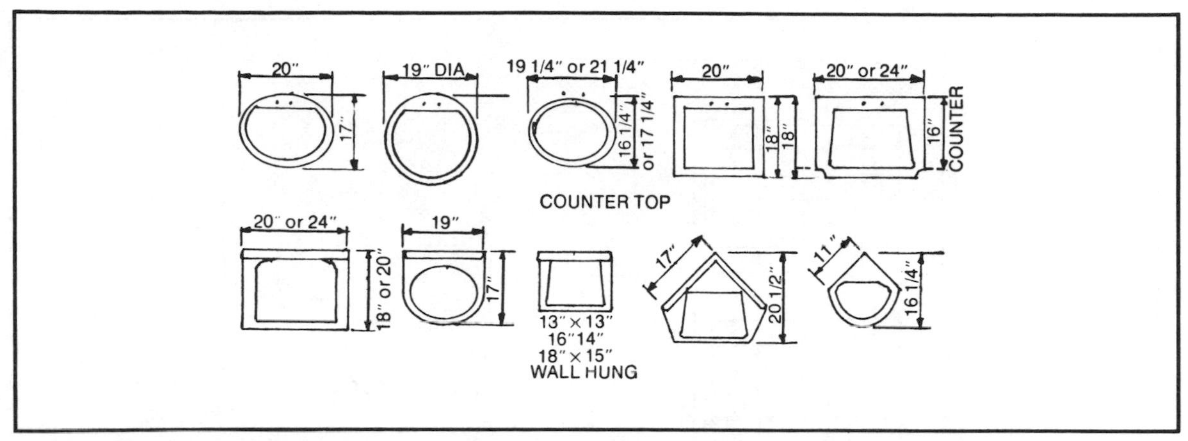

Fig. 13-2. Standard lavatory designs.

various bathroom installations.

Also to be considered is storage for towels and bath supplies, as well as a towel bar for each person using the facility.

A bath (or medicine) cabinet with mirror is usually installed above the lavatory. If there are children, a *lockable* medicine compartment is a safety measure.

A full length mirror on the door is a helpful feature, and so is a built-in ventilated hamper for soiled linens.

BASIC PLANNING CONSIDERATIONS

When planning a bathroom you should remember that toilets need cold water while the other fixtures require both hot and cold water.

In addition to the pipes that carry waste water away, there must also be vent piping running up through the roof. The vent piping, or vent stack, prevents the rush of waste water down a pipe from causing a vacuum that might draw the water that seals the trap below each fixture. (Fixture traps are simply U-shaped sections of pipe below fixtures. Water trapped in the U-shaped sections act as a seal.) This seal prevents gases and odors from entering the house. Plumbing codes have rigid rules as to the distance the various traps can be from a vent. Like the National Electric Code there is a National Plumbing Code, though it is not yet in general use. You must check with your local code authorities.

Particularly when planning homes with two floors and basement, it is necessary to locate first floor partitions so that pipes may be properly concealed.

When there will be more than three persons living in a house, try to include two toilets, two lavatories, one tub, and a separate shower. If possible, isolate the toilets and bathtub in separate compartments.

If there are more than two bedrooms, the usual requirement is a private bathroom for the master, or primary bedroom *and* a general bathroom, accessible from a common hallway from the other bedrooms. If the baths and bedrooms are on a second floor, a toilet or powder room should be placed on the first floor.

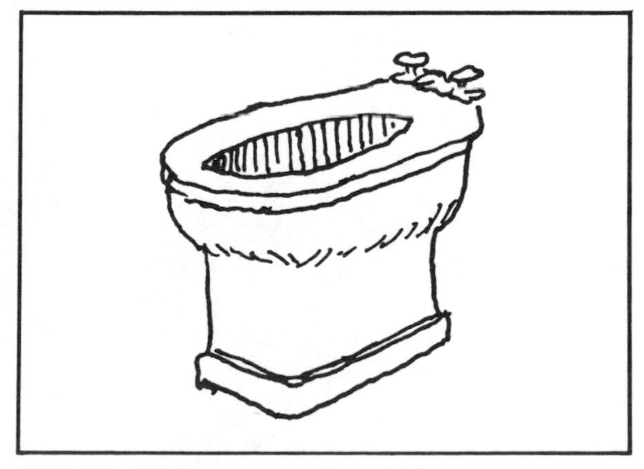

Fig. 13-3. A bidet.

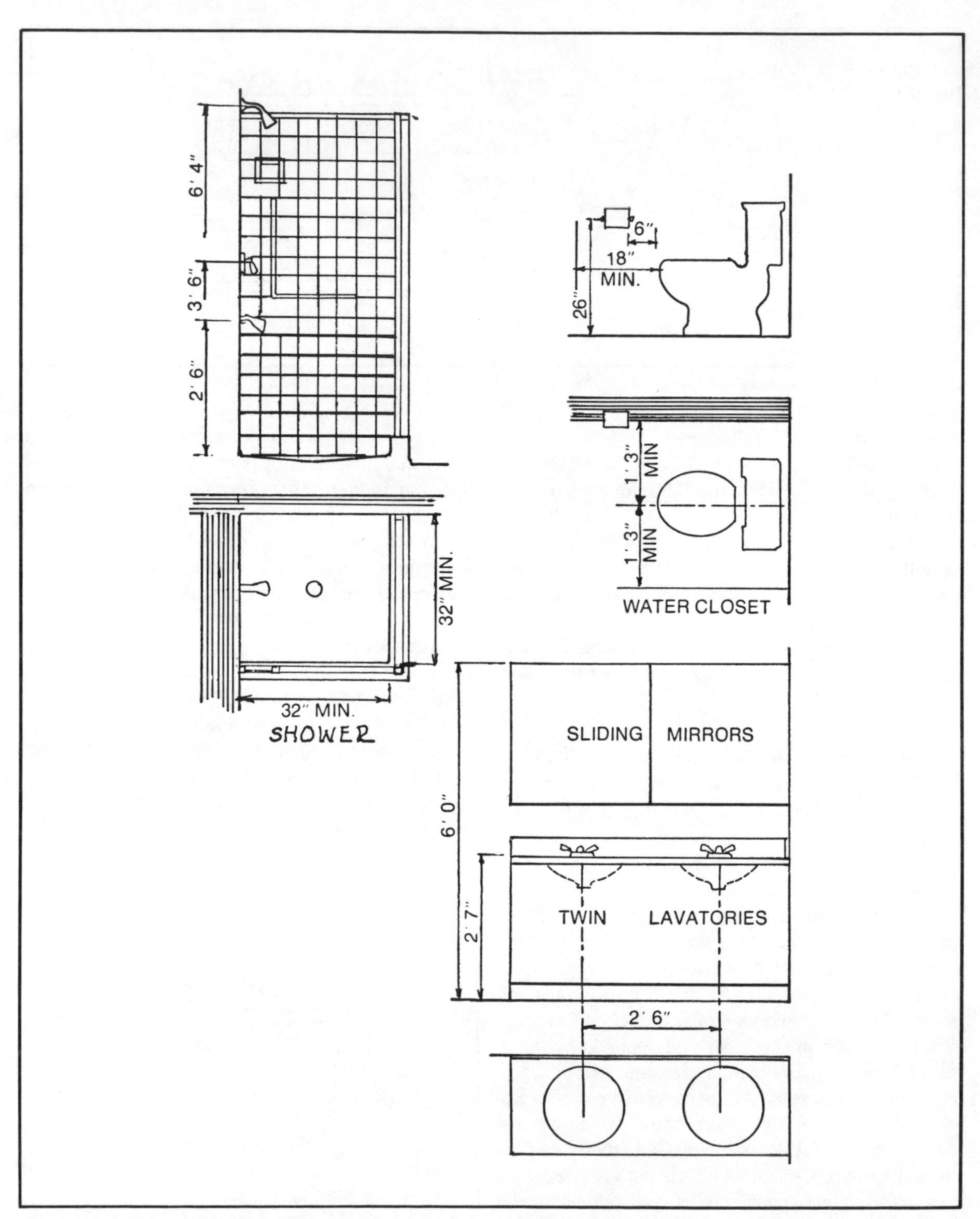

Fig. 13-4. Standard bathroom dimensions.

Privacy is as important as convenience when planning bath facilities. If possible the bathroom should be located so as not to be readily visible when the door is open.

Bathrooms should be ventilated by at least an exhaust fan so that other parts of the house need not be unduly cooled in winter.

Especially in houses of frame construction, it is desirable to insulate water piping. This prevents loss of heat in hot water lines, condensation on cold water lines, and contact with the frame of the house (such contact transmits vibrations from the pipes through the house).

Windows should not be located in a shower or over a tub. But if this is unavoidable, place them as far from the shower head as possible. If a bathroom is on the ground floor, the windows should be obscure glazing.

Safety should be considered. All too many accidents occur in the bathroom. There should be grab bars at the tub and shower with both vertical and horizontal hand holds (Fig. 13-5). Nonskid floors save falls; both ceramic and vinyl floors may be slippery when wet. Locate light switches well away from the tub or shower.

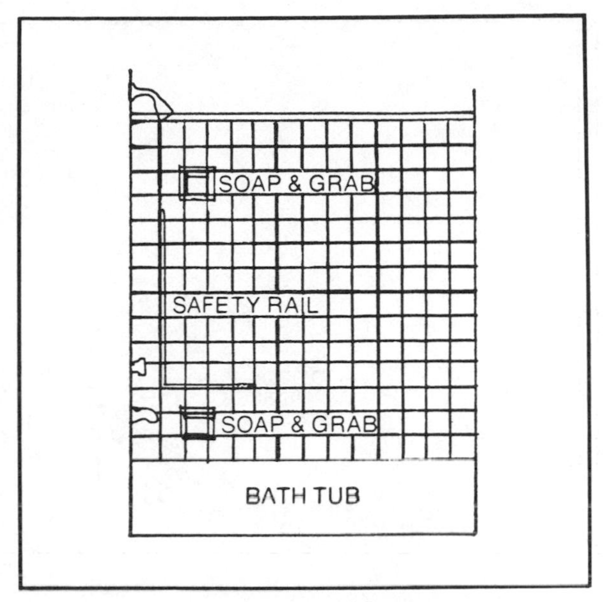

Fig. 13-5. A bathtub, shower, and fixtures.

Because of spray and splash from tubs and showers and high humidity, bathroom walls must be treated differently from elsewhere. Ceramic tile is a favorite covering, especially around tubs and showers. Tile set in cement plaster over metal lath and waterproof paper is a skilled craft and is the desired but more expensive method. Nowadays such tile is more often set in adhesive on walls covered with a fairly waterproof tile backer type of gypsum board. There are also very thin tiles of plastic or aluminum, and there are preformed wall panels of plastic laminate or fiberglass for use at tubs and showers.

While it is desirable to cover bathroom walls with tile to 48 inches above the floor, plaster or dry wall may be painted or finished with moisture resistant-paper or fabric.

The water heater is an important part of any plumbing system and should be placed as close as possible to the point of hot water use. If gas is to be used for water heating, the water heater's location is likely to be governed by the position of a chimney flue. If there is no chimney, some sort of insulated flue will be necessary to carry the fumes away.

THE PLUMBING

You are free in your choice of the visible fixtures for bath and kitchen, but the piping must conform to the plumbing code for your area. A plumbing code may regulate the size and location of pipes—and even what materials pipes can be made of.

The responsibility for meeting code requirements falls on the licensed plumber whom you employ to perform that part of building your home. He is familiar with the requirements, and you will not be permitted to cover or conceal piping until it has been inspected and approved.

Although your plumber will probably be able to connect the various fixtures wherever you locate them on your plan, the better you understand the plumbing code requirements and the problems of piping, the more you can avoid excessive costs.

Waste piping is commonly referred to as *drains*. The main soil, or main waste, pipe starts with a cleanout fitting 5 feet outside the foundation wall (Fig. 13-6). There it will be connected to a public

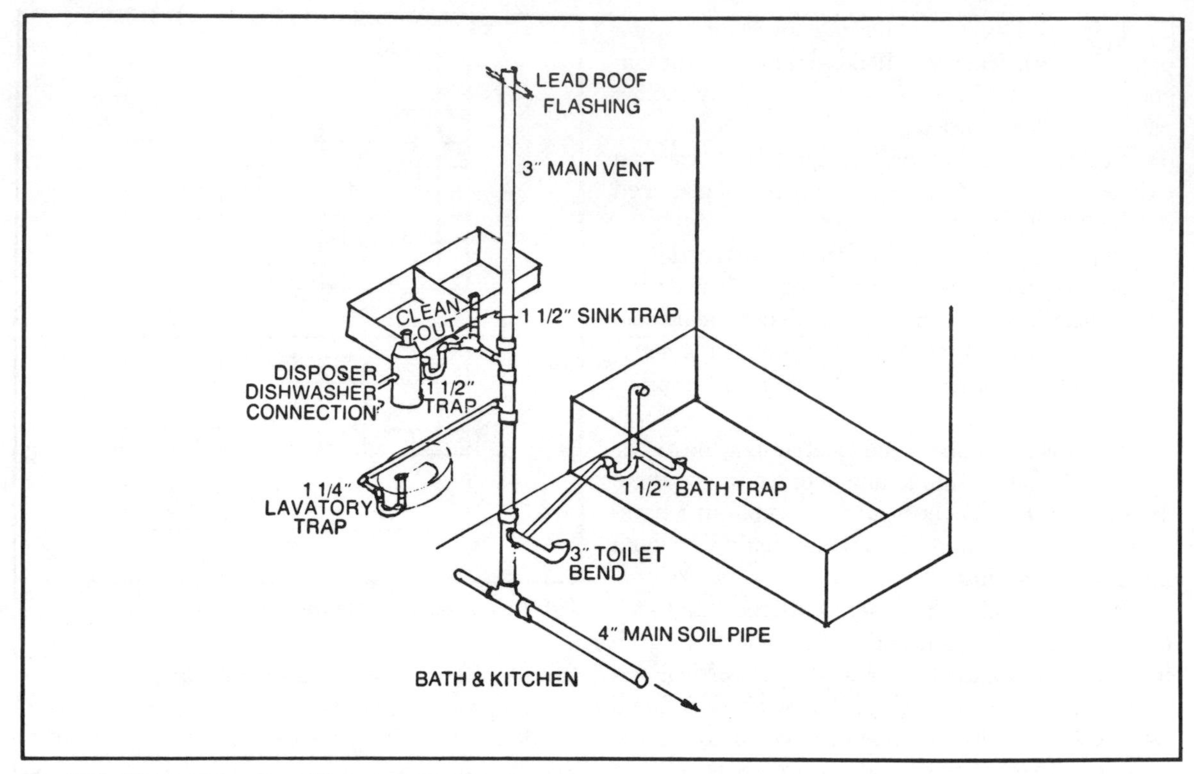

Fig. 13-6. Waste piping for a kitchen and bathroom.

sewer or to a septic tank if you must have your own disposal system. The main soil pipe is usually a 4 inch diameter pipe. The old standby has been cast-iron pipe with joints poured of hot lead. Because of the temperature changes due to the hot and cold water this pipe must carry, there will be some expansion and contraction, which the somewhat flexible lead joints allow without leakage. More recently, piping of plastic material with joints made solid with a bonding material is becoming more common. Such piping takes less labor to install. They are flexible and seem not to have expansion problems.

For water piping, copper piping with soldered joints used to be the first choice. But like waste piping, it is being superseded by plastic piping. Where not concealed, galvanized steel piping is probably least expensive but is more subject to corrosion with some waters or the liming up in hot water lines. Sometimes a combination of all different materials will be least costly. Figure 13-7 shows a typical water piping arrangement for a bathroom and kitchen.

Every plumbing system must have at least one vent stack open-ending 6 inches above the roof. In house construction where the main soil pipe is 4 inches in diameter, the main vent must be 3 inches in diameter. In some cases the 4 inch pipe must be carried full size as a vent up through the roof.

In one-story houses where fixture traps are more than the code-required maximum distance from the main vent (Fig. 13-8), each fixture must be separately vented by running 1 1/2 inch or 2 inch diameter vents through the roof or by running secondary vents from each fixture trap *up* to the main vent (Fig. 13-9). Typical code requirements pertaining to maximum distances of traps from the main vent are shown in Table 13-1.

When planning a house with a basement, it is easy for the plumber to connect the piping to the fixtures on the floor above. In a basementless slab-floor house, you should give the precise locations

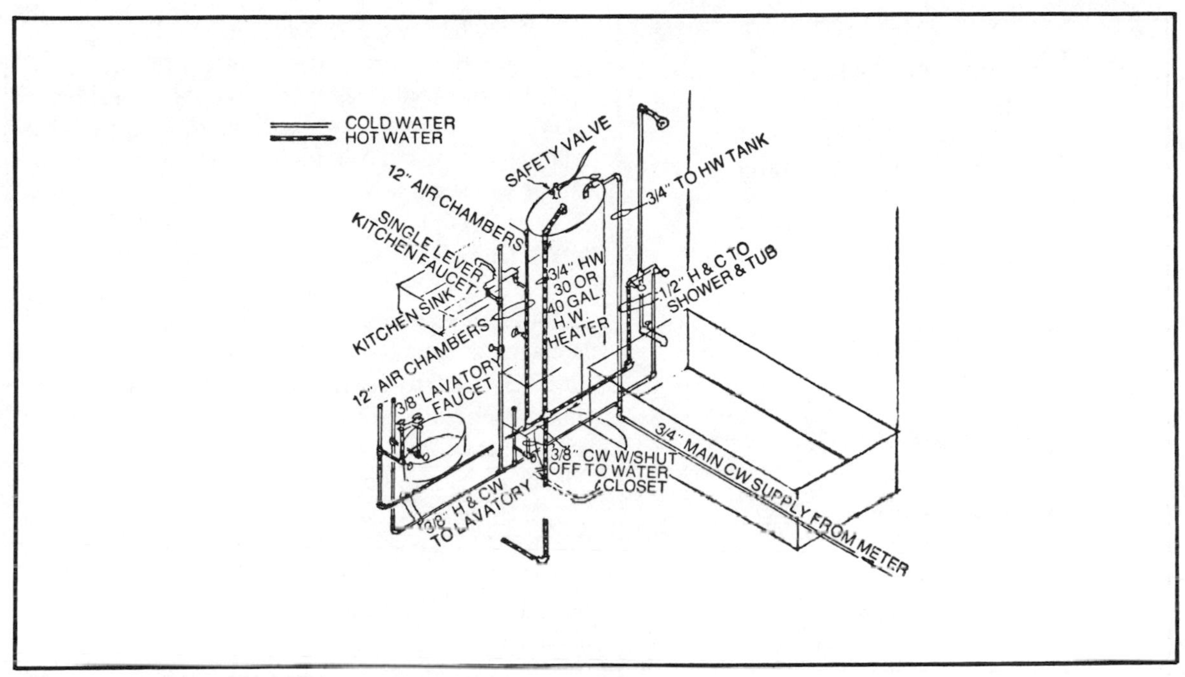

Fig. 13-7. A water piping arrangement for a kitchen and bathroom.

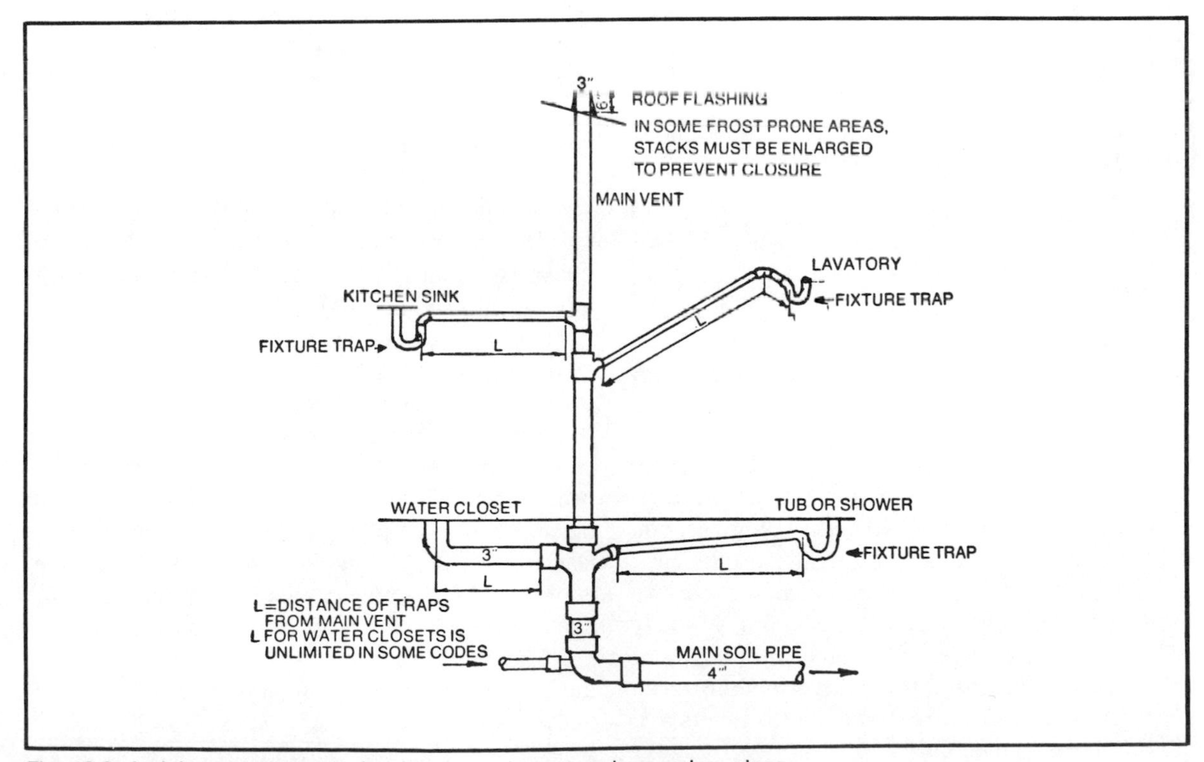

Fig. 13-8. A piping arrangement showing the main vent and secondary pipes.

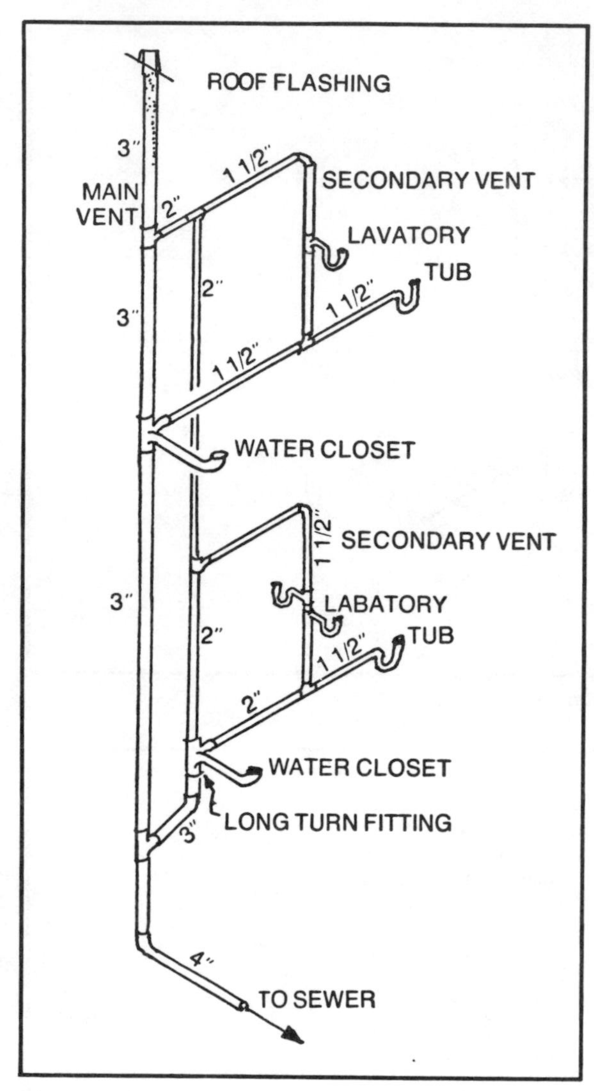

Fig. 13-9. A two-story piping network showing secondary vents.

for the fixtures because piping must come up through the floor exactly at the proper place. If locating such piping is left to a busy plumber, you may find it located in a less than desirable spot.

If there are water problems, such as excessive hardness, iron stains, bad taste, or odors, you may want to install a water softener or other type of treatment equipment. It may be worthwhile to run separate piping for untreated water to toilets and outside hose connections.

If you have your own well, space should be provided for the pump and pressure tank.

The disposal of sewage presents no problem if there is a public sewer available. The plumber simply extends the 4 inch sewer pipe from the house to the place of connection to the main sewer. If your street is not paved, this connection will be a Y-branch. If paved, a branch sewer will have been brought to a point outside the edge of the paving. Your plumber will get the location and depth of the connection from the local sewer authority, though you may wish to get this information yourself and place it on your plot plan. There will probably be a connection charge; in some places this is a considerable sum to be added to the cost of building.

If there is no public sewer available, as is usual in rural and some suburban areas, you must install your own disposal system. Again, check with the local health authorities or the county farm agent as to what you can do.

A common waste disposal system is the *septic tank* with a *leaching bed* (Fig. 13-10). In such a system, sewage leaves the house through the main soil pipe and enters the septic tank, usually located about 5 feet away from the foundation wall. Through bacterial action the septic tank breaks down the solid wastes. The remaining liquid wastes are then piped into a distribution box where they are channeled out into a dispersion field (Fig. 13-11). The pipes that fan out into the dispersion field have open joints, allowing the liquid wastes to seep into the porous soil.

A septic tank of proper capacity connected to an underground leaching bed is satisfactory if the soil is able to absorb waste water. If not, a *sand filter bed* above ground can purify waste water so it can be safely discharged into an open ditch. Such a filter bed, being visible, is best located in the backyard out of sight. A septic tank with an underground leaching bed can be located in the front yard if there is room and if the pipe can be readily connected to a public sewer system when one becomes available.

The installation of a septic tank with a leaching bed usually requires an assessment by a local authority of the ability of your soil to dispose of waste water. This is called a *percolation test*. To

FIXTURE	DIAMETER OF TRAP AND DRAIN	MAXIMUM DISTANCE OF TRAP FROM MAIN VENT	
		*1/4″ SLOPE	*1/2″ SLOPE
LAVATORY	1 1/4″	4′-6″	3′-0″
		2′-0″	1′-6″
	1 1/2″	7′-0″	3′-6″
		6′-0″	3′-0″
FLAT BOTTOM SINK, TUB SHOWER	1 1/2″	10′-0″	5′-0″
		8′-0″	4′-0″
	2″	10′-0″	5′-0″
		6′-6″	3′-6″

Table 13-1. Maximum Unvented Lengths of Drains for Stack-Vented Fixtures.

make such a test, first several small (6 inches or 8 inches in diameter) holes are dug in the area down to the probable depth of the bed. The holes are filled with water once or twice to wet the soil. Then after a short time, they are again filled with water and the rate of its absorption is timed and must meet standard requirements. Sandy porous soils, usually make good leaching beds, but heavy clay soils do not.

A sand filter *above* ground accomplishes about the same thing that a leaching bed does underground. It consists of a low wall of concrete or blocks, containing a system of open-jointed drains installed at a level that will permit the purified water to flow to an open ditch or other place of disposal. The drains are covered with 18 inches or more of gravel and sand. A perforated pipe system laid on boards covers the top of the gravel and sand.

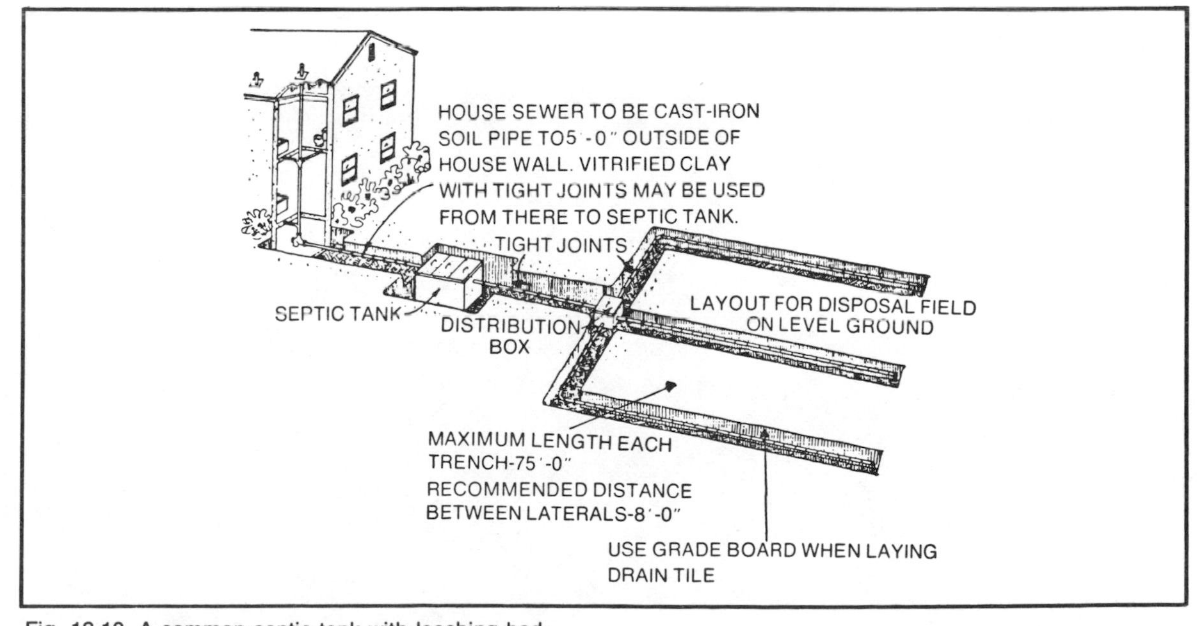

Fig. 13-10. A common septic tank with leaching bed.

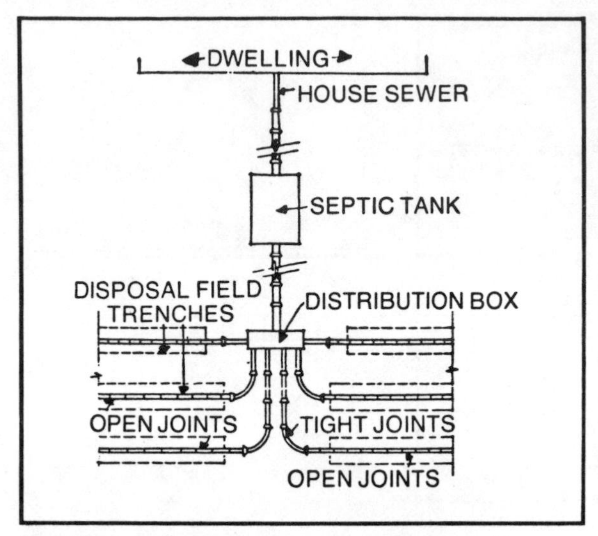

Fig. 13-11. A distribution box and dispersion field.

The perforated pipe system distributes the partly purified water from the septic tank over the sand filter surface. The water then seeps through the sand and into the drains, being purified as it goes. The drains then carry the purified water to a ditch or sewer. A small float-actuated pump at the underground septic tank periodically forces the water uphill to the sand filter. But sewage can flow by gravity downhill to the filter bed, it is necessary to hold it at or near the septic tank where a self-actuating siphon directs it at intervals to the filter, since the sewage cannot be allowed to trickle slowly out.

Your local health authority or county agricultural agent can advise you on such disposal matters and will probably have bulletins and pamphlets of detailed instructions.

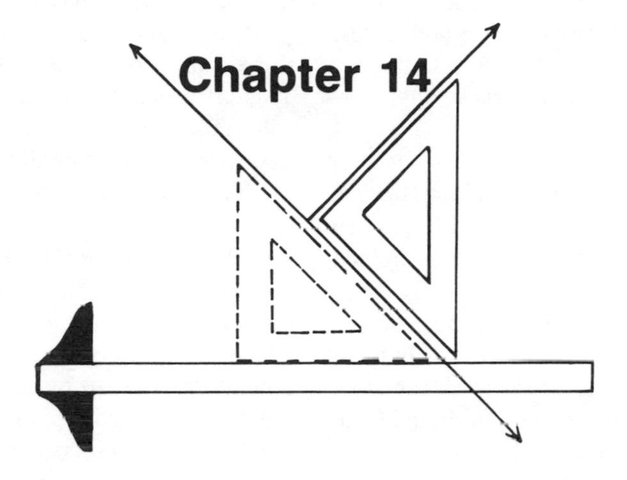

Stairways

Stairways can be ornamental as well as practical. They can also be a source of accidents. Because of this, design standards have been established to make stairs as safe as possible. Here are some of those design criteria:

☐ To avoid head bumps, main stairs should have at least 6 feet 8 inches clearance, which is the same as the height of a door. Seven feet might be better if you have large pieces of furniture to be moved.

☐ Stairs should have at least 2 feet 8 inches between railings, which requires at least 3 feet between walls if the stairs are enclosed. This is also the minimum width for hallways. If possible, stairs *and* halls should be 39 inches or 40 inches wide. A wider hall and stairs is safer.

☐ Climbing stairs takes a lot of energy, and studies have found that if treads and risers are properly proportioned, using stairs becomes easier and more comfortable. Use either of the following formulas to determine correct tread and riser dimensions:

1. The depth of 1 tread plus the height of 2 risers should be between 24 inches and 25 inches.
2. The tread depth multiplied by the riser height should be between 70 inches and 75 inches.

Stock stair treads come in 9 1/2 inches, 10 1/2 inches, and 11 1/2 inches depth. Because of the usual 1/2 inch rabbet on treads, for net 9 inch deep treads, risers can be 6 1/2 inches to 7 inches high. Of course minor variations of these figures are not too critical.

☐ All risers height must be the same for any flight of stairs.

☐ An isolated flight of stairs should have at least 3 risers and not more than 18. A flight having only two risers should be separated by a landing from another flight of stairs.

☐ The width of a landing should not be less than the width of the stairs. The depth of a landing should not be less than 2 feet 6 inches.

☐ The swing of a door opening on a stairway should not overlap the top step.

☐ Handrails should be provided where needed to protect occupants from falls. The handrail should be continuous on at least one side of each flight of more than three risers. Stairs open on both sides, including basement stairs, should have a continuous handrail on one side and a railing on the open portion of the other side.

Sometimes a disappearing or folding stairs is valuable for access to attic space (Fig. 14-1). They can go between joists or trusses spaced on 24 inch centers.

Whether a stair is fully enclosed or open on one or both sides, handrailings are necessary. They can be plain or ornamental to conform with the style of the building. They should be continuous from floor to floor and should be placed 30 inches above the nose of the tread and 34 inches above landings.

Exterior stairs or steps, such as from ground level to the house entrance, are usually designed for smaller riser heights and wider treads. A 6 inch riser and 12 inch tread is a good combination. If there are more than two risers, provide a handrail.

Sometimes a ramp may be preferred if the slope can be kept to 2 inches in 12 inches or less. This criterion is especially important if the ramp will be used by an elderly person. A ramp is a necessity where a wheelchair must be used.

To draw a stairway, proceed as follows:

1. Determine the height from floor to floor. Divide this by 7 1/2 to determine the approximate number of risers. This is not likely to come out an even number, so you must change the riser height a small fraction to make an even number.

2. Check your plans for available space for a stairs run. Decide which of the stock depths of stairs treads is most suitable. As previously noted, stock depth are 9 1/2 inches, 10 1/2 inches, and 11 1/4 inches (or a net depth of 9 inches), 10 inches, or 11 inches). The 9 1/2 inch depth is most commonly used.

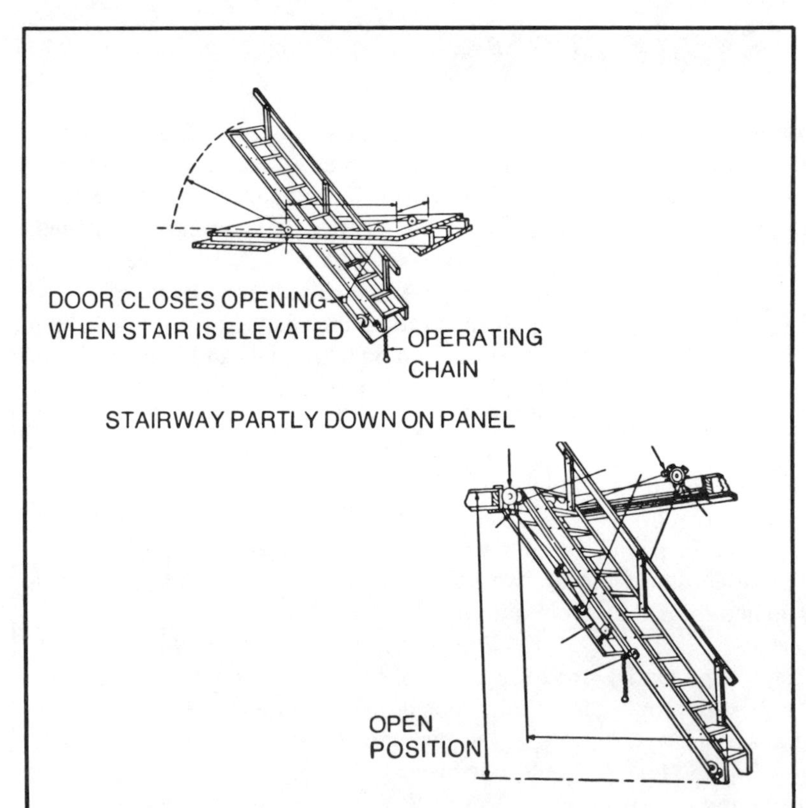

Fig. 14-1. A disappearing attic stairs.

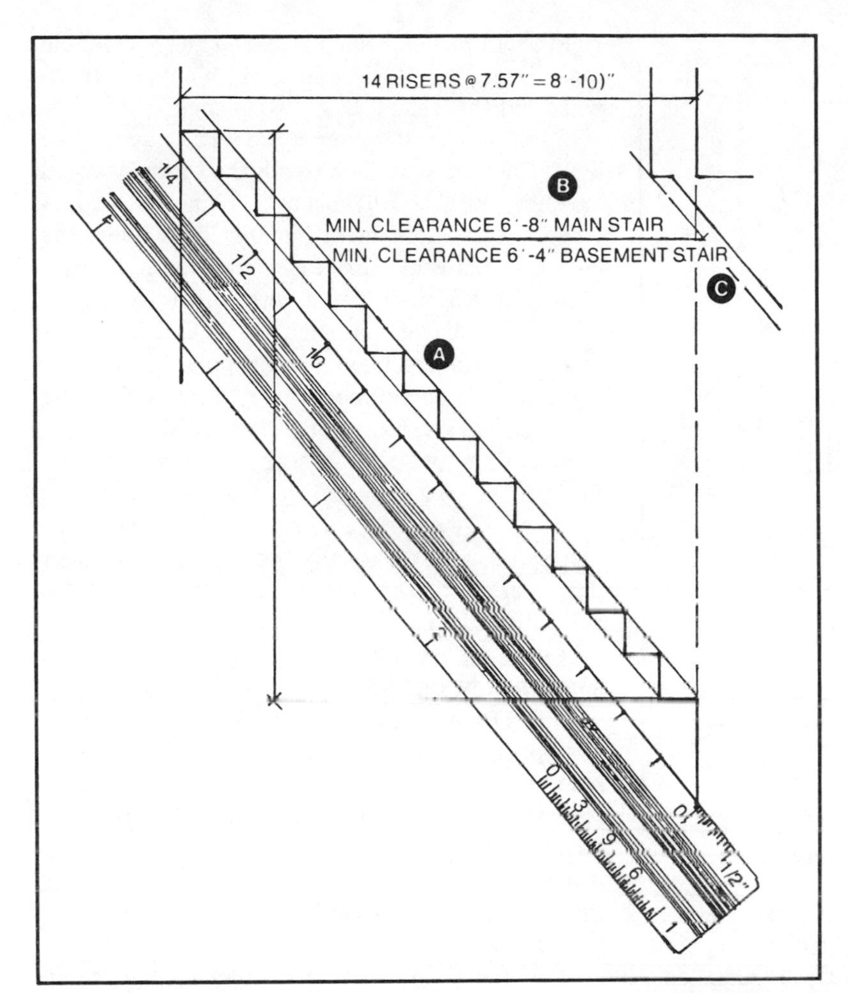

Fig. 14-2. Use your scale to draw stairs.

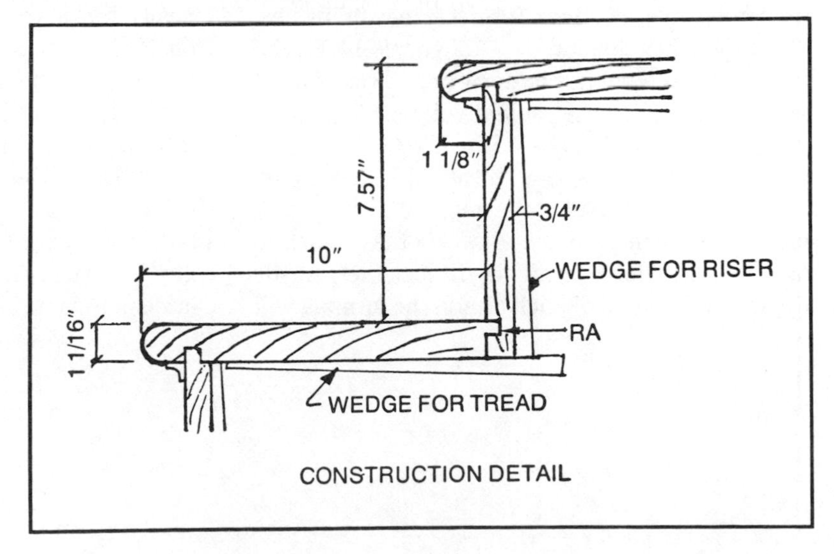

Fig. 14-3. A cross section of treads and risers.

Fig. 14-4. A stairs drawing with railing details.

3. Use the stairs formulas to check suitability. For example, a house with an 8 foot ceiling, 2 × 10 inch second-floor joists, and 1/2 inch rough floors has a total rise of 8 feet 10 inches or 106 inches. The 106 inches divided by 7 1/2 equals 14.13, the number of risers. Thus either 14 or 15 risers must be used. If 12, the riser height will be 7.57 inches. If 15, it will be 7.07 inches. If stock 9 1/2 inch (net 9 inch) treads are used with the 7.57 inch risers, the depth of 1 tread plus the height of 2 risers will equal 24.14 inches, which is *within* the 24 inch to 25 inch limit. However, if the 10 inch (net) treads are used with the 7.57 inch risers, the formula will yield 25.14 inches, which is *outside* the acceptable limit. Using the 15-riser stairs with a 9 inch tread will yield this result: 2 × 7.07 inches + 9 inches = 23.14 inches. Of course, 23.14 is outside the limit. But with a 10 inch tread, the formula would look like: this 2 × 7.07 inches + 10 inches = 24.24 inches. And 24.14 is within the limit. Thus the choice is between 7.07 inch risers with 10 inch treads, or 7.57 inch risers with 9 inch treads.

4. Small fractions such as 7.57 inches (7 9/16 inches) are difficult to lay out on a 1/2 inch scale drawing. It's sometimes easier to use a drawing scale to mark off the total riser height into the number of individual risers, as shown in Fig. 14-2. For example, let's say you want to draw stairs with 14 risers. Use any suitable scale (with 14 intervals) that can be laid diagonally between the two floor lines marked on the plan. Point off the 14 spaces, and with a T square, project the points over to lines representing the slope of the stairs (Fig. 14-2A). There will be one less tread than there are risers.

5. On the drawing, locate the ceiling opening or sloping stair soffit by drawing a dimension representing the required 6 foot 8 inch vertical clearance (Fig. 14-2B). Draw a light line through the top end of the dimension line parallel to the stair nosings (Fig. 14-2C). This determines the size of the stairwell and also the limits of the second-floor framing or header.

6. Draw the outline of the stair framing. Usually there are three supporting stringers cut from 2 × 10s or 2 × 12s. At least 3 inches of solid wood must remain after the stringer cuts are made. Show a cross section of all or a few of the treads and risers (Fig. 14-3).

7. Draw the railings in sufficient detail to show a builder the type of railing you want (Fig. 14-4). If you plan to use stock stairs millwork, you can select a design from a millwork catalog or you can design some other type of stairs and railing.

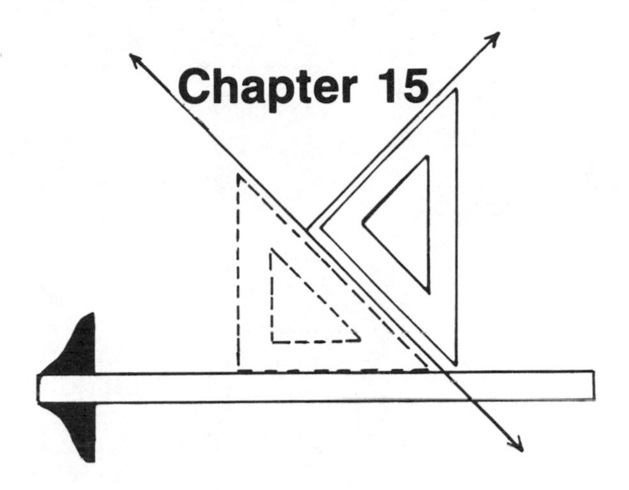

Fireplaces

Until little more than a hundred years ago, all homes had to have a fireplace, but since that time fireplaces have been built more as a decorative center of interest.

Nowadays with the increasing cost of heating oil and gas, many families are seeking security in rural or suburban areas where wood is a self-replenishing product and a fireplace is insurance of comfort in case of power failure or energy shortage.

In the old days there were masons who were experts at fireplace building and knew by experience how to dimension a fireplace to avoid the problem of smoke escaping into the room. This type of artisan is no more, but manufacturers of fireplace equipment have developed tables of dimensions that make it possible for any planner to design a satisfactory all-masonry fireplace (Figs. 15- 1A and 15-1B).

But if you don't want to build an all-masonry fireplace, you can buy modern fireplace liners (metal preformed fireplace shells) and build masonry partitions around them.

Some of these liners have double walls so that warmed air circulates back into the room rather than being lost up the chimney. A plan for a fireplace with one of these heat circulating liners is shown in Fig. 15-2.

Nowadays you can also buy a freestanding fireplace (Fig. 15-3). They don't have to be placed on a firebrick hearth or surrounded by masonry for fireproofing. These units are completely insulated and may be placed directly on a wood frame floor. They come with insulated lightweight flues.

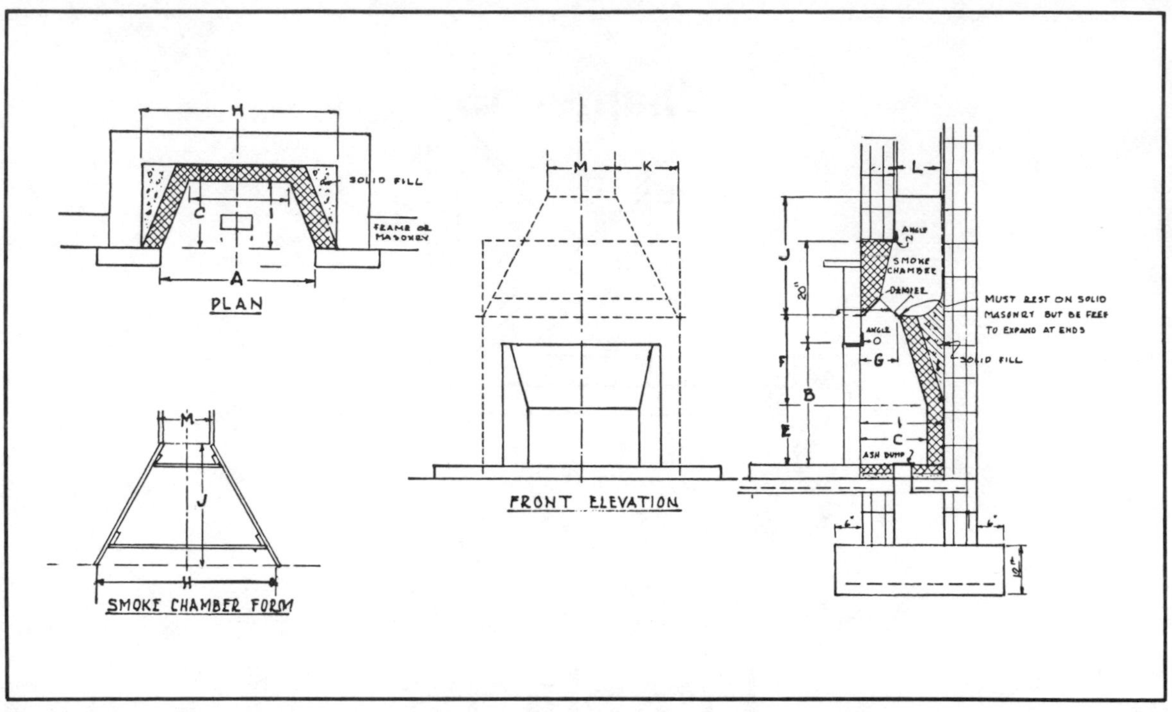

Fig. 15-1A. A plan for an all-masonry fireplace. Dimension A is the key dimension; it's the width of the fireplace opening.

RECOMMENDED DIMENSIONS (IN INCHES)

FINISHED FIREPLACE OPENING							ROUGH MASONRY			FLUE SIZE									DONLEY DAMPER	
										NEW SIZES			OLD SIZES			ROUND	STEEL	ANGLE	ROTARY	POKER
A	B	C	D	E	F	G	H	I	J	K	L	M	K	L	M		N	O	NO	NO
24	24	16	11	14	15	8¾	32	20	19	10	8	12	11¾	8½	8½	8	A-36	A 36	324	224
26	24	16	13	14	15	8¾	34	20	21	11	8	12	12¾	8½	8½	8	A-36	A 36	330	230
28	24	16	15	14	15	8¾	36	20	21	12	8	12	11½	8½	13	10	A 36	A 36	330	230
30	29	16	17	14	18	8¾	38	20	24	13	12	12	12½	8½	13	10	A 42	A 36	330	230
32	29	16	19	14	21	8¾	40	20	24	14	12	12	13½	8½	13	10	A 42	A 42	333	233
36	29	16	23	14	21	8¾	44	20	27	16	12	12	15½	13	13	12	A 48	A 42	336	236
40	29	16	27	14	21	8¾	48	20	29	16	12	16	17½	13	13	12	A 48	A 48	342	242
42	32	16	29	14	23	8¾	50	20	32	17	16	16	18¼	13	13	12	A 54	A 48	342	242
48	32	18	33	14	23	8¾	56	22	27	20	16	16	21½	13	13	15	B 60	B 54	348	248
54	37	20	37	16	27	13	68	24	45	26	16	16	25	13	18	15	B 72	B 60	—	254
60	37	22	42	16	27	13	72	27	45	26	16	20	27	13	18	15	B 72	B 66	—	260
60	40	22	42	16	29	13	72	27	45	26	16	20	27	18	18	18	B 72	B 66	—	260
72	40	22	54	16	29	13	84	27	56	32	20	20	33	18	18	18	C 84	C 84	—	272
84	40	24	64	20	26	13	96	29	61	36	20	24	36	20	20	20	C 96	C 96	384	284
96	40	24	76	20	26	13	108	29	75	42	20	24	42	24	24	22	C 108	C 108	396	296

ANGLE SIZES A= 3x3x3/16 B= 3½x3x¼ C= 5x3½x5/16

Fig. 15-1B. Recommended fireplace dimensions for given widths of fireplace openings.

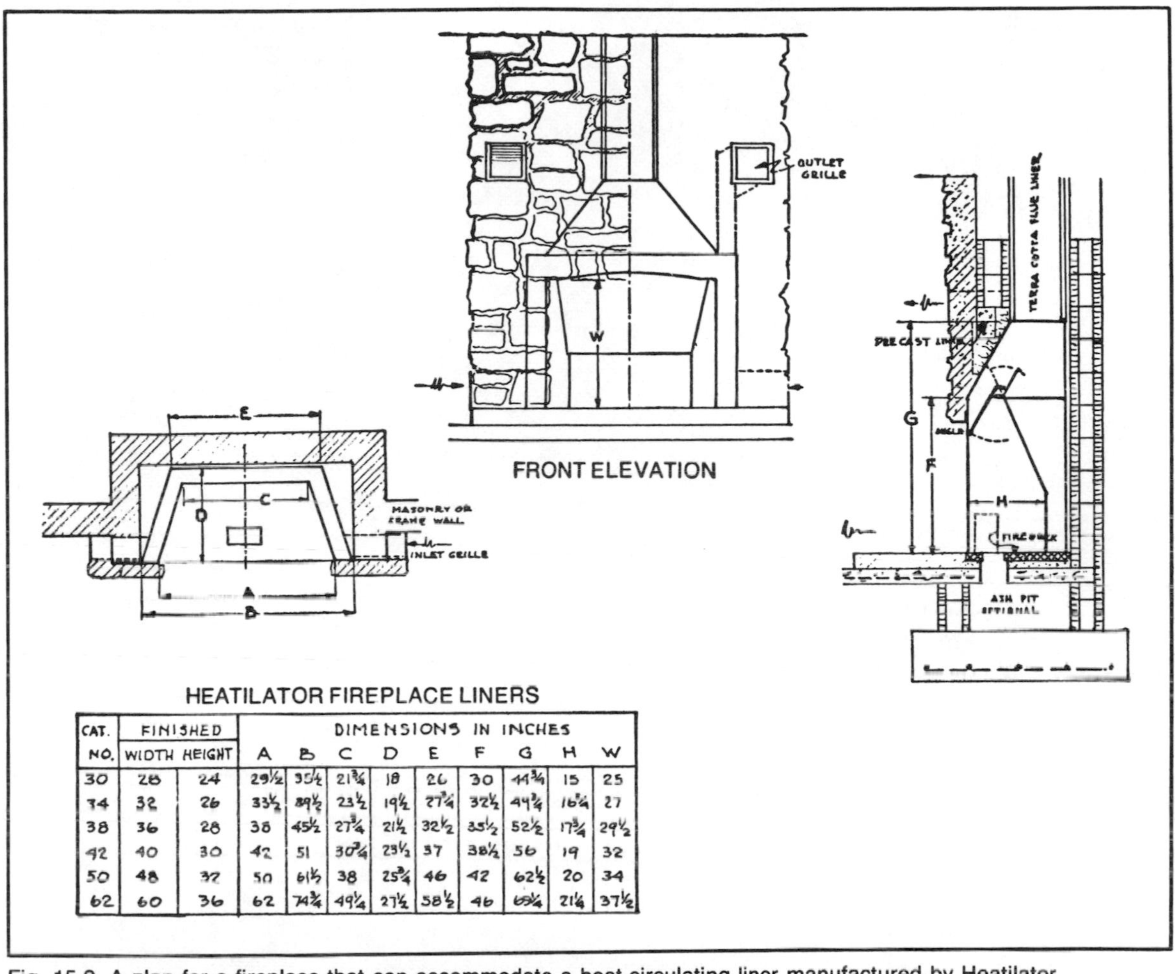

HEATILATOR FIREPLACE LINERS

CAT. NO.	FINISHED WIDTH	FINISHED HEIGHT	A	B	C	D	E	F	G	H	W
			DIMENSIONS IN INCHES								
30	28	24	29½	35½	21¾	18	26	30	44¾	15	25
34	32	26	33½	39½	23½	19½	27¾	32½	49¾	16¾	27
38	36	28	38	45½	27¾	21½	32½	35½	52½	17¾	29½
42	40	30	42	51	30¾	23½	37	38½	56	19	32
50	48	32	50	61½	38	25¾	46	42	62½	20	34
62	60	36	62	74¾	49¼	27½	58½	46	65¼	21¼	37½

Fig. 15-2. A plan for a fireplace that can accommodate a heat-circulating liner manufactured by Heatilator.

Fig. 15-3. A freestanding fireplace.

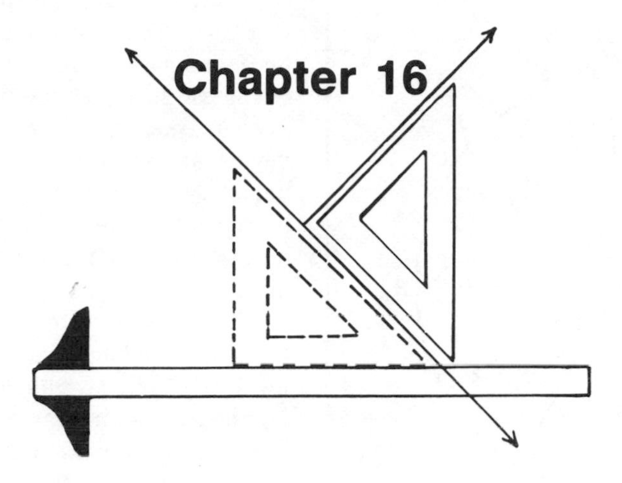

The Plot Plan

Usually the final drawing of a set will be the plot plan, although it's a good idea to draw a simple lot outline at the start of your planning to be sure that your house will fit within the required setbacks (required distances a house must be set back from a street, property line, etc.)

The plot plan is a detailed boundary outline of the piece of land on which your house is to be built. It must show how the house will be located in relation to setbacks and clearances and all other information required by the permit authority.

In many instances the plot plan need only be a small 1 inch = 20 foot or 1 inch = 16 foot scale drawing that can be fitted into a corner of a larger drawing, preferably the foundation plan. Or it can be on a separate 8 1/2 × 11 inch sheet.

The builder will use the plot plan when he stakes out the house before digging the foundation.

BASIC PLOT PLAN NOTATION

The plot plan must be dimensioned in surveyor's language, so you should be able to understand it.

In early times, land was surveyed with a rather primitive magnetic surveyor's compass and a measuring chain 66 feet long with 100 links. Nowadays surveying is done with an accurate *transit* and a 100 foot steel tape that reads to tenths of a foot and can be estimated to hundredths of a foot.

From school days you will remember that a circle is divided into 360°, each degree is divided into 60 minutes, and each minute is divided into 60 seconds. Minutes are indicated by ′ (as in 20′); seconds are indicated by ″ (as in 30″). As far as your drawing is concerned, you can forget the seconds, though a surveyor must use them in his figuring.

The direction along a property line is written in compass language. Thus N 15° 21′ 44″ E indicates a line that is 15° 22′ to the east, or right, of a north-south line (when you're looking north). If you were looking south, the same line would have a bearing of S 15° 22′ W. See Fig. 16-1.

Note that no bearing can be greater than 90°. And the number of minutes (′) or seconds (″) in a bearing cannot exceed 60.

You should also understand that any compass

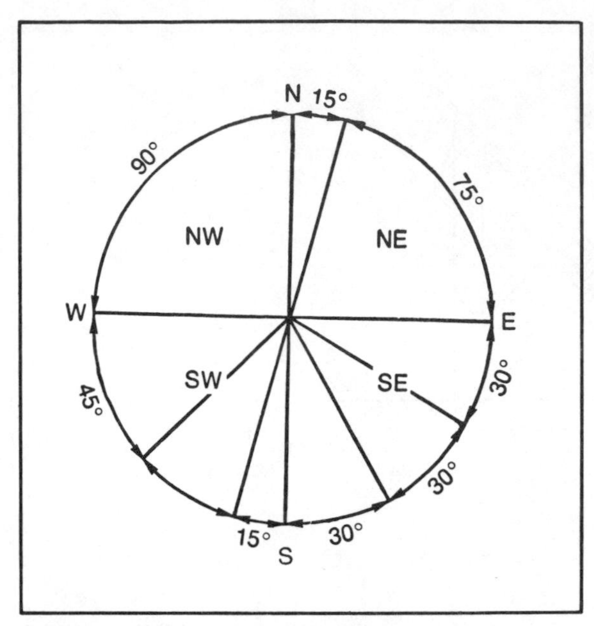

Fig. 16-1. A bearing of N 15 degrees 22′ E is the same as a bearing S 15 degrees 22′ W.

bearing can be written two ways. For example, N 15° E can also be S 15° W. It depends on the direction you are looking. Thus when you are adding or subtracting bearings to get the angle between them, you may change the bearing notation to suit your convenience.

READING A SURVEY PLAT

When you purchase a lot in a developed area, you'll probably receive a *survey plat*. A survey plat is a map of a lot. It shows the boundaries of the lot and its location (indicated in surveyor's notation).

A survey plat can save you a lot of work. You can easily turn it into a plot plan by drawing your house (with setbacks and clearances) on it. Or you can use it as a guide to draw a plot plan on a separate piece of paper.

Take a look at the plat survey in Fig. 16-2. This plat tells you that you should find a cement monument (CM) on the north edge of McKay Creek Drive 265 feet from the centerline of Bluff Drive. The notation N 15° 00′ 00″ E 100.0′ means that if you were to survey this lot with a compass and sight 15° east of true north and measure off 100 feet along that bearing, you should find an iron pipe (IP) corner marker. The plat also indicates that 60 feet from this IP on a bearing of 75°, there should be another iron pipe corner marker. A surveyor would use such plat notations to help him pinpoint every marker on a piece of property. Usually a surveyor starts at one marker and walks all around the boundary line, checking markers as he goes until he returns to the first marker, or place of beginning (POB). If he finds any markers missing, he replaces them.

If you wanted to redraw a survey plat, you would need to know the angles formed by intersecting lines. To find the angles, think of the intersecting lines as being drawn on a circle (like the one in Fig. 16-2). For example, the western boundary of the property shown in Fig. 16-2 has a bearing of N 15° E; the southern boundary has a bearing of N 75° W (which is the same as S 75° E). See Fig. 16-3. Since you know there are 90° in a quarter circle, the angle formed by the intersection of the two boundaries must be 90°.

Note that when opposite sides of a lot have bearings that read a similar number of degrees (even though one reads N-E and the other reads S-W), they are parallel lines.

Another rule to remember is that for every lot with four sides, the sum of the inside angles must be 360°. For lots with more than four sides, the number of sides plus one, multiplied by 90°, is the sum of the inside angles. Thus for a five-sided lot, the sum of the inside angles is $5 + 1 = 6 \times 90° = 540°$. You can check the accuracy of any survey by this rule.

Figure 16-4 shows a completed plot plan developed from the survey plot of Fig. 16-2.

DRAWING AN ENLARGED VERSION OF A CURVED-LINE PLAT

Figure 16-5 shows a survey plat of a lot that has curved boundaries. Siting (planning the position of) a house on such a lot takes a great deal of care. So to draw a plot plan for a house on a curved-boundary lot, first copy the lots small-scale survey plat at a *larger* scale, either 1/8 inch or 1/4 inch to 1 foot. Then draw the preliminary floor plan on this larger plat. It's easier to site a house on the large-scale plat.

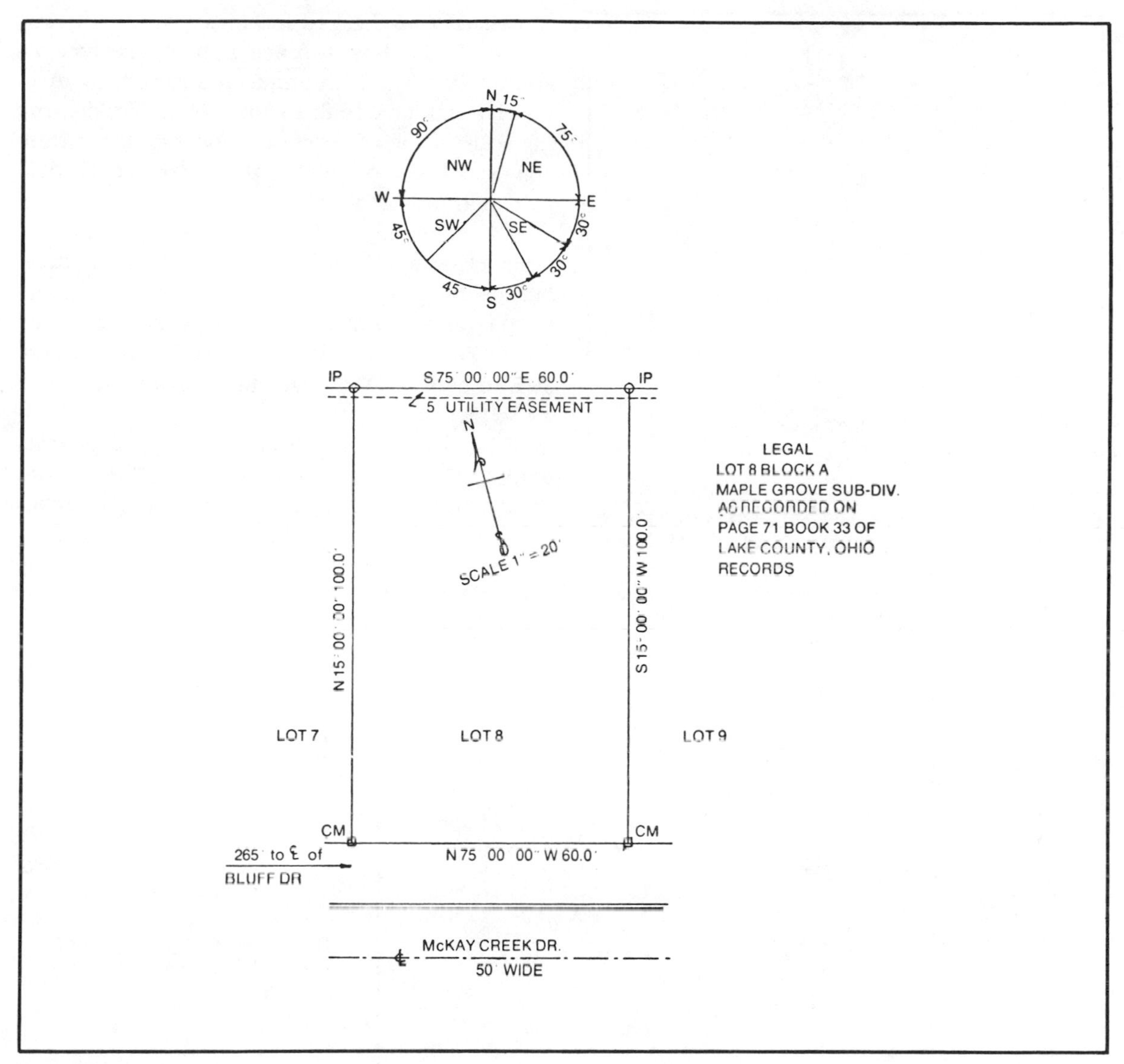

Fig. 16-2. A typical plat survey.

To copy the survey plat in Fig. 16-5 at a larger scale (1/8 inch to 1 foot), proceed as follows:

1. Tack a large sheet of paper to your drawing board and draw an east-west base line near the bottom edge. Let a point near the center of the paper (about 15 inches from the righthand edge of the paper) represent the lower rear corner of the lot (Fig. 16-5A). A vertical line through this point (A) will be your north-south reference line.

Note: A curved line which can be considered part of a circle (like the line from A to B in Fig. 16-5) is called an *arc*. A line between two points on an arc is called a *chord* (Fig. 16-6).

2. Set your adjustable triangle to 22° 48′, which is the bearing of the chord of the curved line from A to B in Fig. 16-5. Draw the chord with the triangle (starting at A). Be sure to draw the line to scale (at 113.63 feet). However, do not try to draw the arc of the chord until the lot outline has been

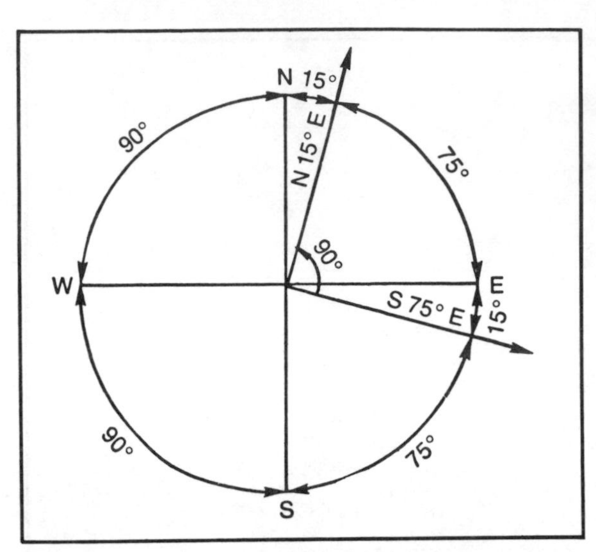

Fig. 16-3. Think of intersecting property lines as being drawn on a circle. You can then determine the angle of intersection.

completed and checked for accuracy.

3. Again, start at A and draw the side boundary line (at 46° 32′) to scale (at 121.10 feet). Actually you'll have to draw the line at 43° 29′, for the scale of the triangle reads only to 45°. After you draw the line to scale, then draw the 25 foot half-width of the street (Fig. 16-5).

4. Note that the line from C to E in Fig. 16-5 is not a true arc. The line from C to D is a true arc, but the line from D to E is a straight line. To find point D on your drawing, draw the C-to-D chord (at 28° 38′) to scale (at 34.29 feet). To find the corner at E, draw the D-to-E line (at 8°34′) to scale (at 19.78 feet).

Note: The distance from the center of a circle to any segment of that circle is called the *radius*. Since arcs can be thought of as segments of a circle, the arcs in your drawing have a theoretical radius.

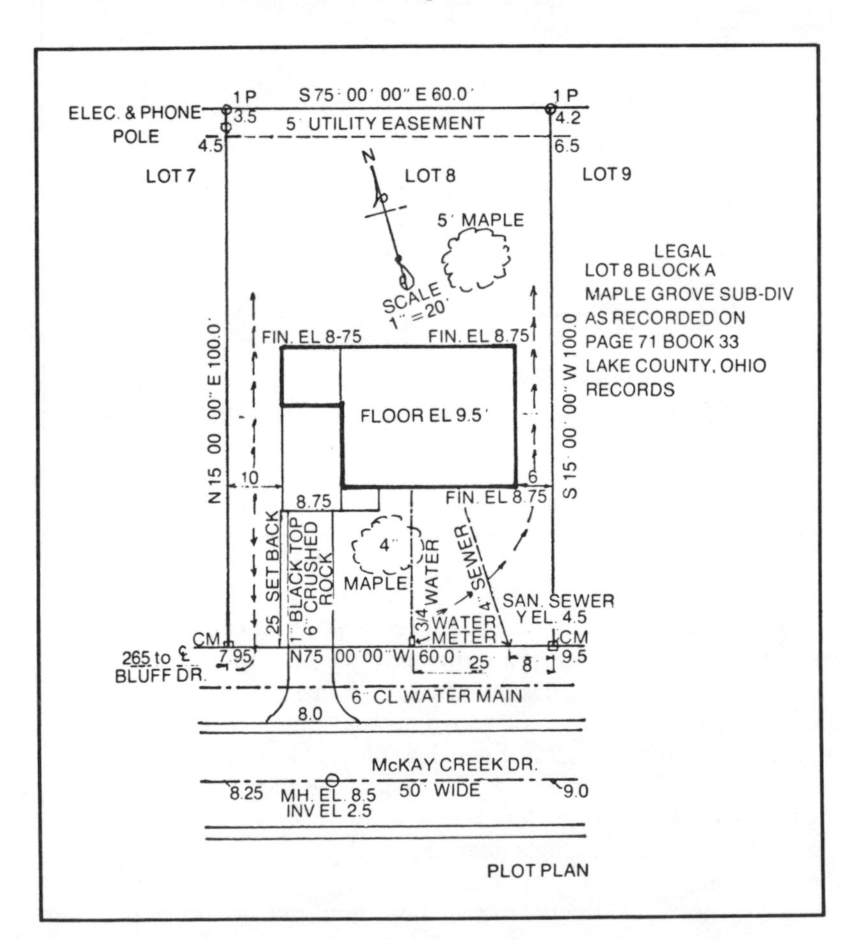

Fig. 16-4. A completed plot plan developed from the plat survey of Fig. 16-2.

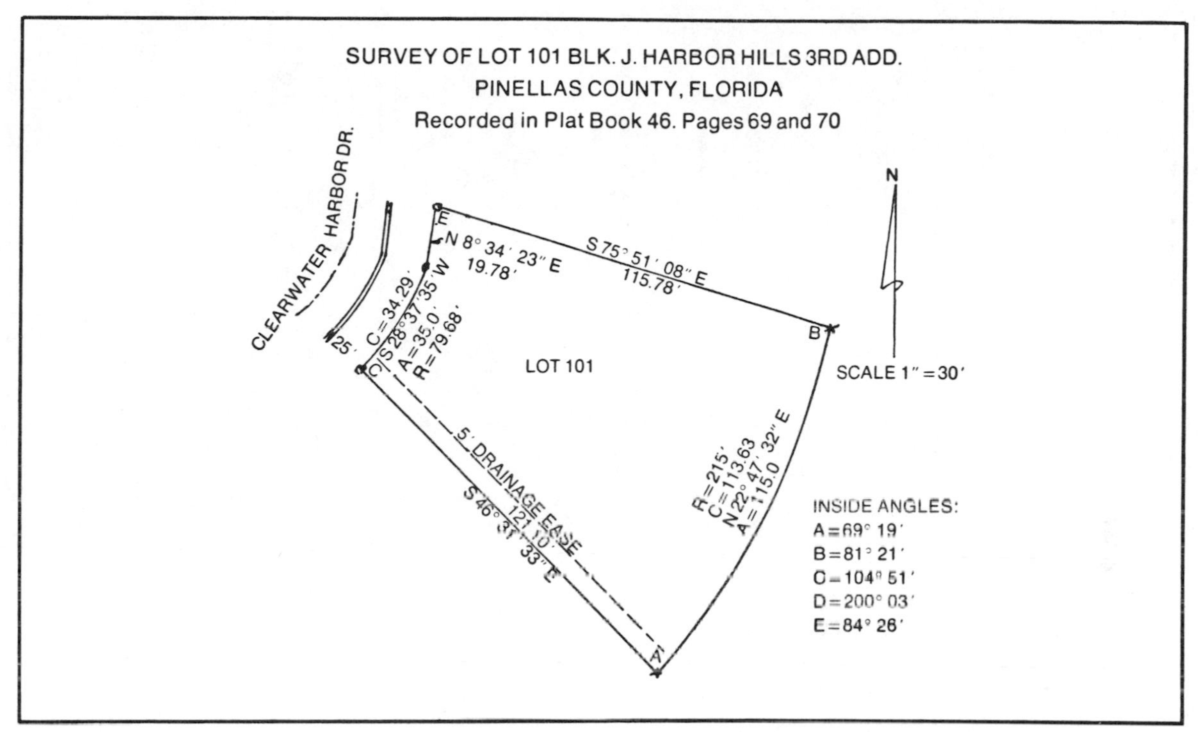

Fig. 16-5. A survey plat of a lot with curved lines.

So to draw any arc, you need to know its radius and the center of the imaginary circle.

5. To draw the arc from C to D, set a drawing compass to the arc's radius (the scaled distance of 79.68 feet). Put the needle point of the compass at C and draw an arc at about where the center of the imaginary circle would be. Do the same thing with the needle point placed at D. The intersection of these two arcs marks the center of the imaginary circle. Then just put the needle point at this center point and draw an arc from C to D. You can draw the other arc (from A to B) using the same technique.

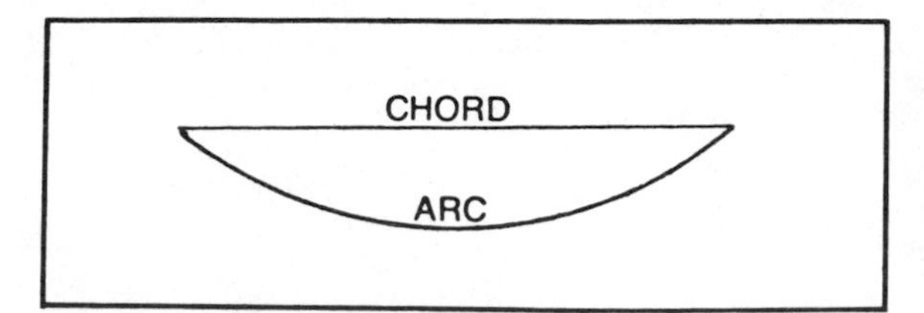

Fig. 16-6. A chord is a straight line that intersects two points on an arc.

Note: If your drawing compass isn't big enough to draw the arcs, you can do equally well with a wooden yardstick (Fig. 16-7). First break off the point of a needle and set it, point out, in the edge

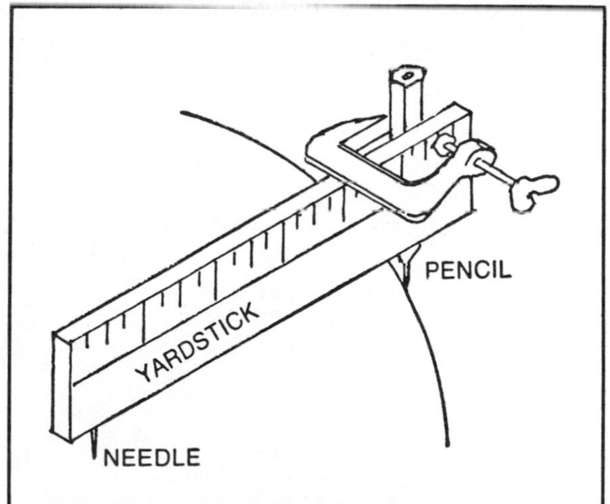

Fig. 16-7. An improvised drawing compass.

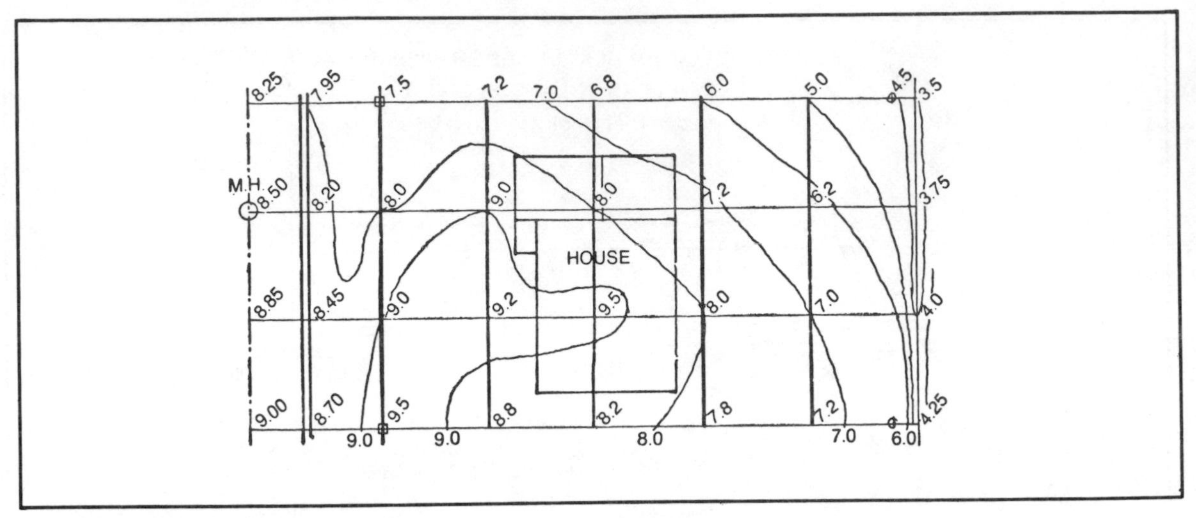

Fig. 16-8. An elevation/contour map.

of the yardstick, near one end. Use a small dime store clamp to hold a pencil to the side of the stick. When you want to draw a large arc, set the needle at the pivot point on the paper and swing the pencil point (set at the appropriate scaled distance along the stick) just as you would a pencil on a standard drawing compass. If your triangle settings and scaled distances have been accurate, you should find the scaled distance between E and B to be 115.78 feet. Actually if you're within 2 feet or 3 feet of that, the outline of the lot will be sufficiently correct for all purposes.

6. If your outline is seriously off, try drawing the upper boundary line (B to E) by starting at the upper rear corner (B). Set your triangle to 14° 09′ (which will give a bearing of S 75° 51′ E) and to the scaled distance of 715.78 feet.

7. Another way to check the redrawn plat is to determine the inside angles (as shown on the plat) and check each corner. This will show if you have made an error in the triangle settings.

8. When you're satisfied with the accuracy of the lot outline, draw the curved street and rear boundaries. Also draw in all clearances and setbacks.

Of course, enlarging a survey plat is just the first step. Then you have to draw in the floor plan of your house. When you've done that, you should have a workable plot plan.

Along with the plot plan, some house planners like to include an *elevation and contour map* like the one shown in Fig. 16-8. Such maps indicate the height (above sea level) of various points on a lot and thus help in the planning of footing depths, landscaping, etc. However, an elevation and contour map is usually unnecessary, for an experienced bulldozer operator can grade a landscape by eye, and footings on lots without severe slopes do not require such detailed plans. But if you must have elevation and contour information, you can ask your surveyor to gather the data for you.

Chapter 17

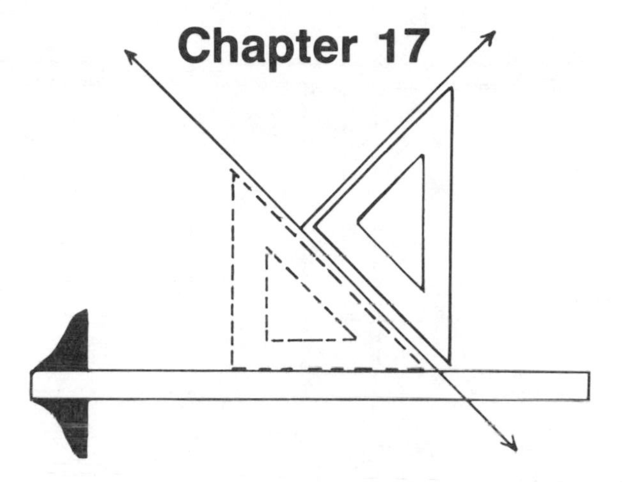

Electrical Plans

The more you know about electricity and its distribution, the better your electrical plan will be. However, do not let lack of electrical know-how discourage you, for planning the electrical layout is usually just a matter of deciding on the location of fixed lights, the switches that control them, convenience outlets, and electrical devices. The engineering of the system can be left to the electrical contractor. He will install the system in compliance with the code requirements of your area (usually the National Electric Code).

An electrical plan is usually just a floor plan with electrical symbols added to show the locations of switches, lights, etc. (Fig. 17-1). The electrical symbols can be drawn directly on the finished floor plan (without dimensions and dimension lines).

HOUSE WIRING

Electric service is brought into a house via three wires, which are either carried overhead on poles, or run through a cable underground (Fig. 17-2). Electric service goes first to a meter which measures (in kilowatts) the amount of service being used. Then it goes to a heavily fused main disconnect (a cutoff switch) in the main service panel.

Thus when you plan, keep in mind where electrical service will enter the house, for there must be a clear, straight run from the outside pole or underground cable to the meter. And there must be a straight run from the meter to the main service panel. This service panel should be in the garage, utility room, or basement. And this may determine where the garage or utility room will be located, for you cannot run service wires over the roof or through the house.

The service panel is a steel box and is usually surface-mounted on masonry block walls but may be flush-mounted in frame construction.

The heavy duty 230-volt circuits to the water heater, range, dryer, air conditioner, etc. are usually protected by cartridge fuses or circuit breakers. The lighter 115-volt lighting and convenience circuits are protected by plug fuses or circuit breakers.

It is recommended that no wire thinner than #12 copper be used and that all circuits be grounded

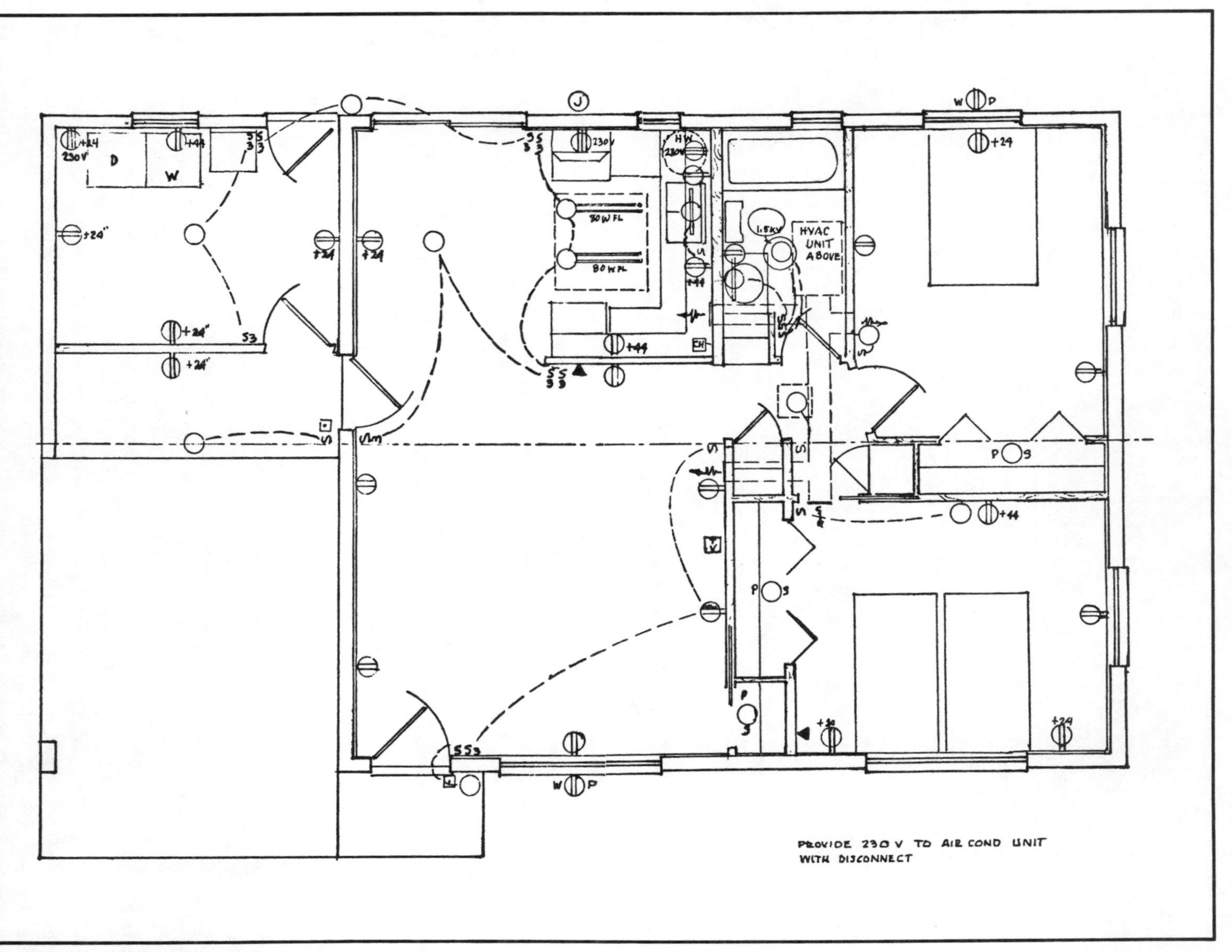

Fig. 17-1. An electrical plan.

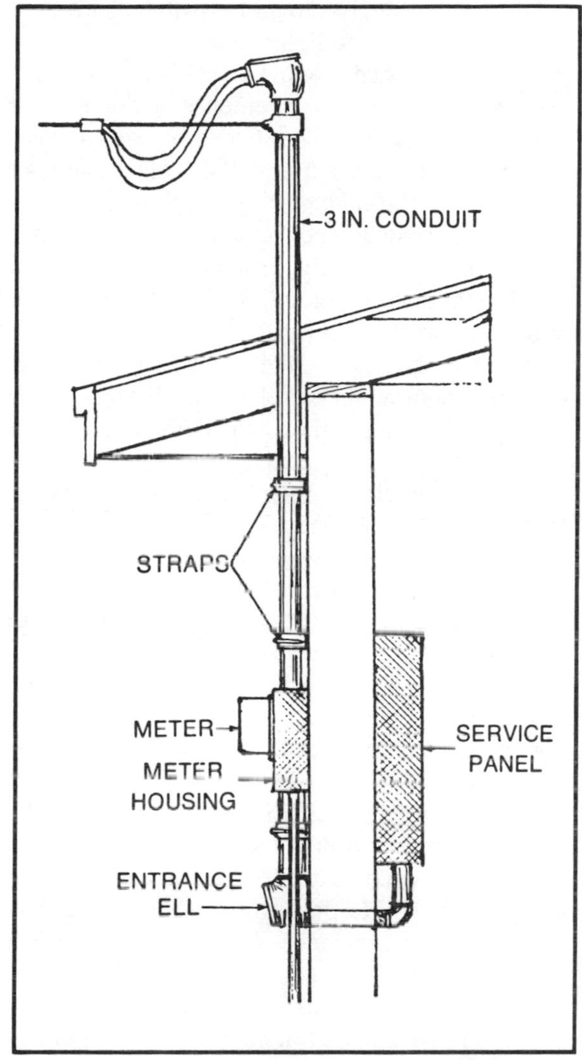

Fig. 17-2. Electric service enters a house via three wires, goes directly to a meter, then to a service panel.

(using three-wire type conductors for maximum safety). There are other safety devices that immediately break the flow of current in case of a short due to defective equipment.

Where wiring is exposed, as in a garage, it must be run in conduit or otherwise protected.

In large homes (especially ones with more than one story) it is often desirable to have secondary distribution panels on each floor so that a blown fuse does not require your going to the main service panel.

THE PLACEMENT OF ELECTRICAL FIXTURES

In drawing an electrical plan, deciding on the placement of electrical fixtures (switches, lights, etc.) will be your primary concern.

Wall switches for controlling lights are usually 48 inches above the floor, but be sure they are accessible and not behind doors. They should be on the latch side where they are most handy. Of course, you must place switches for exterior lights on the inside of the house. And switches for stairway lights should be either at the top or bottom of the stairs or both.

Use multipole switches when you want to be able to control a light from two or more points. For example, a garage light should be controlled from near the garage door *and* from inside the house. Sometimes a pilot-lighted switch will remind you to turn off an unneeded light.

As a minimum, the National Electric Code requires one convenience outlet for every 150 feet of floor area. The code insists that outlets be spaced no more than 12 feet apart.

Install an exterior light at each outside entrance to illuminate the faces of visitors and any entrance steps. It is often desirable to have a post light at the driveway near the street. Floodlights under the eaves to illuminate the yard should be considered. And all these lights should be controlled from *within* the house.

One or more outside weatherproof outlets, 18 inches above the ground, should be installed, one near the front entrance for holiday lighting, others for the use of powered lawn tools. The one at the entrance should be switch controlled from within.

Some means of general lighting should be provided for the living room. It may be ceiling, wall, portable, cove, or valance lighting. Provide lighting outlets (switch-controlled) in appropriate locations. And you may want to provide illumination for a prized painting or a bookcase.

Place convenience outlets (duplex) so that no point along the floor line of any usable wall is more than 6 foot from an outlet. Two convenience outlets or two split outlets should be switch controlled.

You may want to provide TV outlets properly

wired to an outside or attic antenna with a convenience outlet close by.

The dining area should have a switch-controlled light over the dining table. The 6 foot rule for convenience outlets applies here, but if a dining or breakfast table is placed against a wall, provide an outlet 36 inches above the floor.

Consideration should be given to phone outlets. In a small house, it may be desirable to locate such an outlet near the dining table or a chair.

In the bedroom good illumination should be provided by a ceiling or wall light (switch-controlled). A light over a full-length mirror or directly in front of a clothes closet may provide general lighting.

Place a convenience outlet on both sides of each probable bed location and follow the 6 foot rule elsewhere. Keep in mind the family habits such as reading in bed. You may want to provide an outlet 24 inches above the floor at the center of each bed space—so electric blankets can be easily plugged in. If you have a dresser or highboy for which wall space has been planned, locate a convenience outlet at 4 inches above the dresser top.

You may want a second TV or phone outlet in the master bedroom. And in these days when energy saving is being considered, one of the old fashion ceiling fans (switch-controlled) is a good substitute for air conditioning.

Lighting in the bathroom can be provided by lights on each side of the mirror or by a fluorescent lamp directly above the mirror. When an enclosed shower is planned, a vaporproof light should be provided. Of course, the switch for such a light should be well away from the shower entrance.

Consideration should be given to a heater (or a combination heater and light in the bathroom) for use when central heating is not needed. If the bathroom has no outside window, a switch- controlled exhaust fan should be installed.

Install a convenience outlet near the lavatory about 40 inches above the floor. Other convenience outlets may be needed in very large bathrooms.

In the kitchen provide outlets for general illumination. And install lighting above the kitchen sink. All these lights should be switch controlled. Other under-cabinet lights may be locally controlled. Suspended luminous ceilings are often desirable. Provide a convenience outlet for the refrigerator. And install one outlet for every 4 linear feet of counter work surface. Place these about 44 inches above the floor. If there is to be a table in the kitchen, provide a convenience outlet 36 inches above the floor near the proposed table location.

Special purpose outlets include those for the range, dishwasher, garbage disposal, compactor, range hood light, and range exhaust fan. You may want to install a clock outlet. And if you plan to have a food freezer in the kitchen, provide an outlet for it.

Install switch-controlled general illumination for the laundry area. And you may want to install a special light over the laundry sink or laundry appliances. Make sure there's at least one convenience outlet in the laundry area. Special outlets are needed for the washer and dryer.

You may need to install a light in each closet. These lights can be switch controlled from outside the closet, or they can be controlled by pull switches.

In areas of high humidity, a convenience outlet in each closet for a dryer may be desirable.

Lights (switch-controlled) should illuminate the entire hall area. Multiple switches may be needed. Provide one convenience outlet for every 15 feet of hallway (measured along the centerline). In reception hallways and foyers, outlets should be placed so that no point along the floor line of any usable wall is more than 10 feet from an outlet.

Install lights where they can illuminate planned work areas or equipment locations, such as near a furnace, pump, or work bench. if the basement is not partitioned off, install one light for every 150 sq. feet of floor area. If there are closed spaces, provide switch-controlled lighting as needed for proper illumination. Provide appropriate special-purpose outlets for all-purpose equipment.

In the attic install at least one light for general illumination (switch-controlled from the bottom of the attic stairway). If no permanent stairs are provided, this light may be chain controlled if located at the access door. If an unfinished attic is planned for later completion into finished rooms, install a junction box with direct connection to the main distribution panel for future extension of lighting and outlets.

Special-purpose outlets may be necessary if heating and air conditioning will be located in the attic. If this equipment is not easily accessible, a disconnect should be provided in a suitable location.

Each porch, breezeway, or similar roofed area of more than 75 sq. foot floor area should have a switch-controlled light. Large areas may require more than one light. Multiple switch control should be installed where there is more than a single entrance. Install one convenience outlet (weatherproof if exposed to weather) for every 15 feet of wall bordering on the porch or breezeway.

In a garage there should be at least one ceiling light (wall switch-controlled). If the garage is to be used for purposes other than car storage, lighting appropriate to the various uses should be employed. Install at least one convenience outlet, and if there is to be a garage door opener, an outlet for this purpose should be provided.

LOW-VOLTAGE SWITCHING

A variation of conventional light wiring is low-voltage relay switching. In this system the 120-volt current goes *directly* to the lights where there are relay switching devices controlled by 24-volt wall switches. Light bell type wire connects the wall switch to the relay switch at the light.

With low voltage at the switches, there is no danger of shock, and there can be a savings in the cost of wire. Such a system enables you to control every light in your house from a central control at the head of your bed. This central control can be connected to a burglar alarm system so that all the lights will turn on if an intruder attempts to break in.

Many small electrical contractors seem unfamiliar with low-voltage switching and hesitate to bid on it. It is actually not complicated and nearly as easy to install as bell wiring.

Every electrical fixture and appliance you install in your house will require a specific amount of electrical service. The electrical system you plan must accommodate *all* these service requirements. This means that you must estimate your house's *total* service requirements and plan a service panel that will meet these demands. (The power company will also need to know how much electrical service your household will use.)

Current is a measure of electrical service and is rated in amperes. Thus a 100-ampere service requirement for your house means that the power company must supply 100 amperes of service to operate your electrical fixtures and appliances. Sometimes the service requirements for fixtures and appliances are expressed in watts, but there is a simple relationship between amperes and watts. If you know the number of watts, just divide by the voltage of the appliance, which will be either 115 volts or 230 volts. For example, a 115 volt 460-watt hairdryer uses 460 ÷ 115 = 4 amperes. Similarly, if you know the number of amperes, multiply the amperes by the voltage in order to find the number of watts required.

Table 17-1. Wattage for House Appliances.

Range	8,000 to 13,500 watts
Cook top	7,200 watts
Wall oven	4,000 watts
Microwave oven	1,500 watts
Dishwasher	1,500 watts
Compactor	400 watts
Clothes dryer	4,500 to 6,000 watts
Washer	700 watts
Water heater	1,600 to 5,500 watts
Hand iron	1,090 watts
Air cond. (room)	800 to 5,000 watts
Air cond. (house)	3,000 to 10,000 watts
Heat pumps	4,000 watts
Blender	720 watts
Coffee pot (1) cup)	650 watts
Coffee urn	1,500 watts
Coffee maker	1,000 watts
Food frezzer	350 watts
Fry pan	1,500 watts
Mixer (small)	150 watts
Mixer (large)	1,000 watts
Refrigerator	300 watts
Roaster (portable)	1,500 watts
Broiler oven	1,200 watts
Toaster (2 sl.)	600 watts
Toaster (4 sl.)	1,600 watts

The National Electric Code requires a minimum of 60 amperes of service for any house with a living area less than 1000 sq. feet. But usually small houses require considerably more service.

The service requirements of your house can be divided into three distinct categories: (1) general lighting and convenience outlets, (2) electric ranges and ovens, and (3) fixed appliances such as clothes dryers, air conditioners, etc. Your house's *total* service requirement is the sum of these three categories.

This is how you calculate your house's service requirement in each category:

1. For general lighting and convenience outlets the code requires a 20-ampere 115-volt circuit (that's 20 × 115 = 2300 watts) for each 500 sq. feet of living space, or part thereof (this excludes garage, porches, and unfinished space). In addition for kitchen-dining areas, there must be two 20-ampere circuits for small 115-volt appliances. To provide for a future addition, there must also be an unused 15-ampere circuit installed. Add all these requirements (in watts). Now take the first 3000 watts of this total and add to it 35% of the remaining number of watts. This will give you the service requirement for general lighting and convenience outlets.

2. Now for the second category (electric ranges and ovens), add 8000 watts for an electric

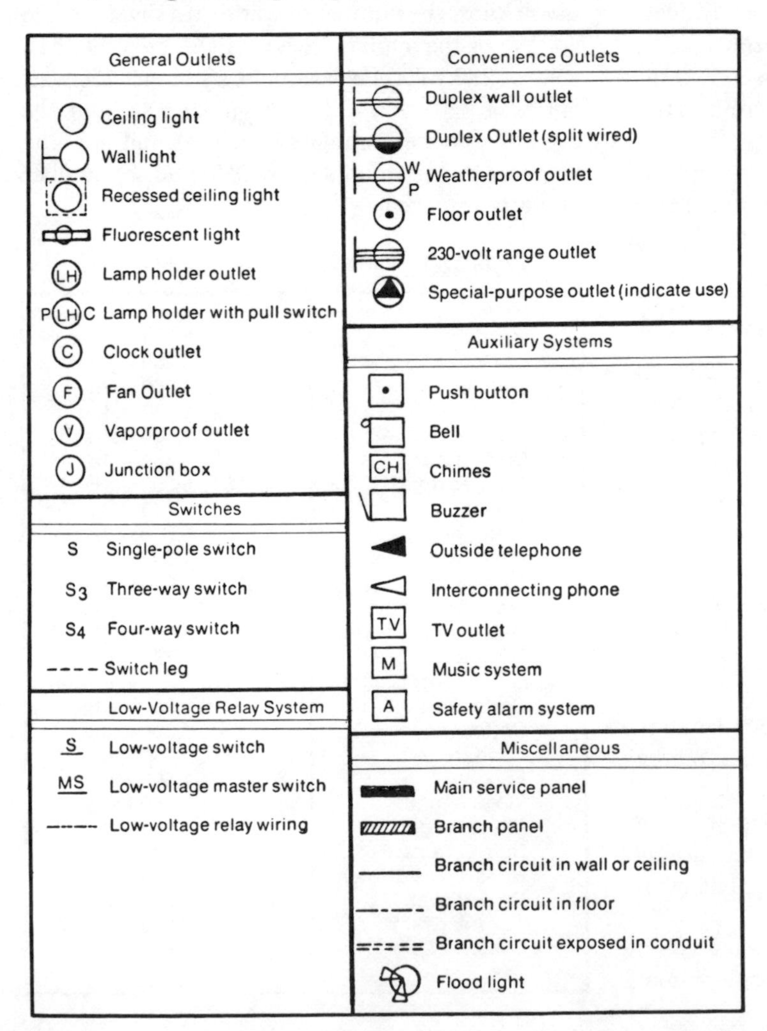

Table 17-2. Electrical Symbols for Residential Wiring.

range of not more than 12-kilowatt rating. The code permits a wall oven plus a separate cook top to be treated as a single range.

3. For the fixed appliances, add together the rated wattage of all fixed appliances to be served by individual circuits not previously accounted for in the calculation. Take 75% of this total. This total will be the service requirement for your fixed appliances. (Usually, the rated wattage of a fixed appliance is shown on an attached plate. Table 17-1 lists wattages for several common household appliances.) If both air conditioning and electric heating are to be used, the rating in watts of only the larger of the two connected loads need be included in the total because both are not operated at the same time.

4. Add the service requirements calculated in 1, 2, and 3. Then divide by 230 volts (for 115/230-volt 3-wire single-phase service) to get your house's total service requirement (in amperes).

ELECTRICAL SYMBOLS

Table 17-2 is a list of common electrical symbols and their meanings. Such symbols are the shorthand of electrical plans.

On your plan, always draw a switch leg (Table 17-2) between lights and their switches. Use appropriate symbols if a light is to be controlled from more than one switch.

The symbol for a fluorescent light should include a notation of wattage. A 20-watt tube is 24 inches long; a 40-watt tube is 48 inches long. A pair of 40-watt tubes should be marked 80 watts.

Chapter 18

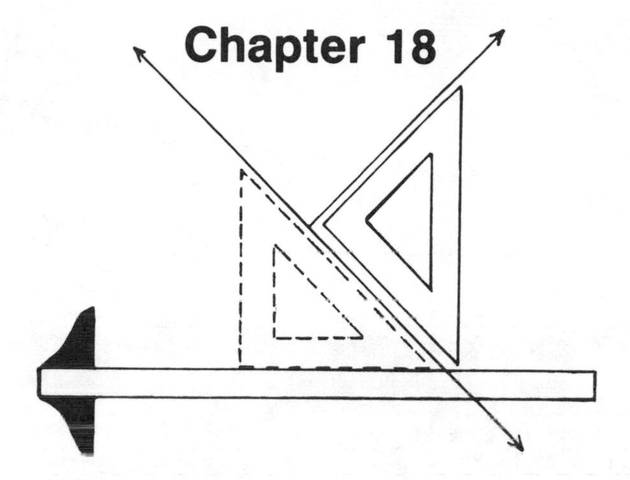

Drawing Plans for a Small House

This chapter will tell you how plans were drawn for the small two-bedroom masonry house shown in Fig. 18-1. It's a simple low cost structure with a car port. The basic steps that were involved in producing the plans are generally the same for *any* house—regardless of structural complexity.

STEP ONE: THE FLOOR PLAN

First, a rough freehand sketch of the floor plan was drawn on cross-ruled paper (with 1/4 inch squares). Each square represented 2 feet, which was a scale of 1/8 inch = 1 foot. The sketch is shown in Fig. 18-2. Such a sketch is a think-on-paper exercise that helps to clarify planning ideas.

Next, a *preliminary* floor plan was drawn. Most planners start by making such preliminary drawings on a 24 × 36 inch sheet of inexpensive paper. When all changes and corrections have been made, tracings are made from the preliminary drawings. Blueprints are made from the tracings.

The first step in drawing the preliminary floor plan was to tape the drawing paper to the drawing board, making sure that the bottom edge of the paper was aligned with the T square. Then a base line was drawn about 1 1/2 inches above the bottom edge of the paper to allow space for dimension lines (see 1 in Fig. 18-3). This line represented the front edge of the car port. Then with the 1/4 inch edge of the triangular scale, a second line was drawn above the base line to represent the front wall line (2 in Fig. 18-3). On the drawing, the distance between line 1 and line 2 represented 4 feet.

Next, the house's center ridge line was drawn (3), and the distance between line 2 and line 3 represented 13 feet, 4 inches. Then a line representing the rear wall line was drawn (4), and of course the distance between line 3 and line 4 also represented 13 feet, 4 inches. So the drawing showed that the rear wall and front wall were 26 feet, 8 inches apart.

Next, lines were drawn to represent the overall width of the house (44 feet), the width of the main body of the house (32 feet), and half the main body width (16 feet). Then vertical lines were drawn at the ends of these lines to represent the left car port/storage wall (5), the left house wall (6),

Fig. 18-1. A small, two-bedroom masonry house.

the house centerline (7), and the right house wall (8).

The *inside* wall lines were drawn next, and the distance between the inside and outside wall lines represented a wall thickness of 8 inches. The left storage wall was drawn *past* the center ridge line (3). In the completed house, such a wall extension would help support the ridge girder.

Next came the *interior* wall lines and some interior details. It was decided to build the house with exposed rafters and a ridge girder, which requires a support at the center of the 32 foot width of the house. This meant that it was desirable to place the living room and bath-kitchen wall on the house centerline (7 in Fig. 18-3).

To draw the interior wall lines (as well as many other features), the coordinate system was used (as discussed in Chapter 5). After the wall lines were drawn, it was possible to render the door and window openings. Then other details were added—kitchen appliances, bathroom furnishings, construc-

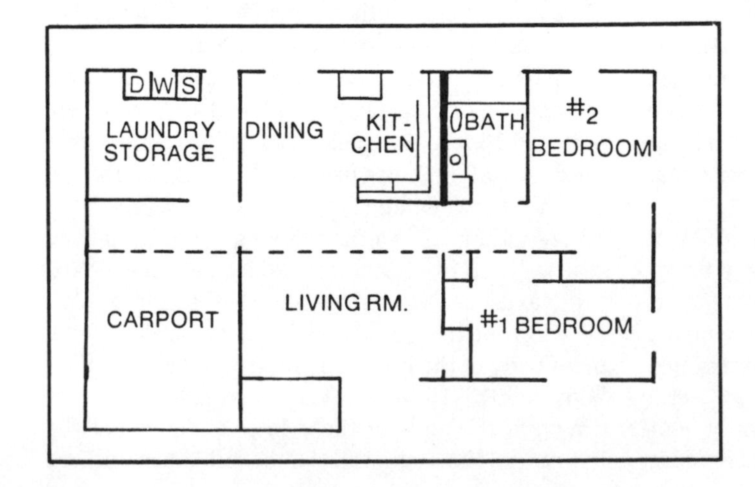

Fig. 18-2. The floor plan sketch.

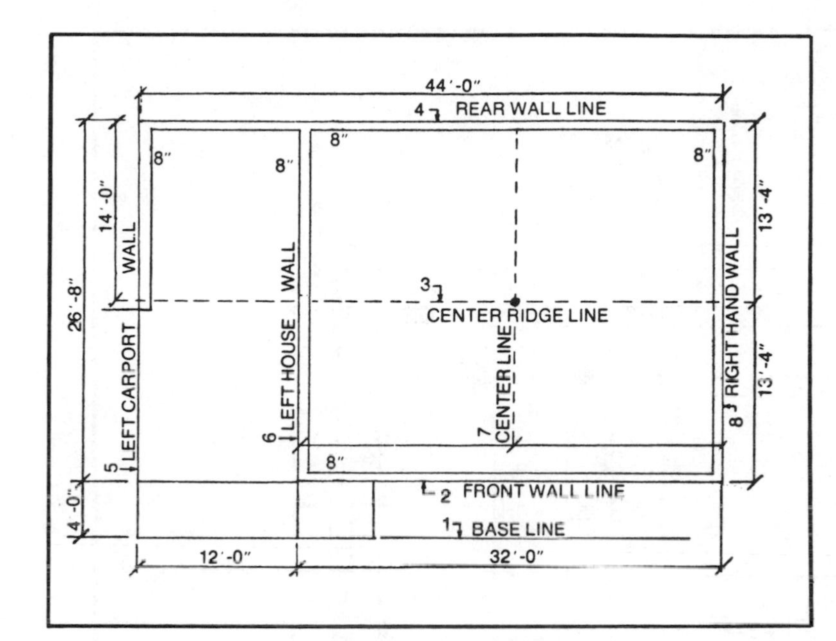

Fig. 18-3. The preliminary floor plan.

tion materials symbols, etc. Of course, all pertinent dimensions were included. The end product of all this is shown in Fig. 18-4.

For the most part, windows were drawn so they would have equal amounts of wall space on each side of them. The principle of modular planning was adhered to wherever possible.

Construction sections had to be made later on. So in the floor plan, short parallel lines were drawn through the walls at certain points, indicating the slices of wall that would later be shown in construction sections. Each pair of short lines was identified by two numbers in a circle, numbers which were keyed to the corresponding construction section.

STEP TWO: THE PLOT PLAN

After the floor plan was completed, the plot plan was drawn. The procedure was very much like that explained in Chapter 16. The lot for the house was 60 feet wide and 100 feet deep, with a 25 foot front setback.

STEP THREE: THE CONSTRUCTION SECTIONS

Construction sections share equal importance with the floor plan. Drawn to a larger sale (1/2 inch = 1 foot), they show relevant cross sections and name the basic building materials.

The main construction section for our small house is shown in Fig. 18-5. It was drawn to scale, using construction materials symbols. It specifies that the footings should be poured in 16 inch wide trenches dug in the ground and that the floor level of the house should be about 12 inches above natural ground level so that finish grading can be sloped away from house walls. The section further indicates that the top of the 8 inch deep footing must not be above ground level.

Other specifications in the construction section include:

☐ The 16 inch wide 8 inch deep concrete footing should be reinforced with two #5 (5/8 inch diameter) steel bars.

☐ The soil below the floor slab should be poisoned to protect against termites and covered with a moisture barrier (0.004 inch thick polyethylene).

☐ The floor slab should be reinforced with welded wire mesh 6 × 6 10/10 (10 gauge wires welded into a mesh of 6 × 6 inch squares).

☐ The exterior walls should be masonry block, 10 blocks high (6 feet, 8 inches).

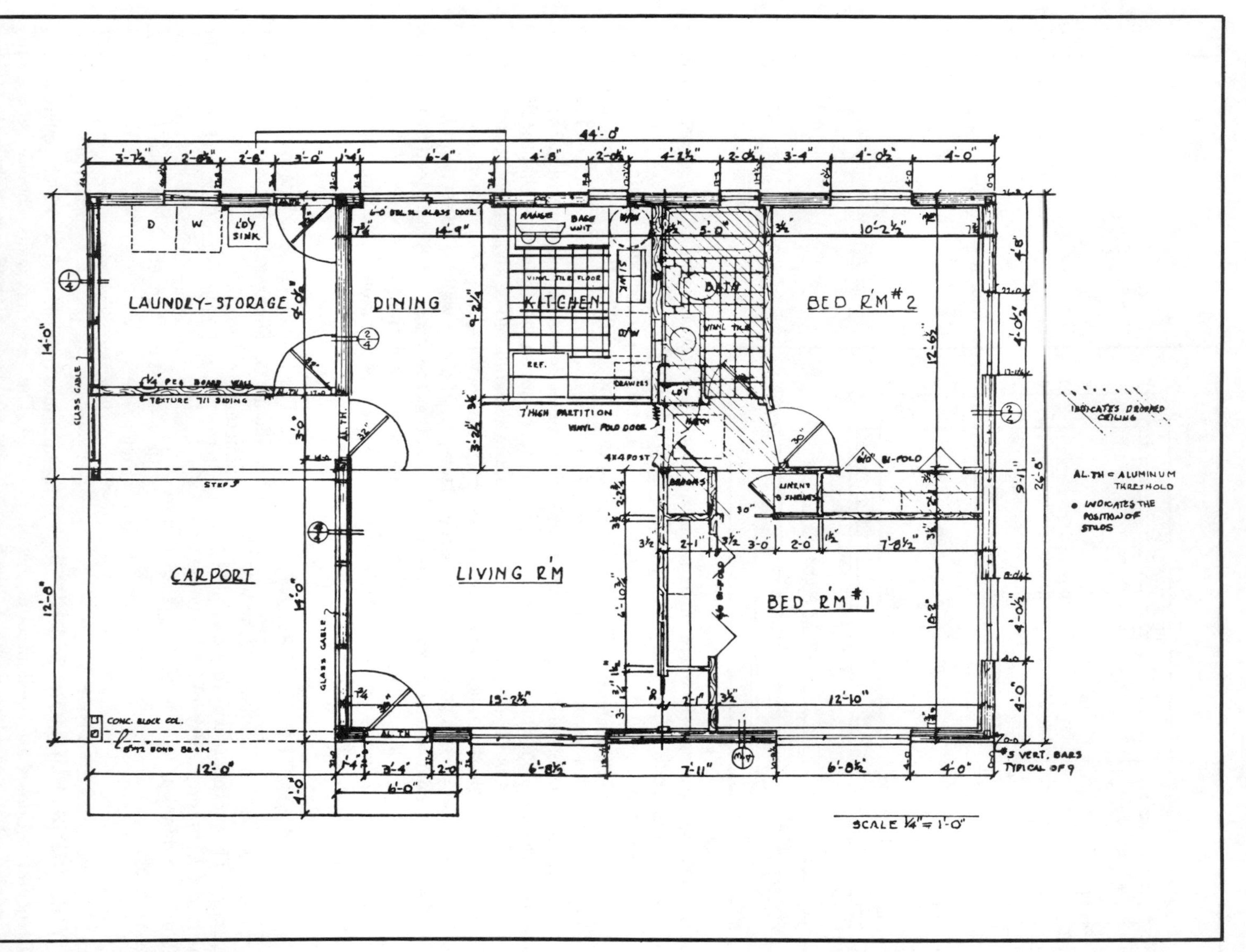

Fig. 18-4. The final floor plan.

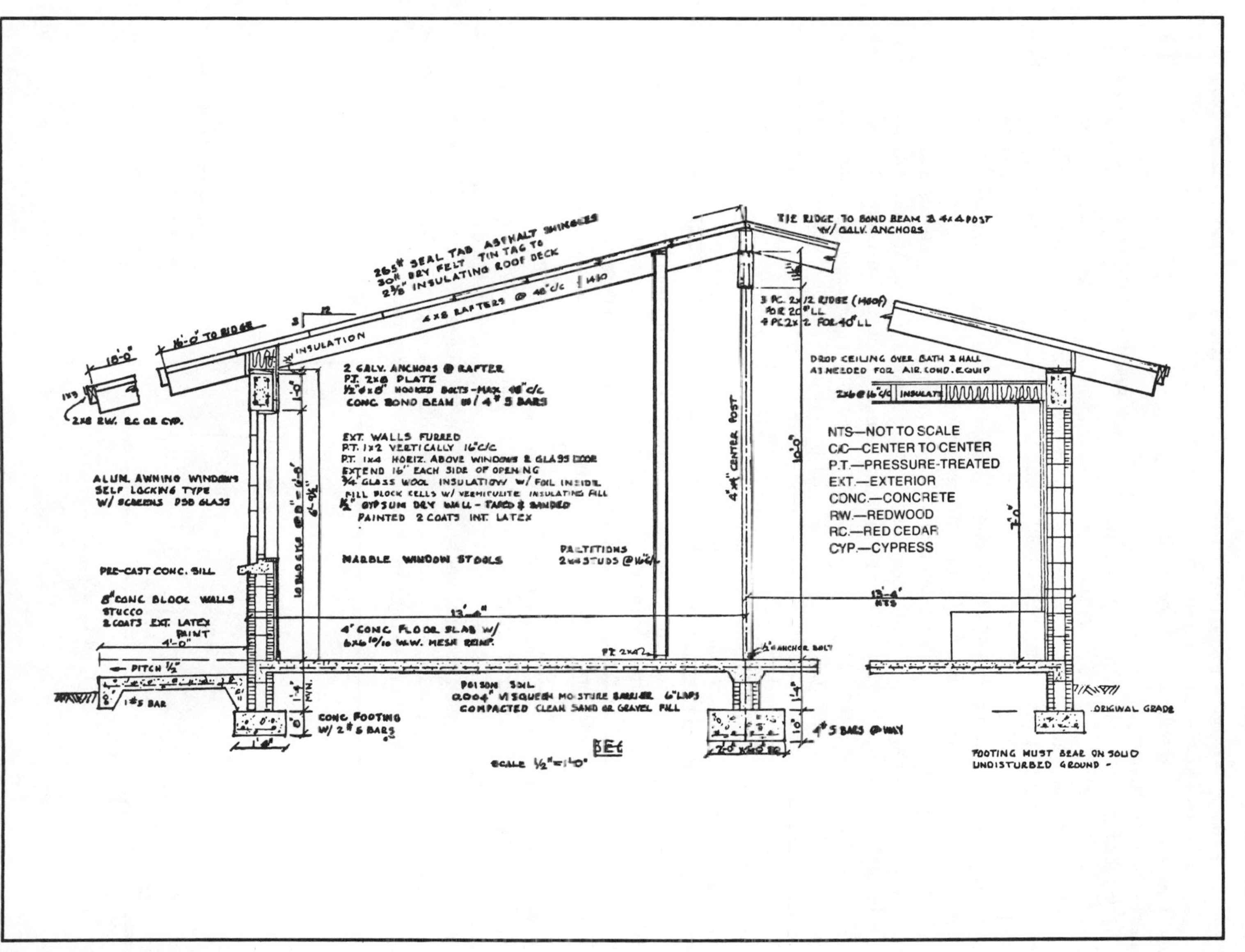

Fig. 18-5. The main construction section.

☐ There should be a concrete bond beam (concrete slab) running along the top of the exterior walls. The bond beam should be reinforced with 4 #5 bars placed horizontally.

☐ Hooked anchor bolts (1/2 × 8 inches) should be set in the top of the bond beam about every 5 feet. (The bolts are used to fasten down 2 × 8 inch wall plates that rest on top of the bond beam.)

☐ The roof should have a pitch of 3/12 (3 units rise for every 12 units of run), or 14°. (The pitch of a roof is determined, in part, by the kind of roofing materials used. A roof covered with materials that shed water well doesn't need a steep pitch. Recommended roof pitches for various roofing materials are shown in Fig. 18-6.)

☐ The roof rafters should be 4-×-8s.

Rendering the construction section of Fig. 18-5 was no harder than creating the floor plan. First, a horizontal line was drawn to represent the floor line. Then the 8 inch walls and the ridge centerline were drawn. Next came the footing and foundation wall, the 4 inch thickness of the floor slab, the 10-block wall height, and the bond beam. After the 2 × 8 inch wall plates were drawn, the 14° roof pitch was determined with an adjustable triangle. The roof was then drawn, and all other details were added.

Figure 18-7 shows detailed construction sections of the walls of the house. Their number designations are keyed, of course, to those in Fig. 18-4. These sections were drawn *after* the main construction section was completed.

STEP FOUR: THE FOUNDATION PLAN

The foundation plan for the house is shown in Fig. 18-8. Essentially it's just a tracing from the floor plan (Fig. 18-4). It indicates the position of the footings, the elevation of the various concrete slabs, the finishes for the slabs, and other pertinent information.

Diagonal dimension lines were drawn to make it easier to accurately lay the foundation slabs later on. The positions of hose connections and plumbing pipes were noted. Special footings (like the one for the center post) were drawn in and described. (The notation 2/0 × 2/0 × 10 inches means 2 feet × 2 feet × 10 inches.)

STEP FIVE: THE ELEVATIONS

Elevations for our small house are shown in Fig. 18-9. They were all drawn at the same scale.

Usually front and rear elevations are drawn together—vertically aligned, one above the other. Right end and left end elevations are made in the same fashion.

Drawing the elevations of Fig. 18-9 was a fairly simple process. First, the base lines were drawn, then the floor lines, then most of the vertical lines. The roofs were next, then the other details, like the doors, windows, materials specifications, etc.

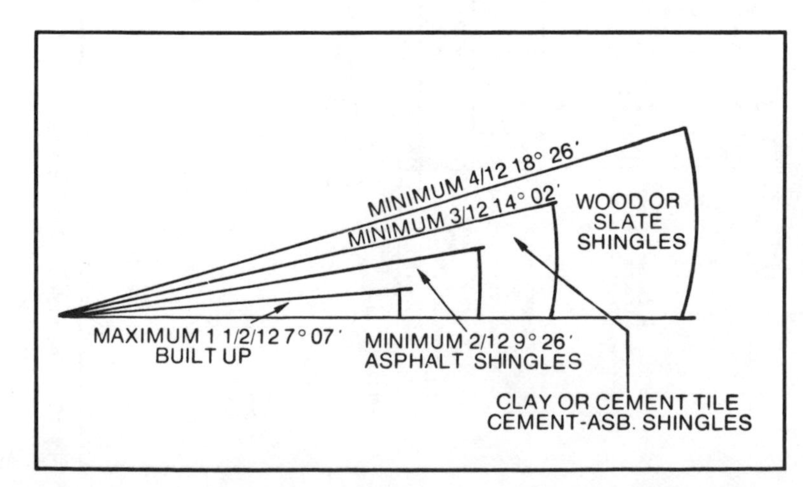

Fig. 18-6. Roof pitches for various roofing materials.

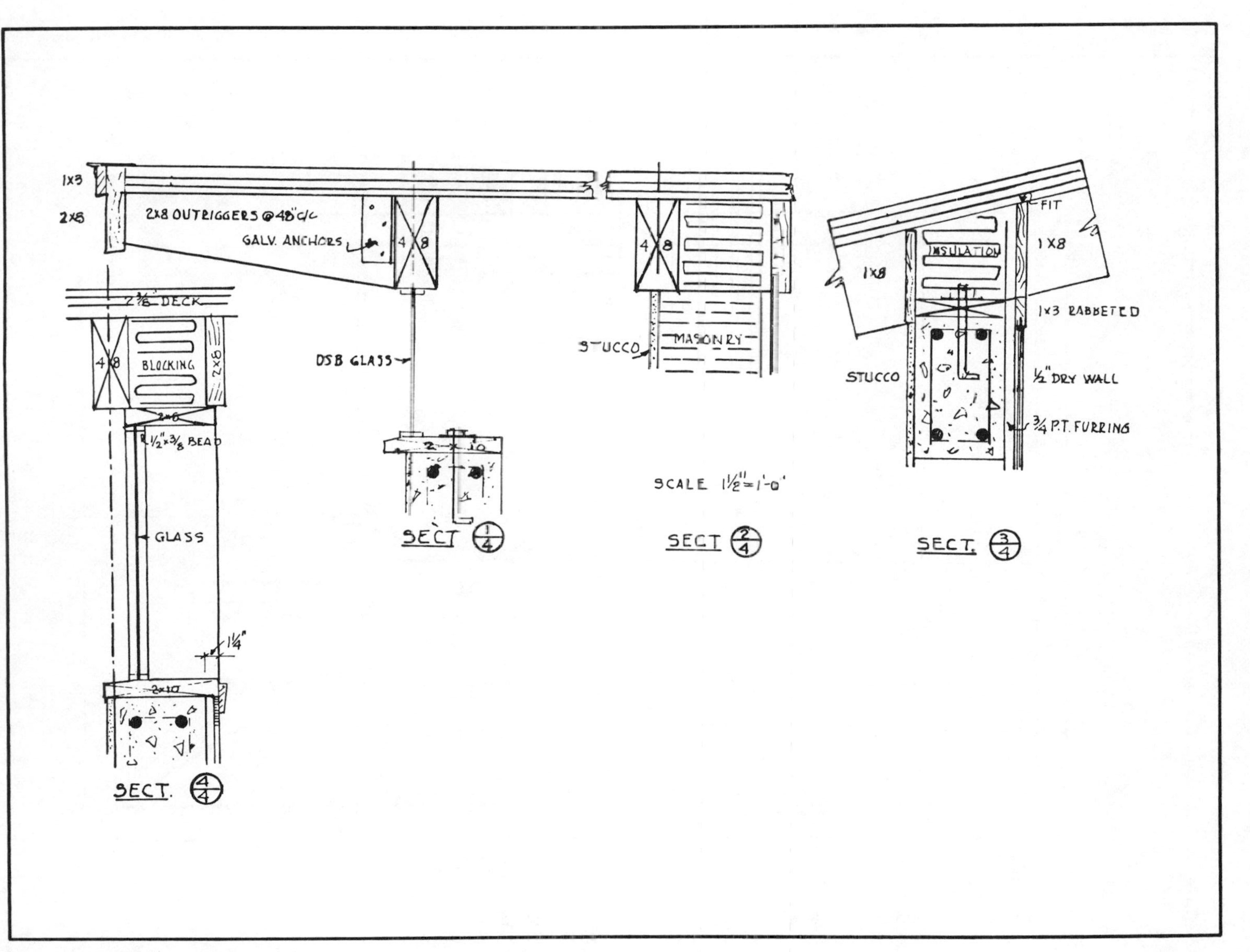

Fig. 18-7. Construction sections of the walls.

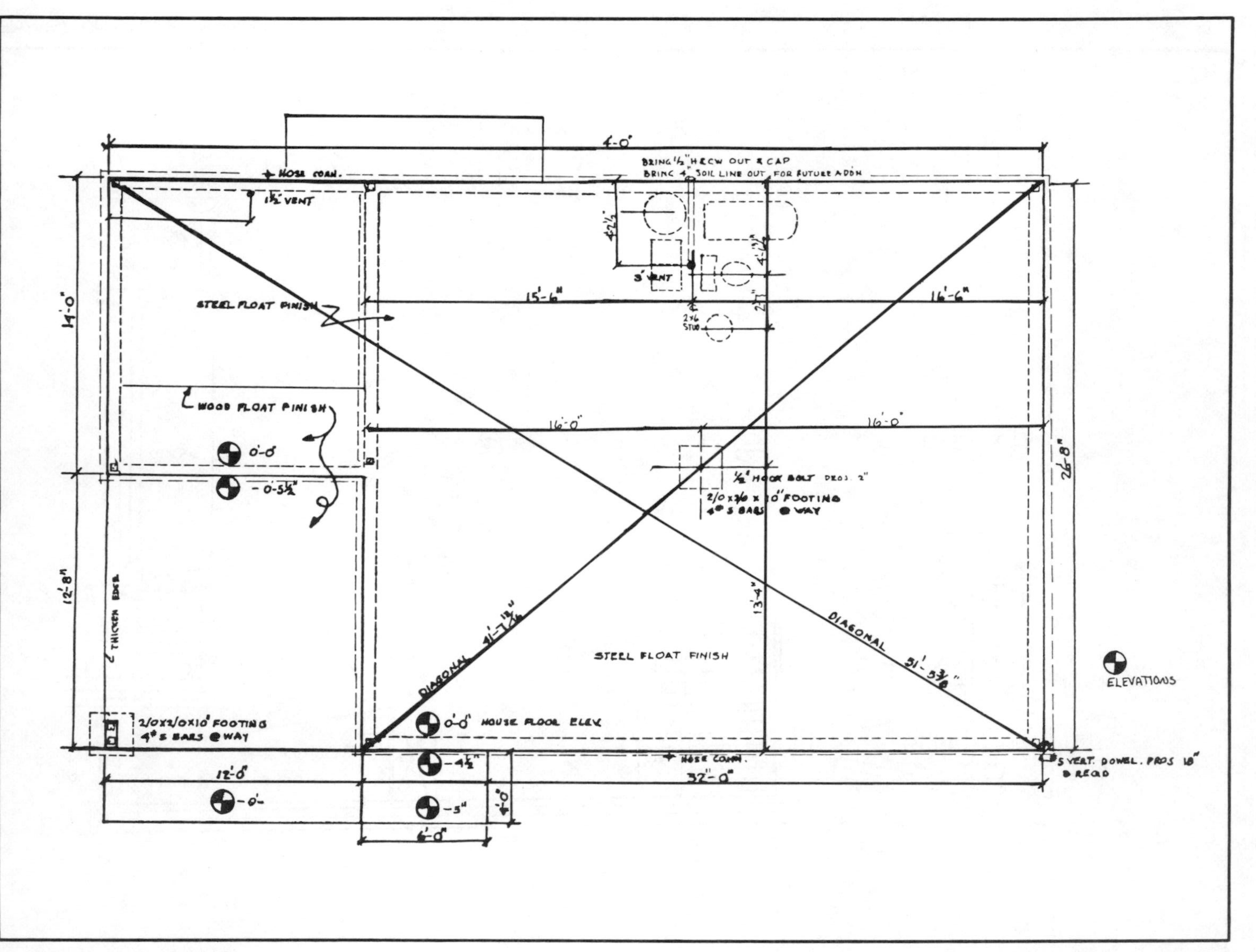

Fig. 18-8. The foundation plan.

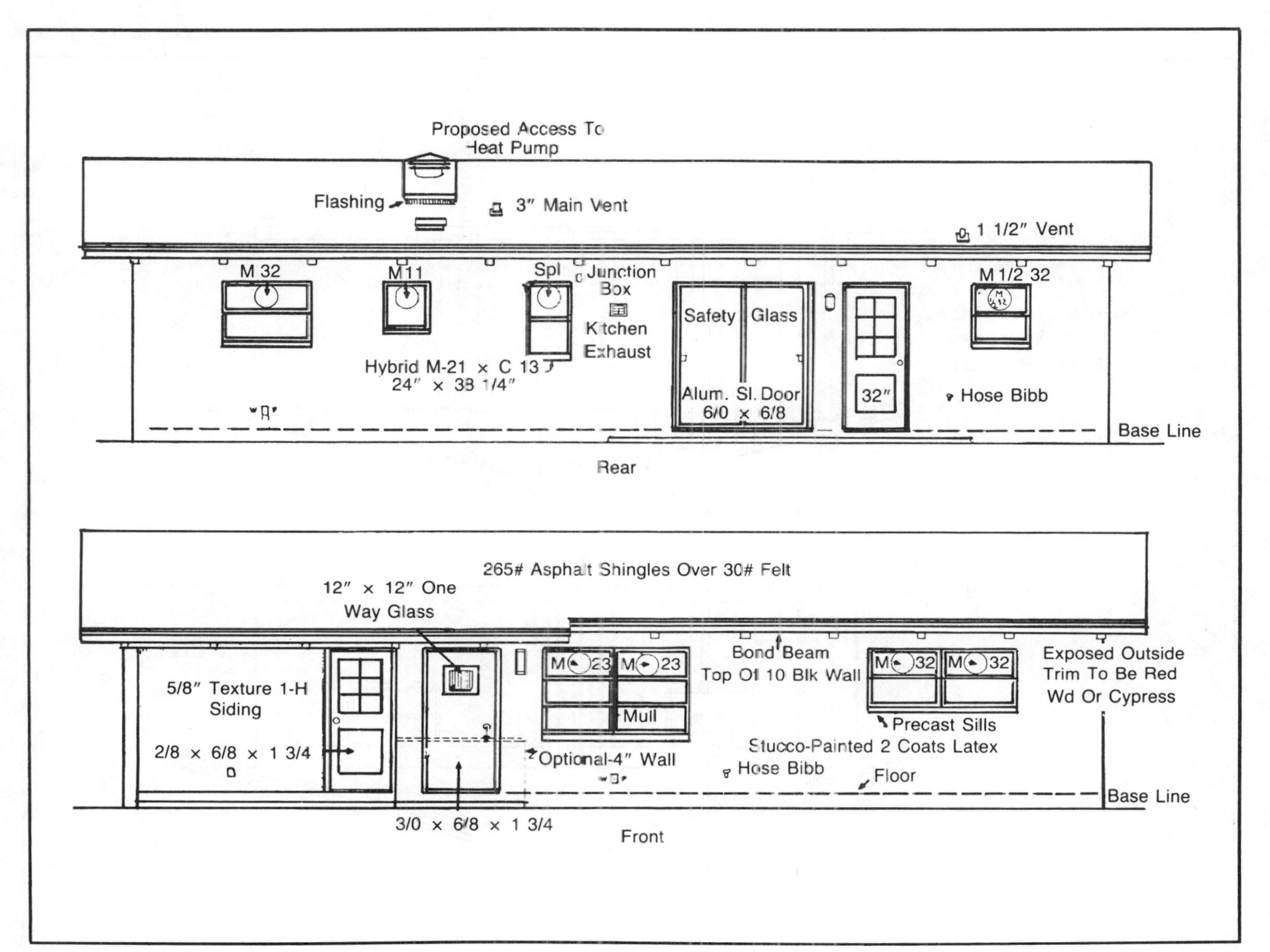

Fig. 18-9. The elevations (continued on next page).

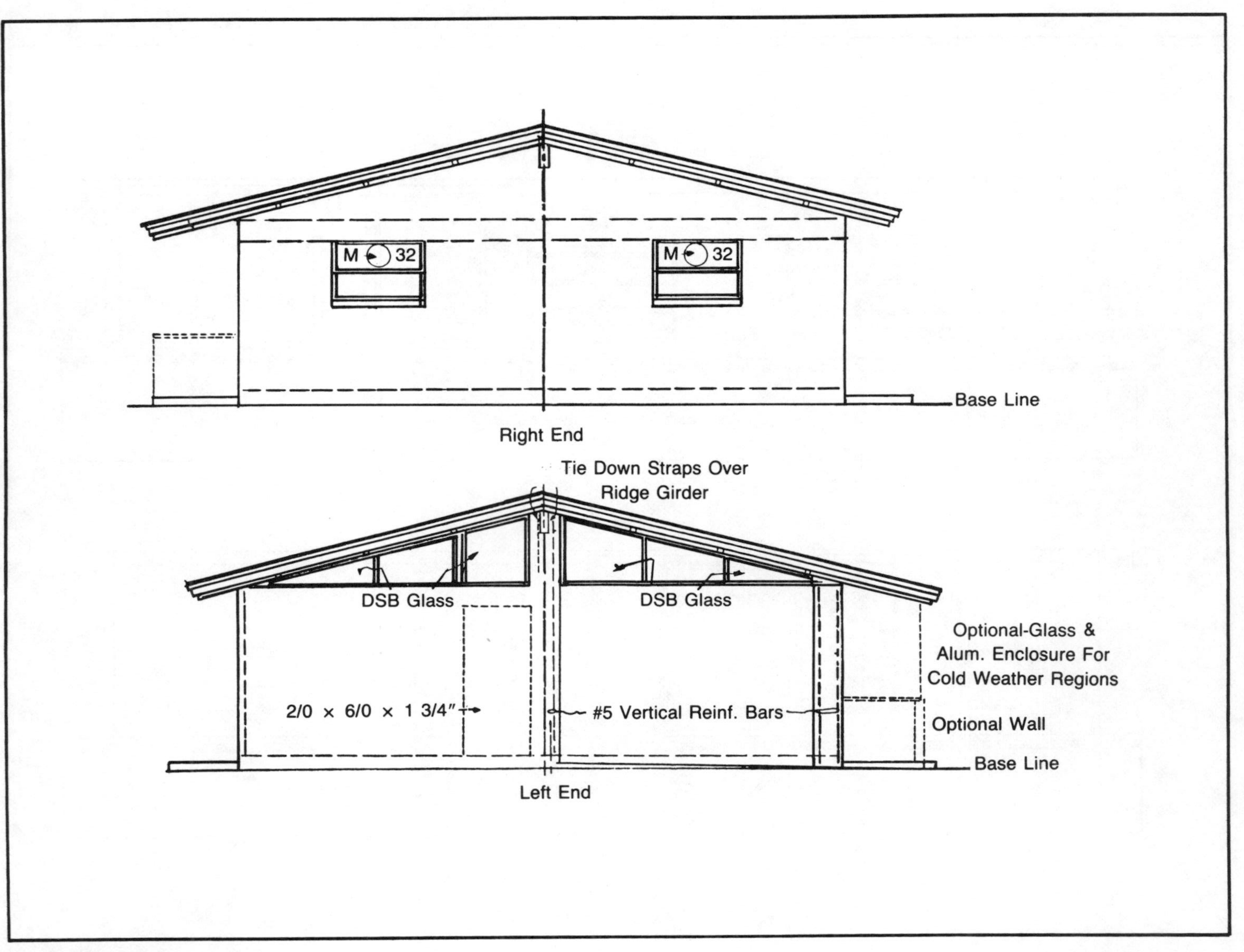

Fig. 18-9. (continued).

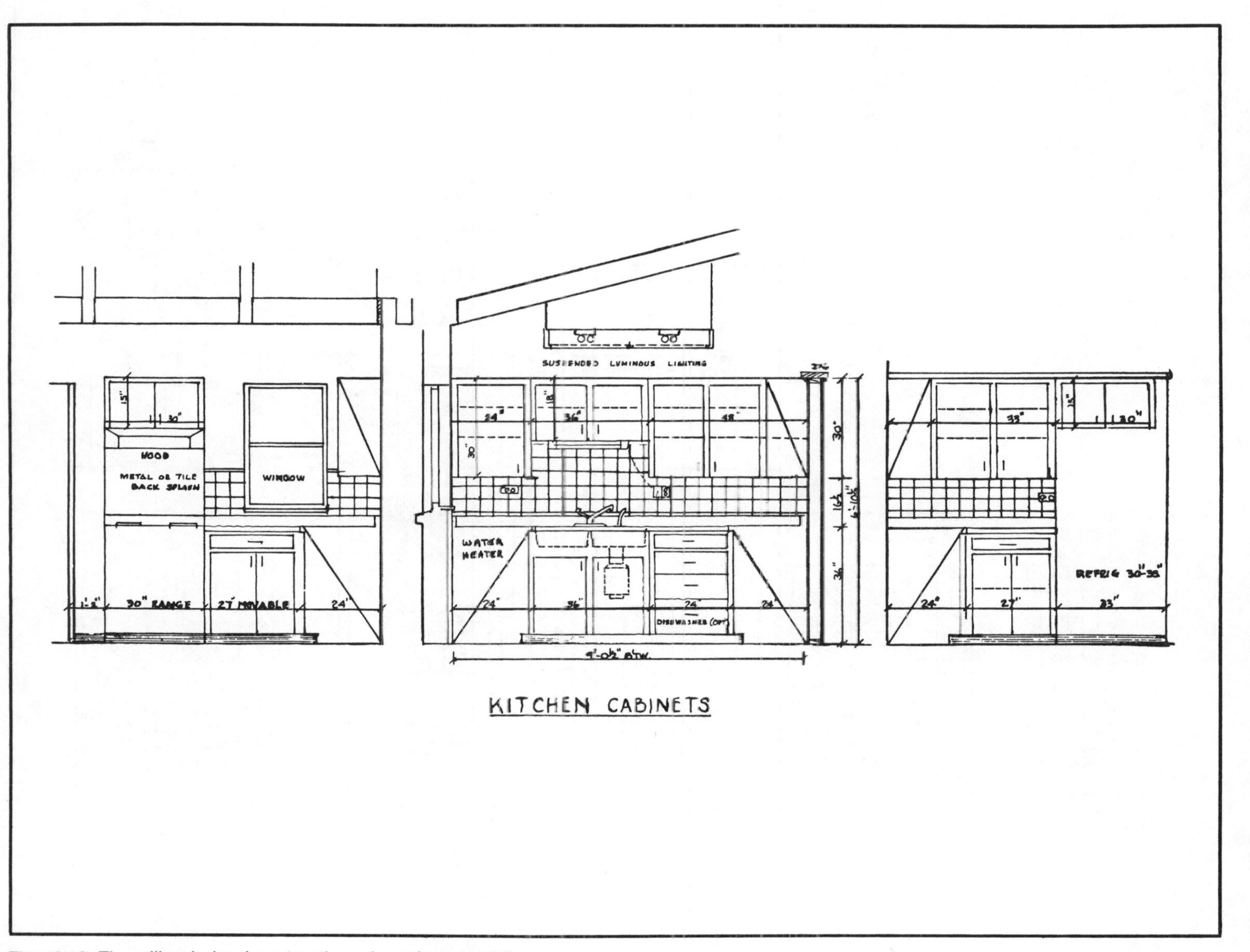

Fig. 18-10. The millwork drawings (continue through page 193).

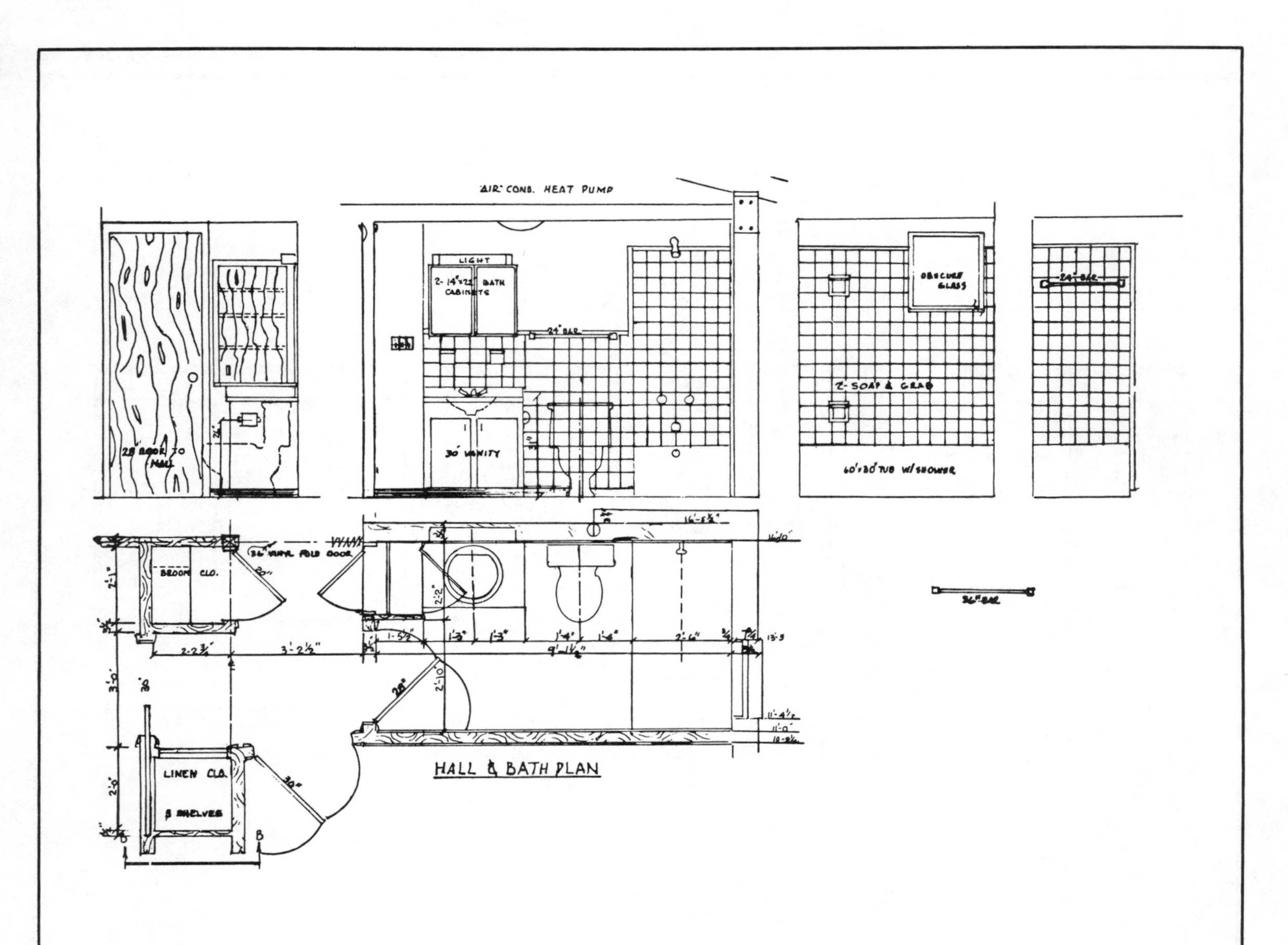

AIR CONB. HEAT PUMP
LIGHT
2-14"x22" BATH CABINETS
24" BAR
30" VANITY
28" DOOR TO HALL
OBSCURE GLASS
2-SOAP & GRAB
60"x30" TUB W/SHOWER
24" BAR
36" BAR
36" VINYL FOLD DOOR
BROOM CLO.
LINEN CLO.
5 SHELVES
HALL & BATH PLAN

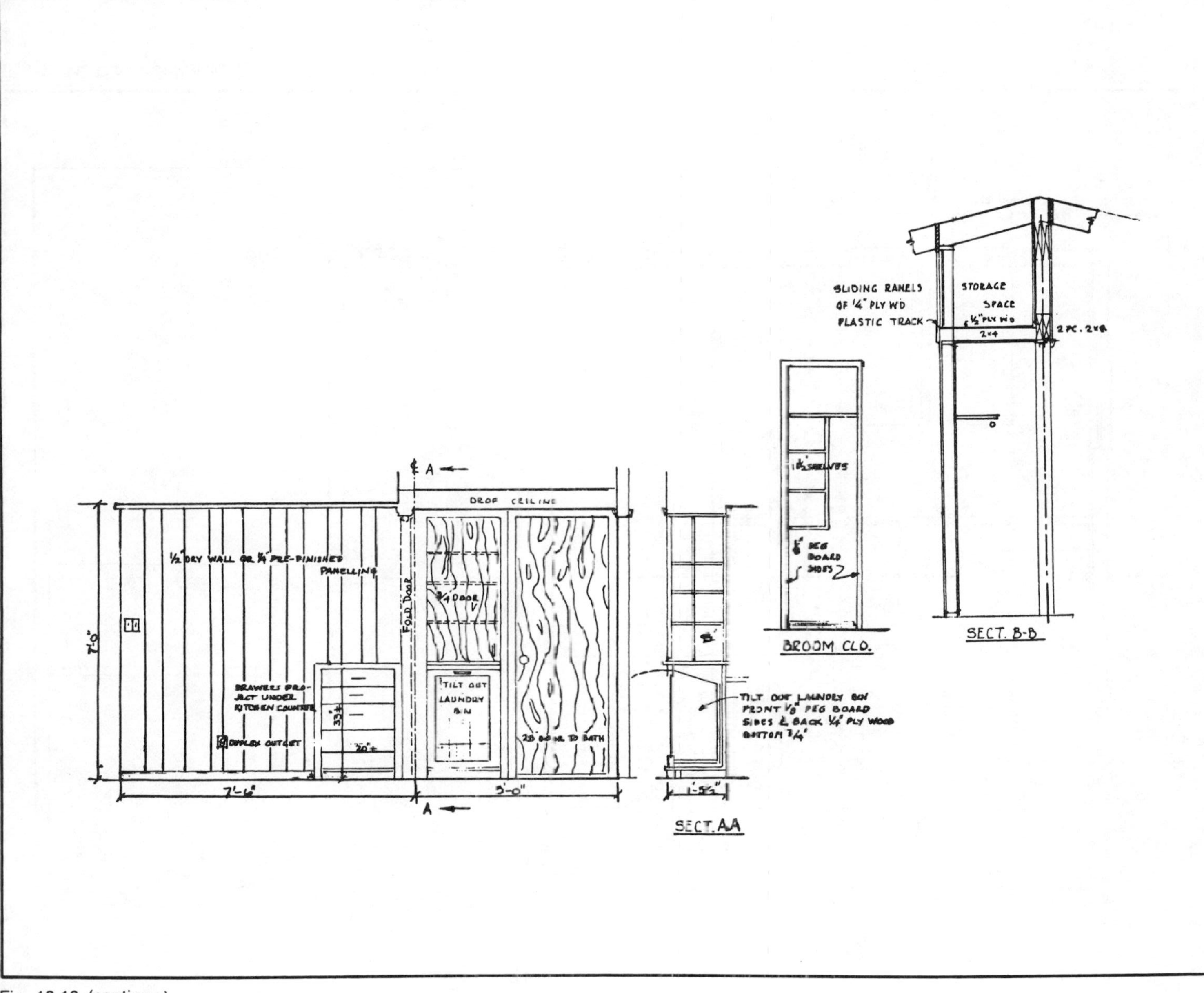

Fig. 18-10. (continue).

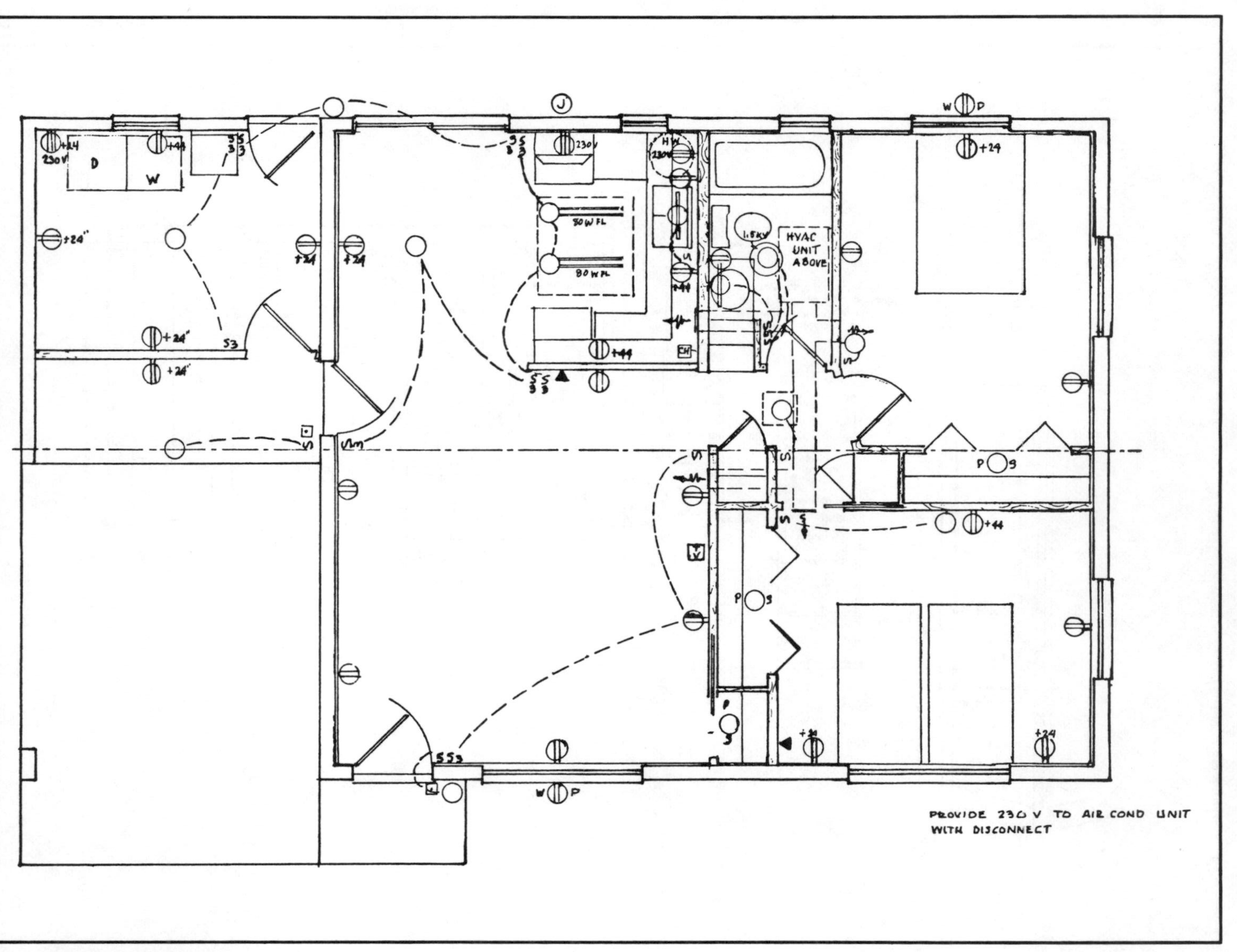

Fig. 18-11. The electrical plan.

STEP SIX: THE MILLWORK DRAWINGS

The millwork drawings are shown in Fig. 18-10. They were all drawn to the same scale.

In some cases it was necessary to include two views of the same room (as in the bathroom plan). Some furnishings required construction sections (e.g., the tilt-out laundry bin).

STEP SEVEN: THE ELECTRICAL PLAN

The electrical plan is shown in Fig. 18-11. After the millwork drawings were completed, it was a relatively simple matter to decide on the locations of electrical fixtures.

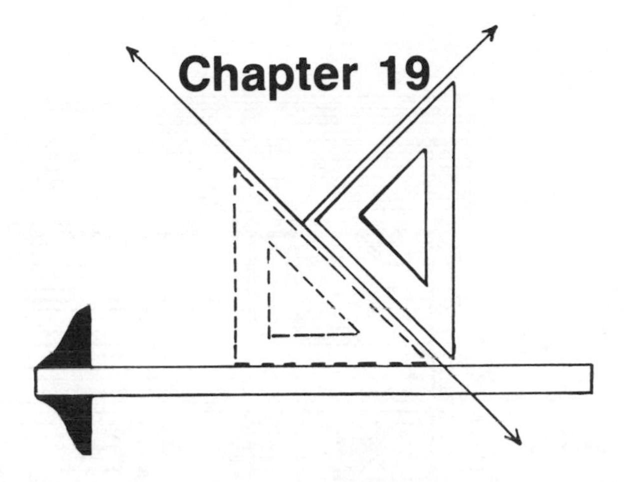

Drawing Plans for a Two-Story House

Planning a two-story home with a basement is a lot more complicated than planning a one story. There must be stairways, partitions to support the upper floor, and provisions made for the plumbing and heating contractor. But if the job is tackled in a logical step-by-step sequence, it will not seem nearly as complex.

Figure 19-1 shows a shaded elevation of a two-story frame house. It has three bedrooms, a living room, a dining area, a kitchen, a garage, a screened porch, and two and a half baths. In the following pages I will tell you how I drew the plans for such a house.

THE PLOT PLAN

For this house, the plot plan came first. The lot for the house was 70 × 100 feet and sloped evenly from the street line to the rear. The street had a 20 foot wide pavement with 6 inch concrete curbs. There was a 4 foot wide sidewalk 1 foot from the property line. The completed plot plan is shown in Fig. 19-2.

To draw this plan, I first drew a line a couple inches above the bottom of a sheet of paper to represent the front wall line of the house. To make sure of vehicle access to the backyard, I wanted 12 feet or more clearance at the garage side of the house. I knew if I allowed 10 feet for clearance on the other side of the house, there would be 50 feet for the overall width of house and garage.

I decided on the more common 16 × 7 foot garage door. So with 2 feet 8 inches each side of the 16 foot door and 4 inches for a stud wall, the garage has to be 21 feet 4 inches, which leaves 28 feet 8 inches for the width of the house.

With a few unit cost calculations, I was able to determine *how much* house could be afforded. I settled on a front-to-back house depth of 27 feet 8 inches. Then I worked out the garage and porch dimensions.

I then finished the scaled outline of the house and drew in the lot boundaries, the front setbacks, the sidewalks, and the clearances. Next, I added all the necessary survey data.

Fig. 19-1. A shaded elevation of a two-story frame house.

THE FLOOR PLAN

Then I started on the preliminary floor plan (Fig. 19-3). I left 8 inches at the left side of the paper for the garage and drew the 28 foot 8 inch width of the house. I wanted side walls to be 2 × 4 inch studs, so I pointed off 4 inches for the inside wall lines.

I planned to use brick veneer on the front of the house. With brick veneer, the front wall would be about 8 inches thick (4 inches brick, 1/2 inch sheathing, 3 1/2 inch studs). So I drew the front wall scaled 8 inches above the front wall line I drew first.

Next, I drew the sides of the house (27 feet 8 inches long), the rear wall, the porch outline, and the garage walls.

Next, I considered the stairway, for its size and location affects the size of many of the inside walls. To determine the length of the stairs I had to know the rise to the second floor (the first floor ceiling would be the usual 8 feet). I also had to know the thickness of the second-floor itself. After I knew these things, I could calculate the number and sizes of treads and risers, as explained in Chapter 14. I computed that the stairs would consist of 14 risers and 13 9 1/2 inch treads—a total stairway length of 10 feet 3 1/2 inches. But then I had to consider how far from the inside front wall the stairs should start. The entrance door had to be 3 feet wide, and I made an additional allowance of 12 inches *beyond* this 3 foot door swing—a total of 4 feet. I added the 8 inch thickness of the front wall. Thus the stairs had to start 4 feet 8 inches from the front wall line. And, of course, a 10 foot 3 1/2 inch stairs (with a 3 foot landing) that had to be 4 feet 8 inches from the front wall meant that the proposed dining room would be 9 feet 1 inch wide.

Next I had to figure out how much of the 28 foot inside width of the house should be given to the stairs. The code says that a main stairs must be at least 32 inches clearance between hand railings. Assuming 2 inches each side for railings, the stair had to be at least 36 inches wide. Because this stairs would be the first feature seen when one enters the house, a 36 inch stairs would appear narrow and cramped. So I made it 40 inches. I added 4 inches for each of the side walls; thus the stairs

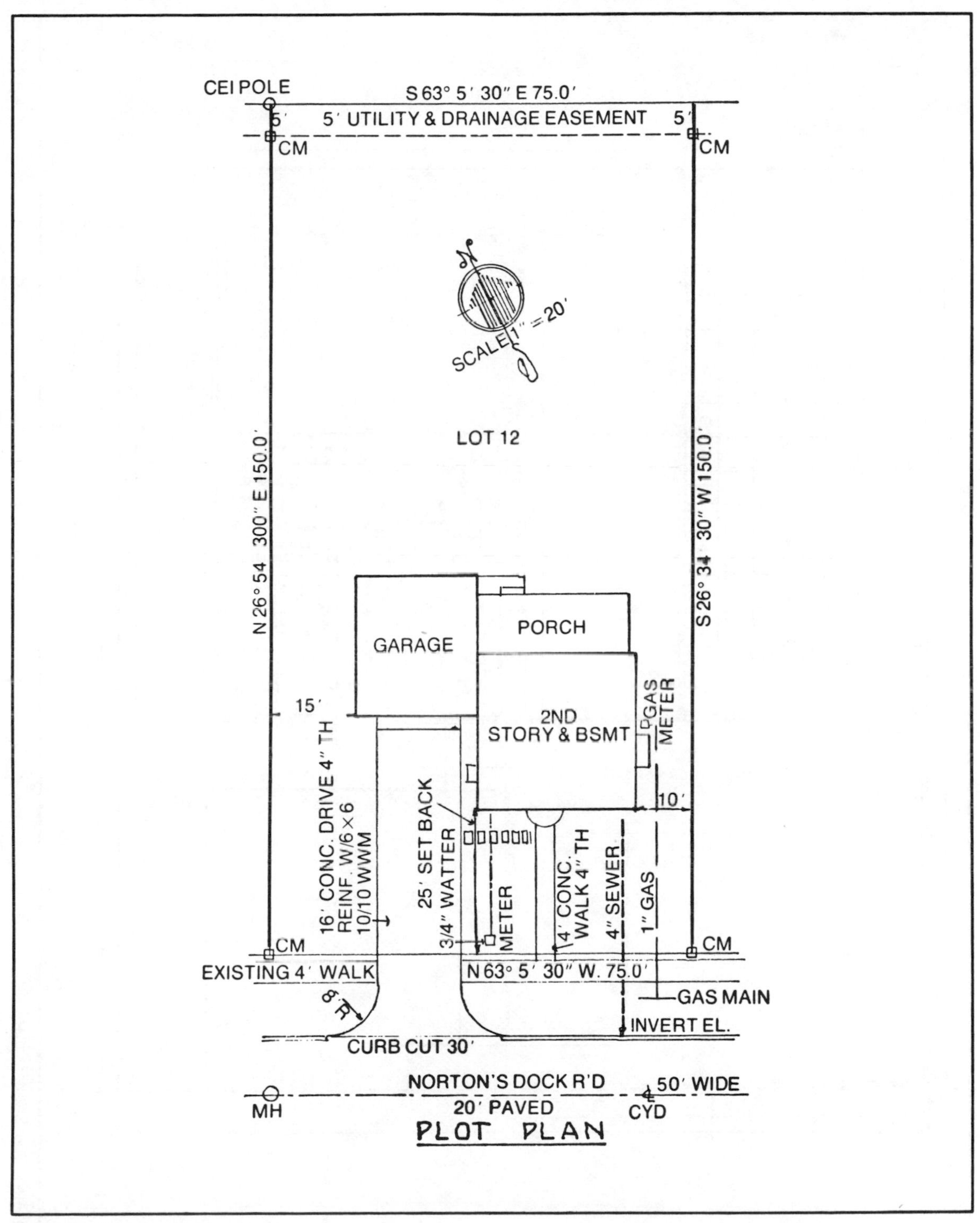

Fig. 19-2. The plot plan.

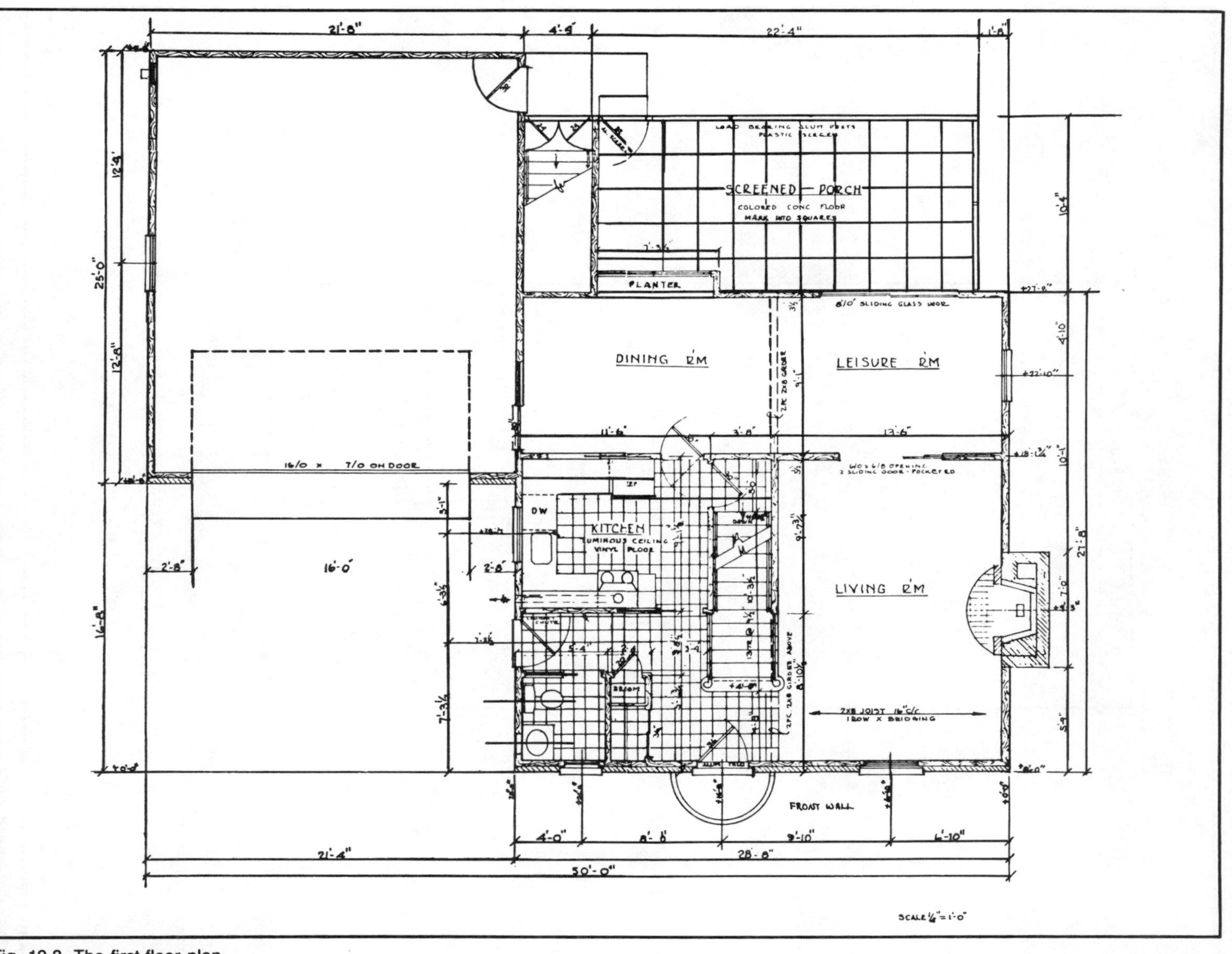

Fig. 19-3. The first-floor plan.

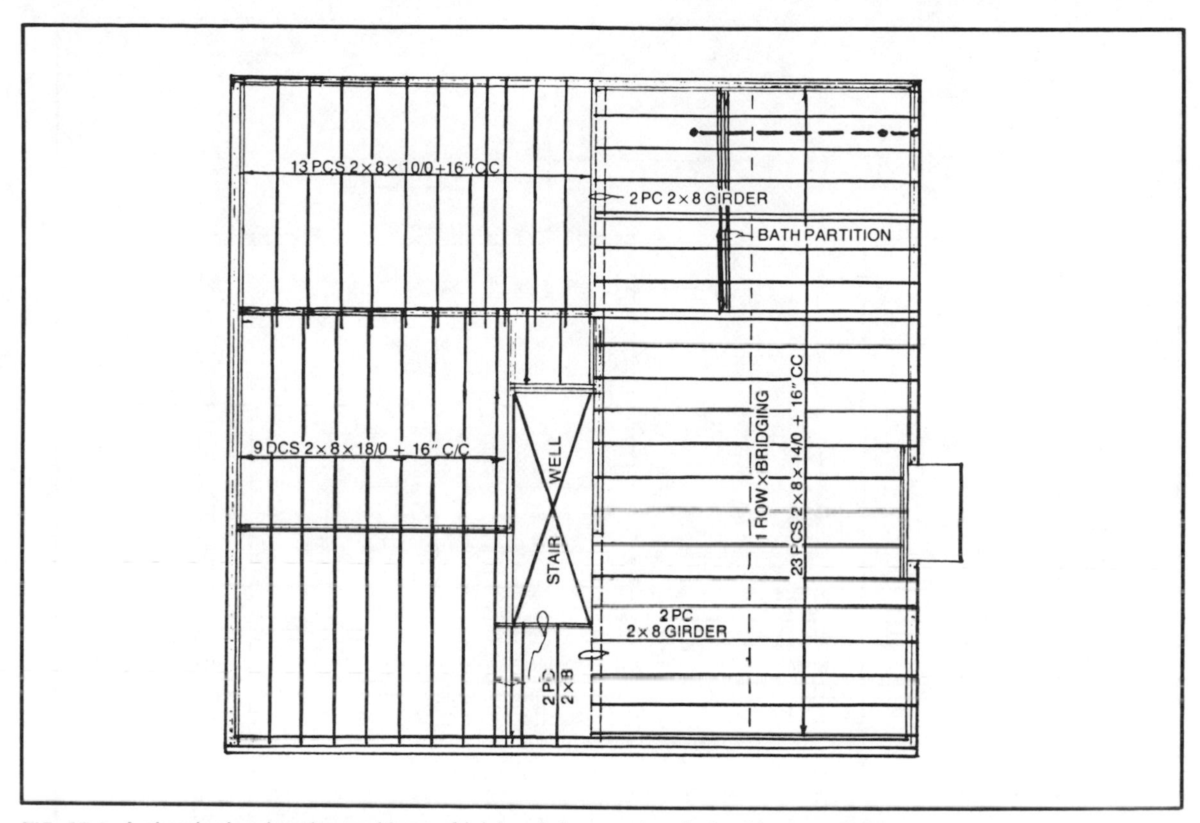

Fig. 19-4. A sketch showing the positions of joists, girders, and walls for the second floor.

would take up 4 feet, leaving 24 feet to be divided between the living room and the kitchen. So I was able to draw in the stairs.

I wanted the kitchen to be U-shaped with a 24 inch wide counter at the left end. Doorways each side were to be 30 inches wide, or 36 inches when trim was included. There had to be at least 30 inches of wall space for a refrigerator at one side and a range at the other. These fixed spaces totaled 7 feet 6 inches (24 inches + 36 inches + 30 inches = 90 inches or 7 feet 6 inches). Thus if I made the kitchen 11 feet long, there would be a 42 inch space for working counters each side. This exceeds the VA-HUD minimum standards for kitchens. This then would make the living room 13 feet wide.

I was then able to draw in the living room wall and a kitchen wall.

Then I had to consider the size of the powder room and wraps closet near the front entrance. Making the stairway a major decorative feature with an ornamental railing required that the first stairs tread have rounded ends about 4 inches wider than the 4 foot width of the stairs. That meant that there had to be 36 inches clearance for walking past this extended tread. This left 7 feet 8 inches for the closet and powder room. The closet had to be 2 feet wide inside with 4 inch walls. That left 5 feet for the width of the powder room. I drew in these walls.

The powder room had to have a window. This required that the toilet fixtures be arranged along the left side wall. I drew in the fixtures and the remaining powder room wall.

I wanted to plan for a secondary exit door near the powder room. I drew the door 32 inches wide with 3 inches each side for trim. I was then able to draw in the front wall of the kitchen.

I then finished off the kitchen by drawing the outlines of the shelves, counters, range, refrigerator, sink, and dishwasher.

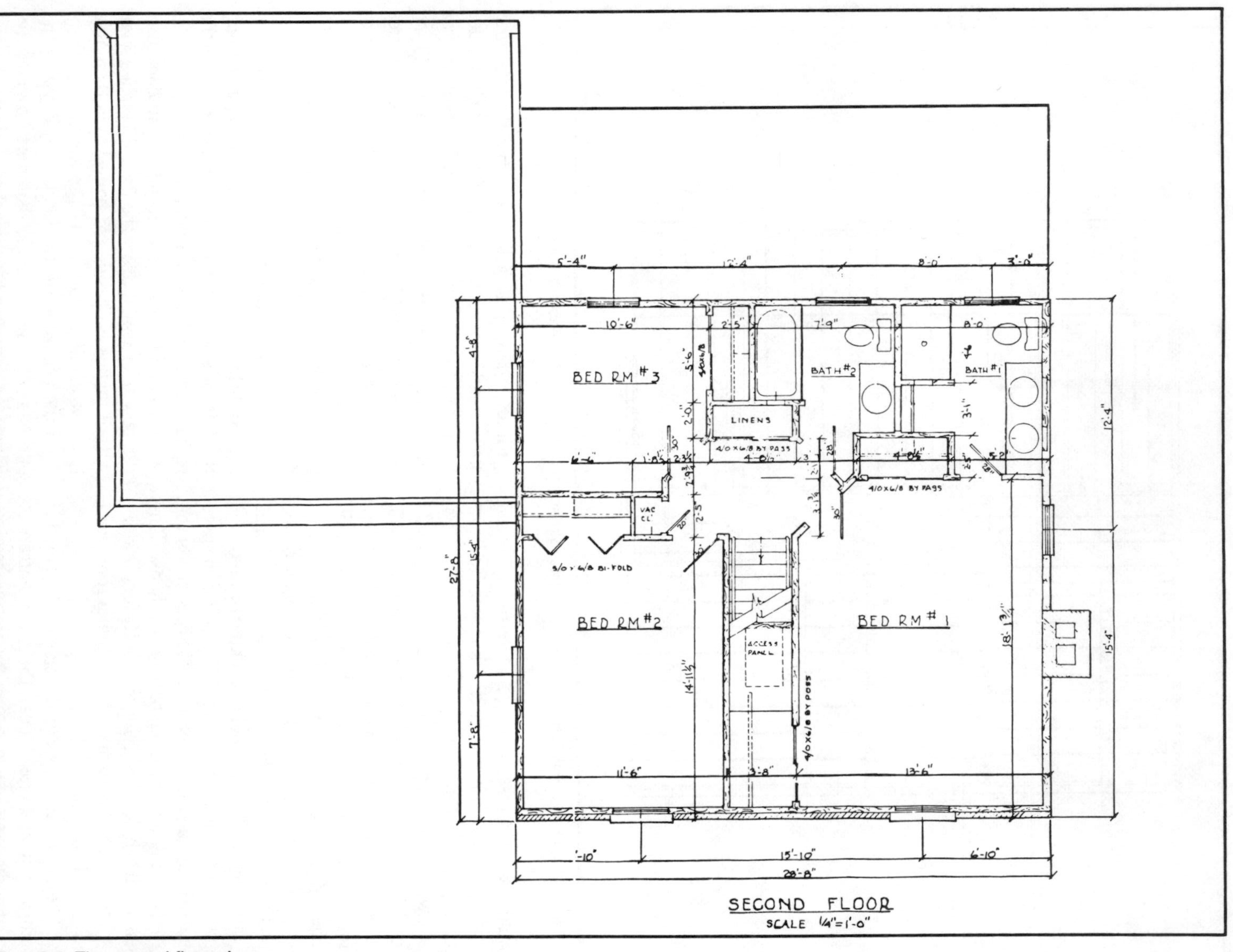

Fig. 19-5. The second-floor plan.

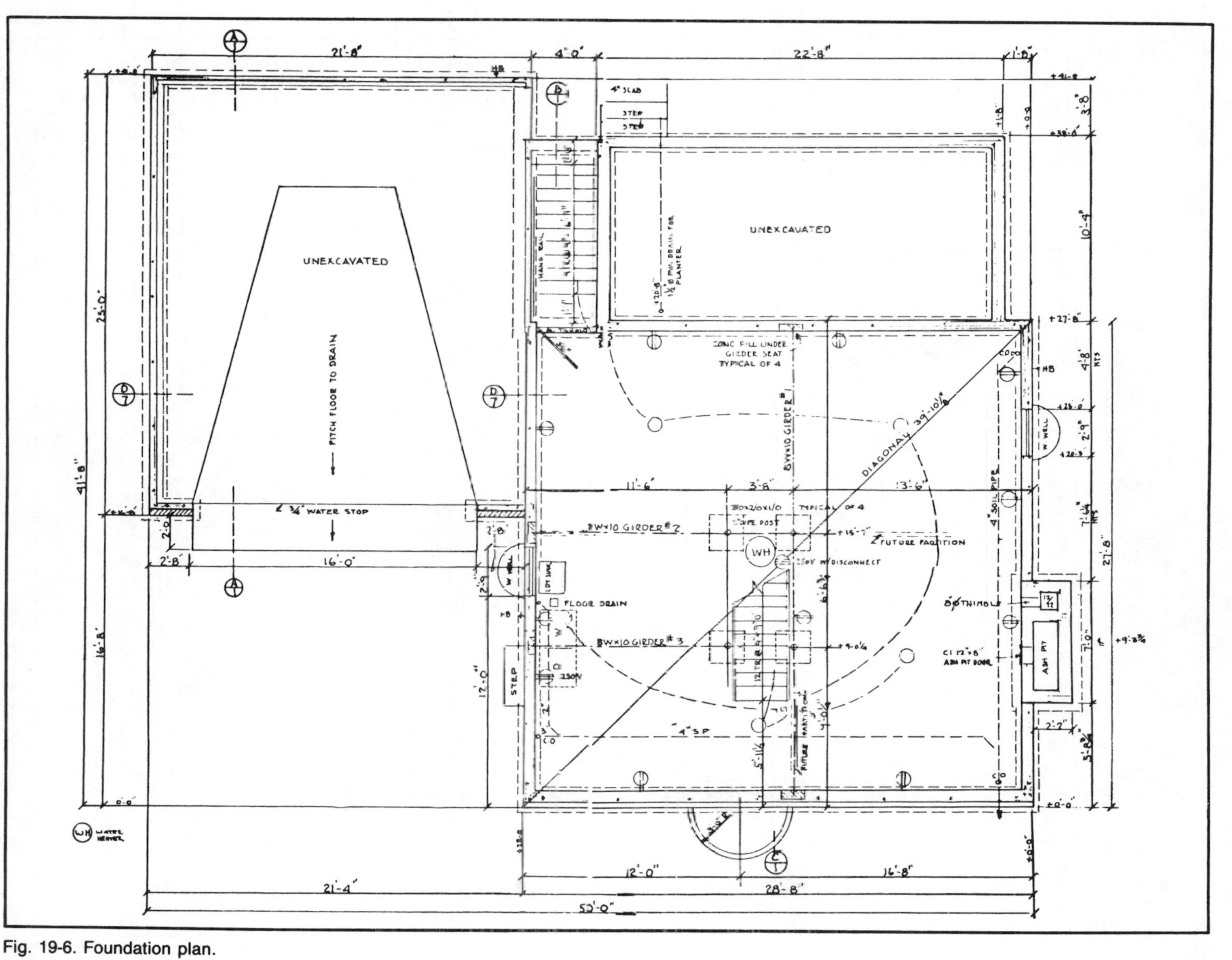

Fig. 19-6. Foundation plan.

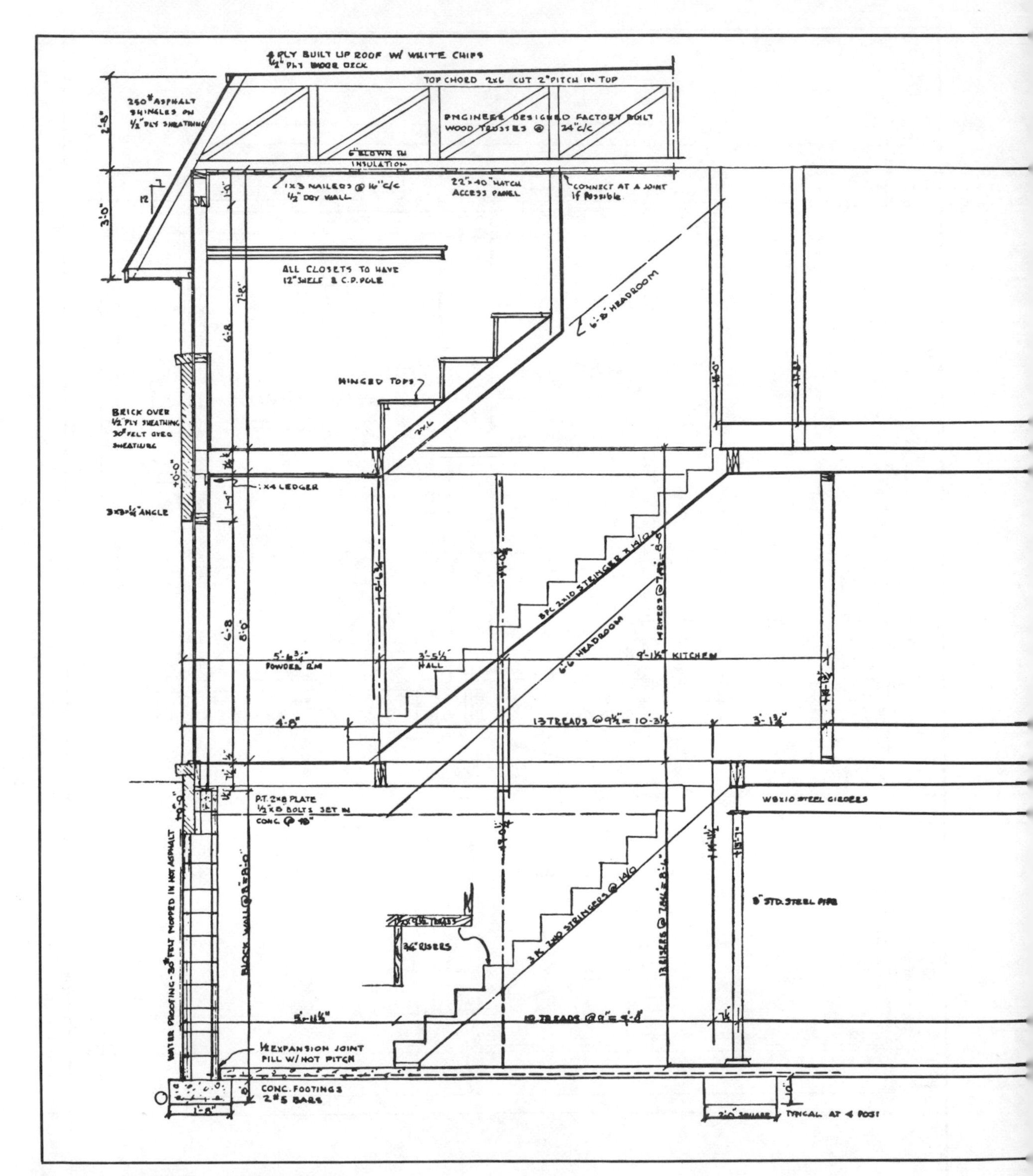

Fig. 19-7. The main construction section.

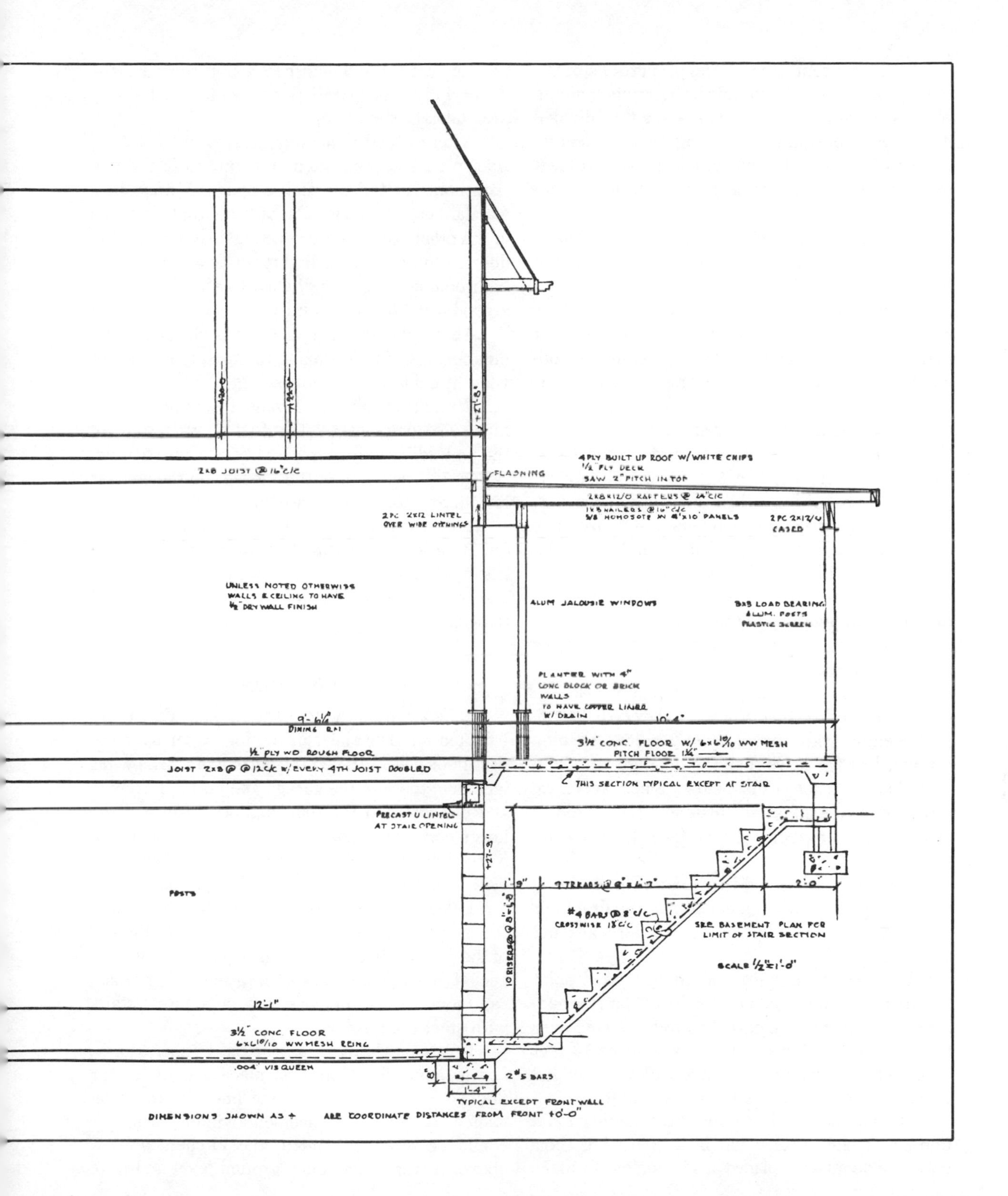
2x8 JOIST @ 16" c/c
FLASHING
4 PLY BUILT UP ROOF W/ WHITE CHIPS
1/2" PLY DECK
SAW 2" PITCH IN TOP
2x8x12/0 RAFTERS @ 24" c/c
1x3 NAILERS @ 16" c/c
5/8 HOMOSOTE IN 4'x10' PANELS
2 PC 2x12 LINTEL OVER WIDE OPENINGS
UNLESS NOTED OTHERWISE WALLS & CEILING TO HAVE 1/2" DRY WALL FINISH
ALUM JALOUSIE WINDOWS
ALUM. POSTS
PLASTIC SCREEN
PLANTER WITH 4" CONC BLOCK OR BRICK WALLS TO HAVE COPPER LINER W/ DRAIN
9'-6 1/4"
DINING RM
10'-4"
3 1/2" CONC. FLOOR W/ 6x6 10/10 WW MESH
PITCH FLOOR 1/4"
1/2" PLY WD ROUGH FLOOR
JOIST 2x8 @ 12" c/c w/ EVERY 4TH JOIST DOUBLED
THIS SECTION TYPICAL EXCEPT AT STAIR
PRECAST U LINTEL AT STAIR OPENING
POSTS
1'-9"
2'-0"
#4 BARS @ 8" c/c
CROSSWISE 18" c/c
SEE BASEMENT PLAN FOR LIMIT OF STAIR SECTION
SCALE 1/2" = 1'-0"
12'-1"
3 1/2" CONC FLOOR
6x6 10/10 WW MESH REINF
.004" VISQUEEN
8"
1'-4"
2 #5 BARS
TYPICAL EXCEPT FRONT WALL
DIMENSIONS SHOWN AS + ARE COORDINATE DISTANCES FROM FRONT +0'-0"

I then concentrated on the proportions of the living room area. I extended the kitchen-dining room wall over to the living area and thus divided it into two rooms, a formal living room (13 feet by 17 feet 4 inches) and a smaller leisure room (13 feet by 9 feet). I centered the fireplace on the longer outside wall.

After that I filled in all other pertinent details on the first floor drawing—the porch details, the garage window, etc.

I didn't plan the location of most of the doors and windows on the first floor until the second floor plan had been worked out. Of course, in positioning various features on the drawing, I used the coordinate system.

In order to draw a good second floor plan, I had to know something about joists, girders, and loads. Joists and girders help to ensure structural stability; they bear most of the weight (load) of a house. Girders, the main structural members, rest horizontally on central support posts that extend from the foundation to the roof. The joists must rest on either girders, exterior walls, interior load-bearing walls, or some combination of these (Fig. 19-4).

Most of the time the position of second floor interior walls doesn't matter structurally: these walls usually don't carry attic and roof loads—a central girder does that. But when they do help support weight from above, their positions become critical. In such cases they should align vertically with load-bearing walls below. Since a central girder was to support the attic and roof, I didn't have to bother about second-floor load-bearing walls.

I started the second-floor plan (Fig. 19-5) by tacking a piece of tracing paper over the first floor plan. I traced the outline of the exterior walls and the stairs.

The top of the stairs was to be a fixed point opening into a hallway. Obviously, the largest bedroom (with its private bath) had to be located over the living room. This bedroom was to be 13 feet wide. The remaining upstairs space was divided among the other two bedrooms and bath.

I found that the floor areas of the number 1 and number 3 bedrooms could be increased if their entrance doors were placed at 45° angles. To make certain that there would be enough room for the doors, I drew a separate floor plan for the hallway area (at a larger scale).

After the bedrooms were drawn on the second floor plan, I concentrated on trying to fit the two bathrooms in the remaining space. I drew in a standard 60 inch tub in each bathroom and sketched in the other bathroom furnishings. By placing the door to the number 2 bathroom at an angle, the bathroom could be lengthened to afford room for a good sized lavatory-vanity.

Above the stairs, I drew in a closet for the number 1 bedroom. I made the closet door 4 feet wide (it could be no wider).

With some shifting of walls, it was possible to make a second closet with 4 foot doors in some of the unused bathroom space. There was space between the back wall of this closet and the shower for storage for both bathrooms.

The next job was to fill in all remaining details (other closets, dimension lines, etc.). Then I made tracings of the finished preliminary floor plans (first and second floors). The tracings were my finished floor plans.

THE FOUNDATION PLAN

Next, I started on the foundation plan (Fig. 19-6). I tacked another sheet of tracing paper over the first-floor plan. I traced the outline of all the walls including those of the garage and porch. I drew in and dimensioned the four support posts and the lower part of the basement stairs. With a rise of 8 feet 4 inches, this stairs was to have 13 risers of 7.69 inches each with 9 inch plank treads.

I also planned a basement exit to the outside. It seemed that the best place for it was to the left of the porch, adjacent to the garage. I made it wide enough so large objects, such as a sheet of plywood, can be taken in. This was to be a concrete stairs with steel reinforcing.

I planned the positions of the washer, dryer, and laundry sink. The best place was under the kitchen where a 2 inch waste line was to be installed. I drew in a basement window where it would be directly beneath the kitchen window above. It was to be below ground level and have

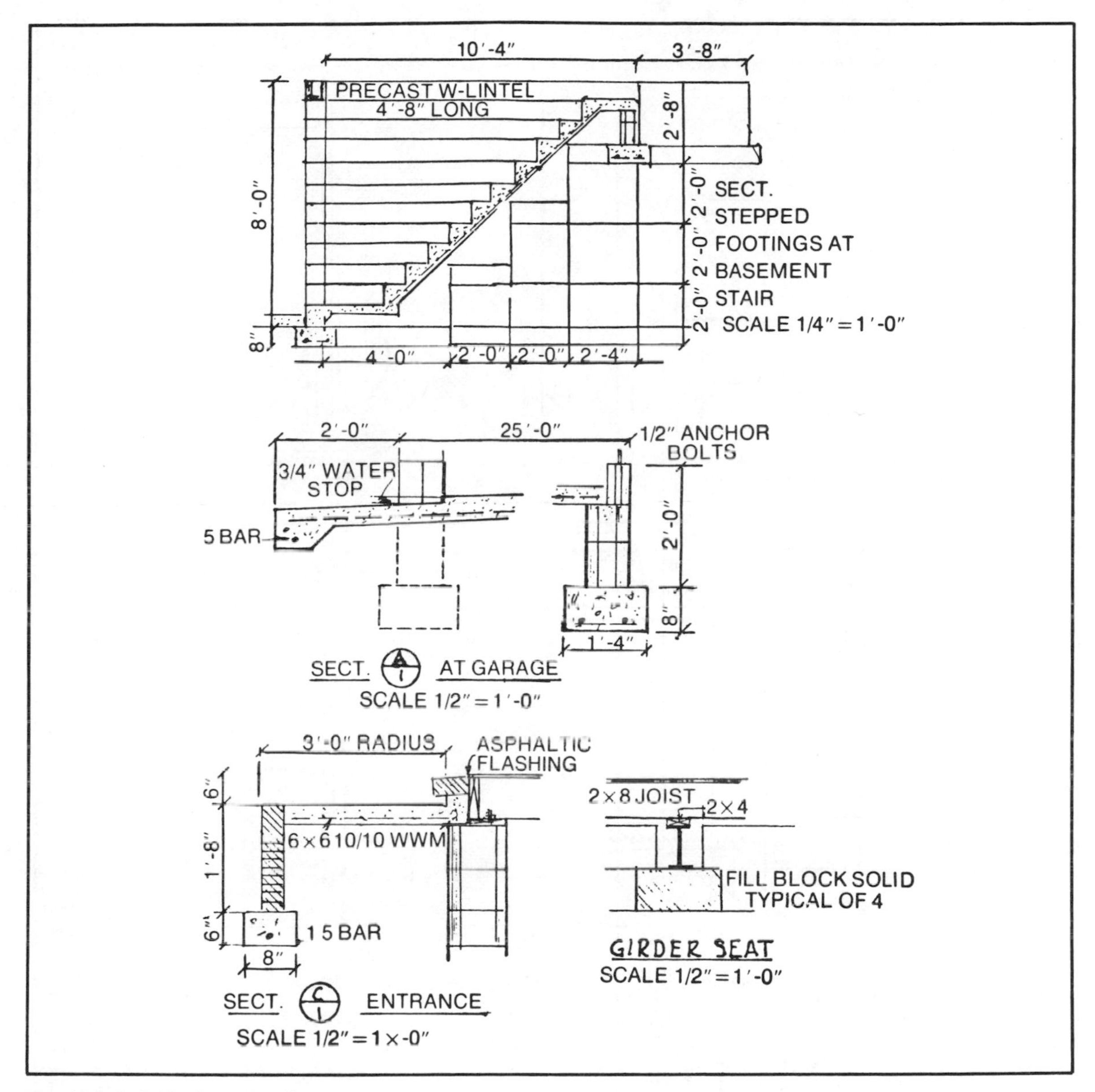

Fig. 19-8. Individual constructions.

a light well and grating.

Then I drew in all necessary electrical symbols, dimension lines, construction section notation, and other details.

THE CONSTRUCTION SECTIONS

Next came the construction sections. I first drew a full section of the whole house. It included details of the stairs, walls, roof, foundation, and floors (Fig. 19-7).

Then I drew individual sections that were keyed to the section notation used in the foundation plan (Fig. 19-8). I also drew a window section, garage section, and basement window-well section (Fig. 19-9).

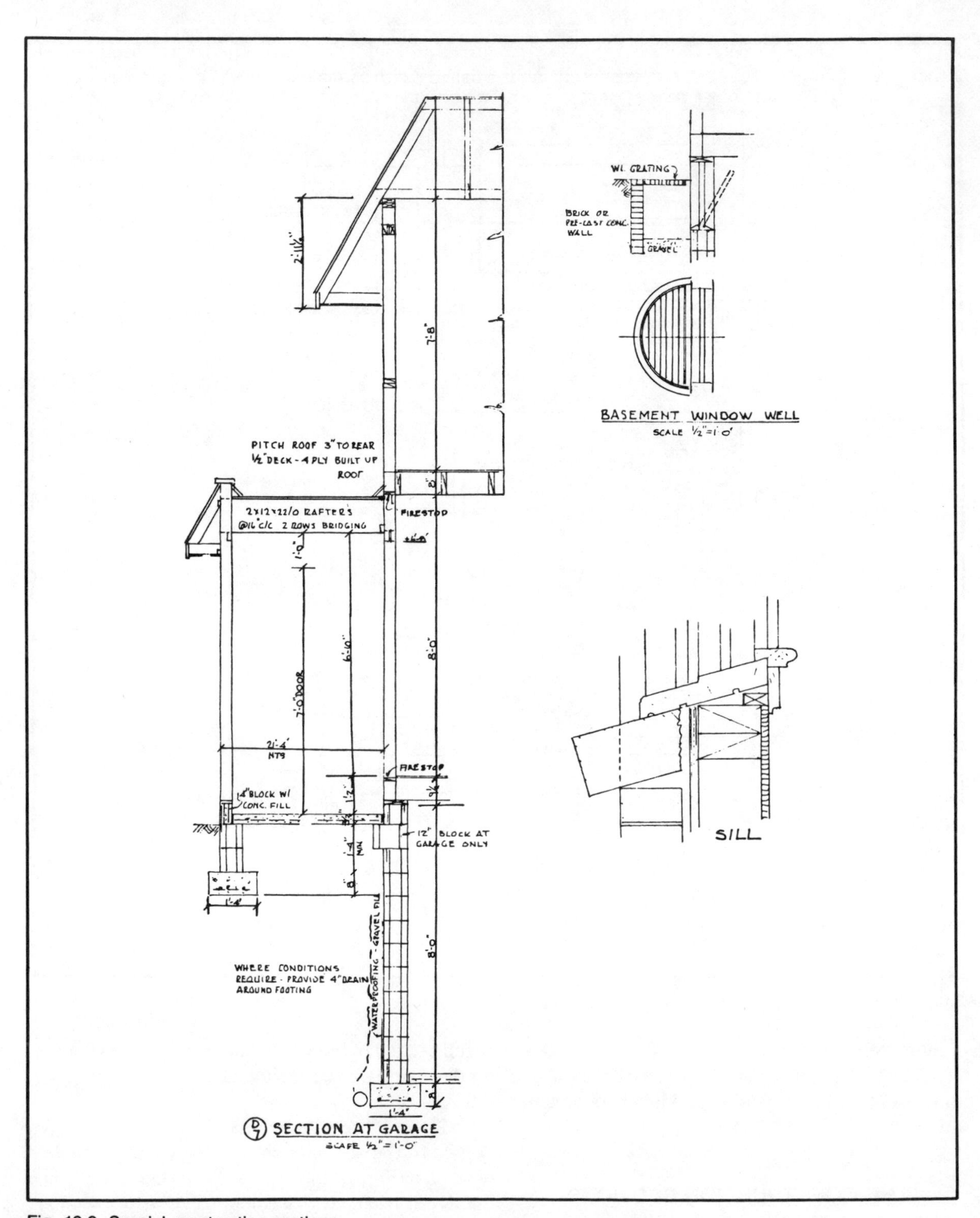

Fig. 19-9. Special construction sections.

THE ELEVATIONS

I then drew four elevations—front, right side, and left side (Fig. 19-10 and 19-11). The plans discussed thus far are for a frame house with comparatively plain window and door millwork. However, to illustrate an alternative architectural style, the elevations of Figs. 19-10 and 19-11 have been embellished with stucco-on-block characteristics, including windows with fancy millwork.

To determine the placement and size of windows on the elevations, I first drew two or three different windows on a small sheet of paper and

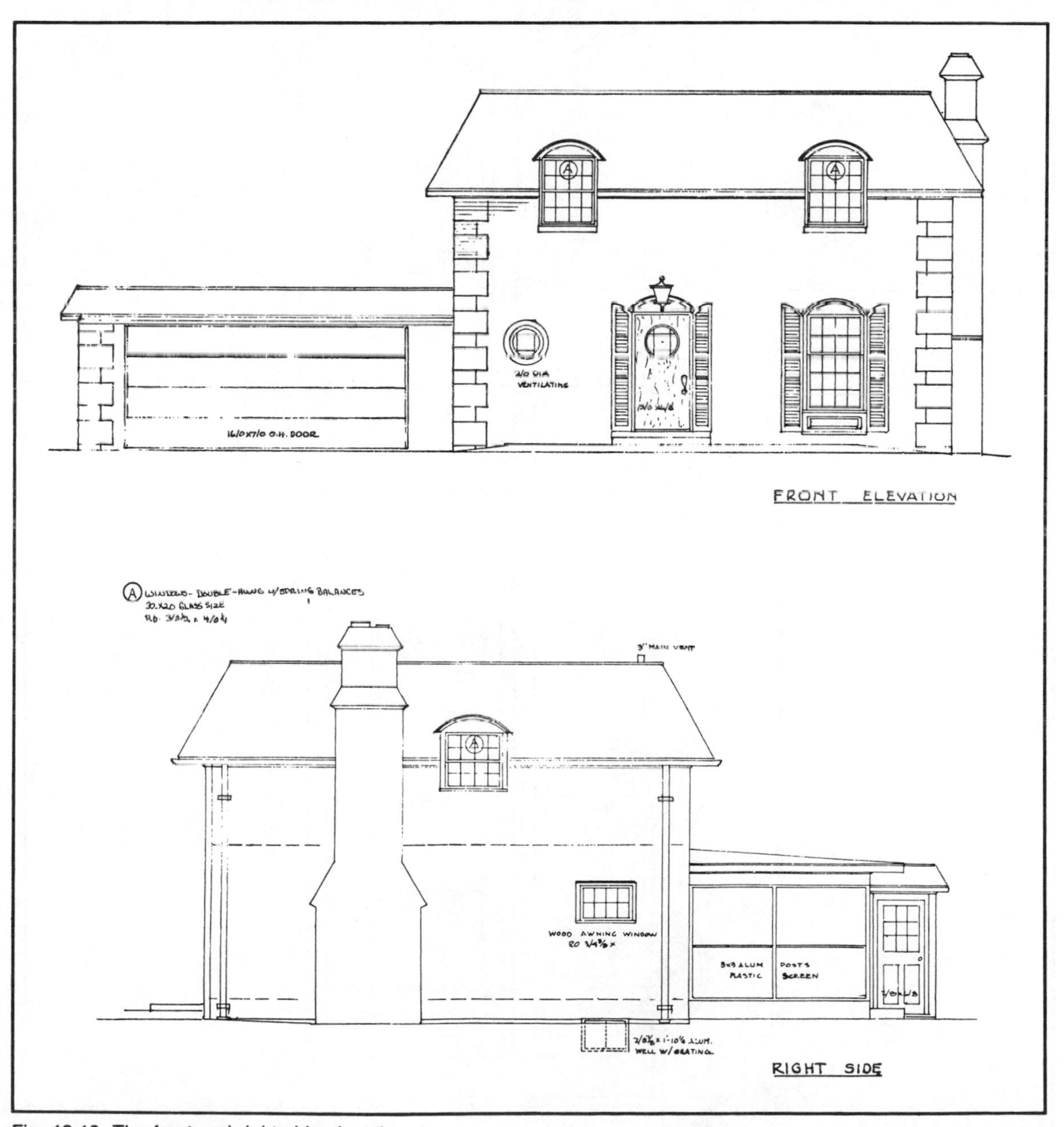

Fig. 19-10. The front and right side elevations.

SCREEN ENCLOSURE IS NOT SHOWN ABOVE
REAR ELEVATION
8/0 x6/8 DS GLASS DOORS
2-2/0x6/8

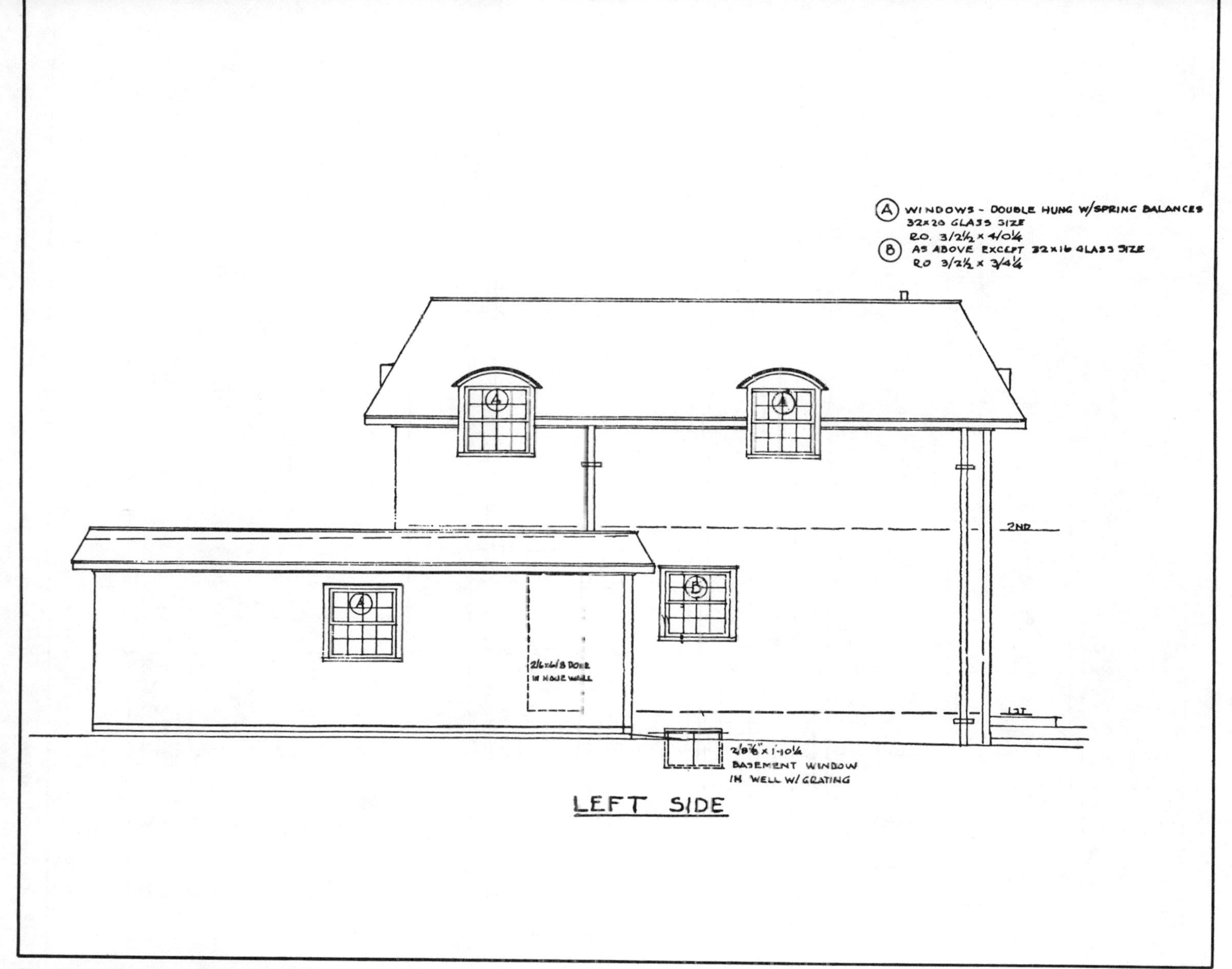

Fig. 19-11. The rear and left side elevations.

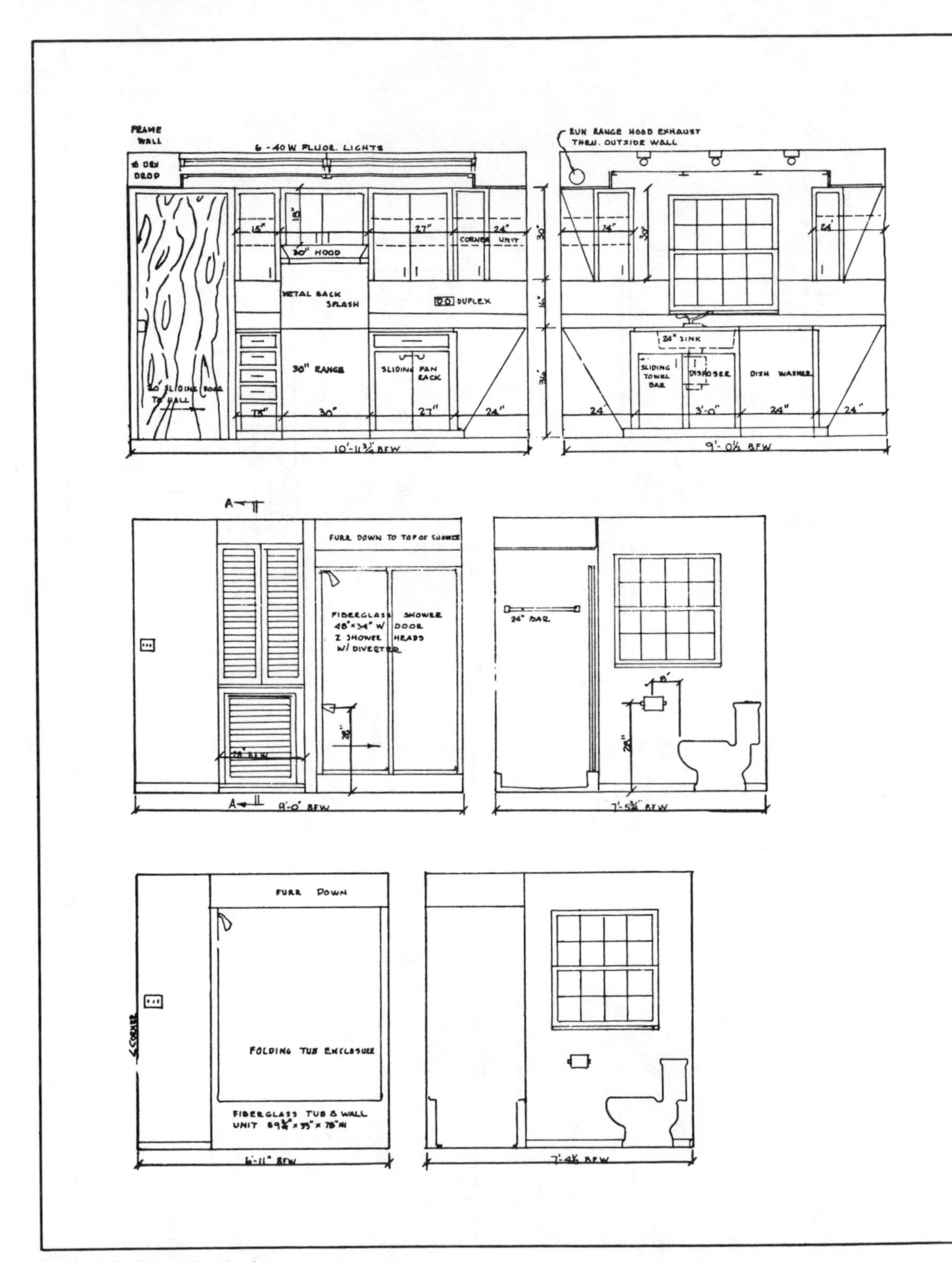

Fig. 19-12. The millwork plans.

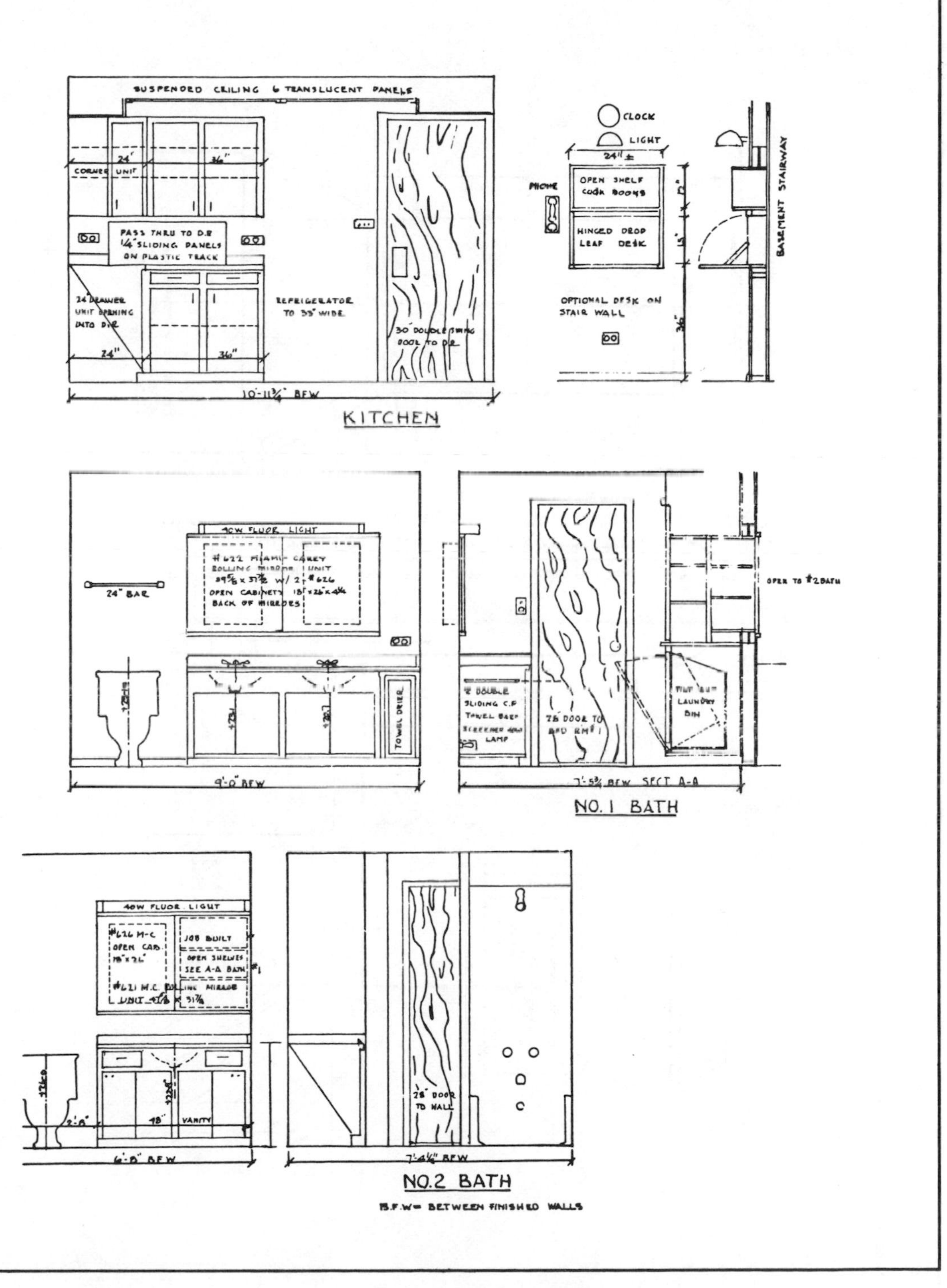
SUSPENDED CEILING 6 TRANSLUCENT PANELS
CORNER UNIT
PASS THRU TO D.R
1/4" SLIDING PANELS
ON PLASTIC TRACK
24" DRAWER UNIT OPENING INTO D.R
REFRIGERATOR TO 33" WIDE
30" DOUBLE SWING DOOR TO D.R
10'-11 3/4" BFW
KITCHEN
CLOCK
LIGHT
PHONE
OPEN SHELF COOK BOOKS
HINGED DROP LEAF DESK
OPTIONAL DESK ON STAIR WALL
BASEMENT STAIRWAY
40W FLUOR. LIGHT
24" BAR
TOWEL DRIER
9'-0" BFW
OPEN TO #2 BATH
LAUNDRY BIN
7'-5 3/4" BFW SECT A-A
NO. 1 BATH
40W FLUOR. LIGHT
JOB BUILT
OPEN SHELVES
SEE A-A BATH #1
VANITY
6'-8" BFW
28" DOOR TO HALL
7'-4 1/2" BFW
NO.2 BATH
B.F.W= BETWEEN FINISHED WALLS

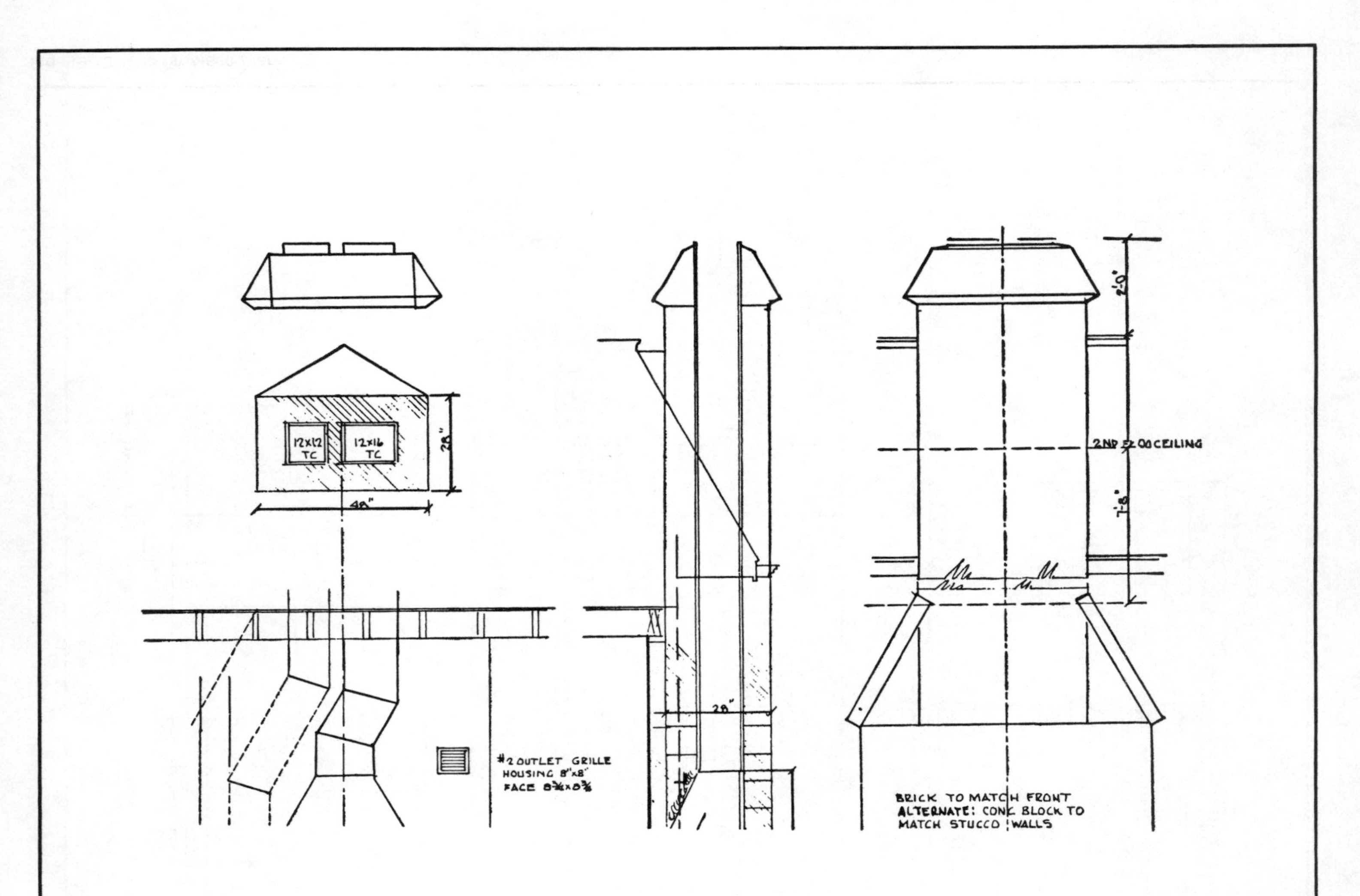

12x12 TC
12x16 TC
28"
48"
28"
2'-0"
2ND FL OO CEILING
7'-8"
#2 OUTLET GRILLE
HOUSING 8"x8"
FACE 8¾x8¾
BRICK TO MATCH FRONT
ALTERNATE: CONC BLOCK TO
MATCH STUCCO WALLS

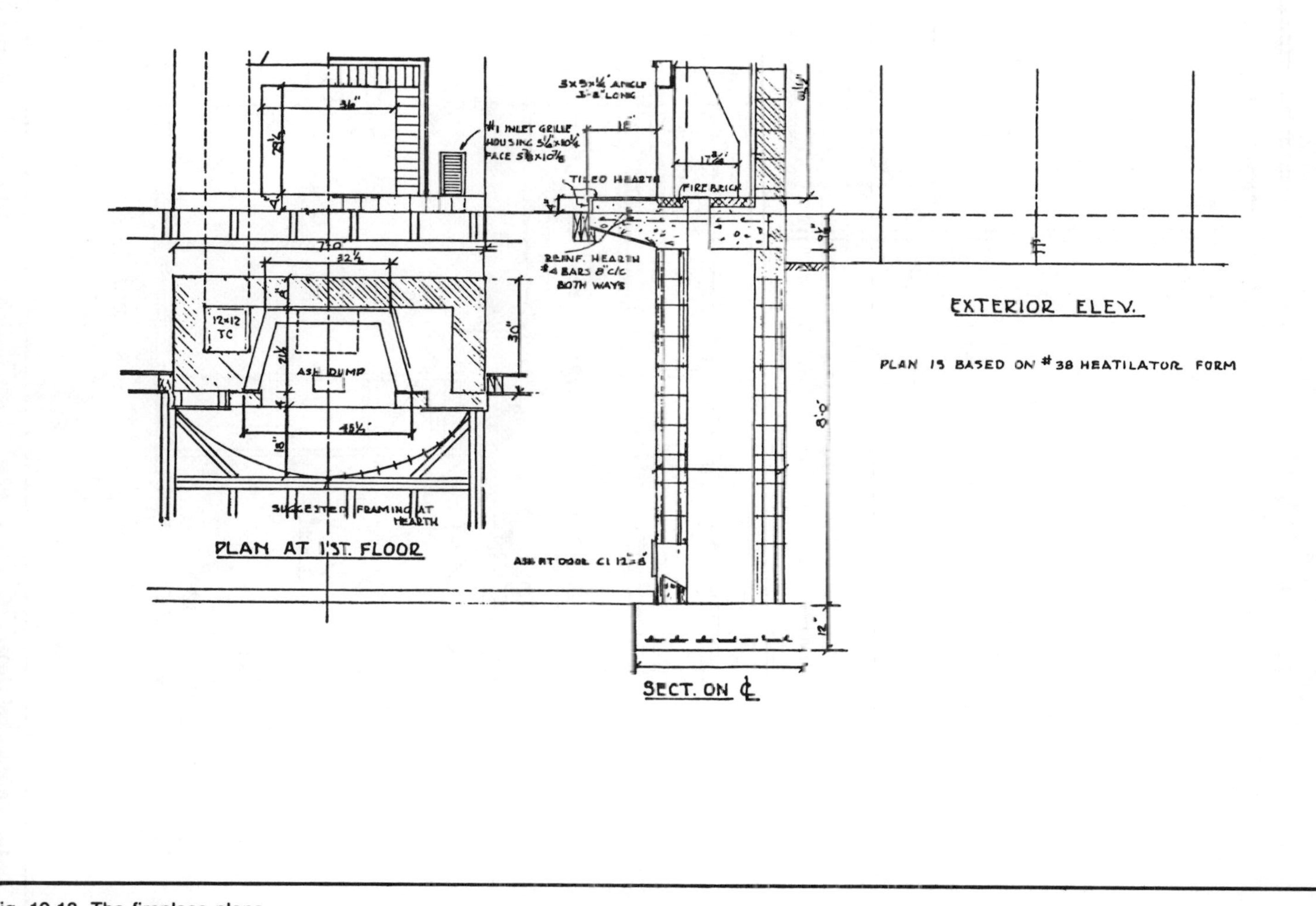

Fig. 19-13. The fireplace plans.

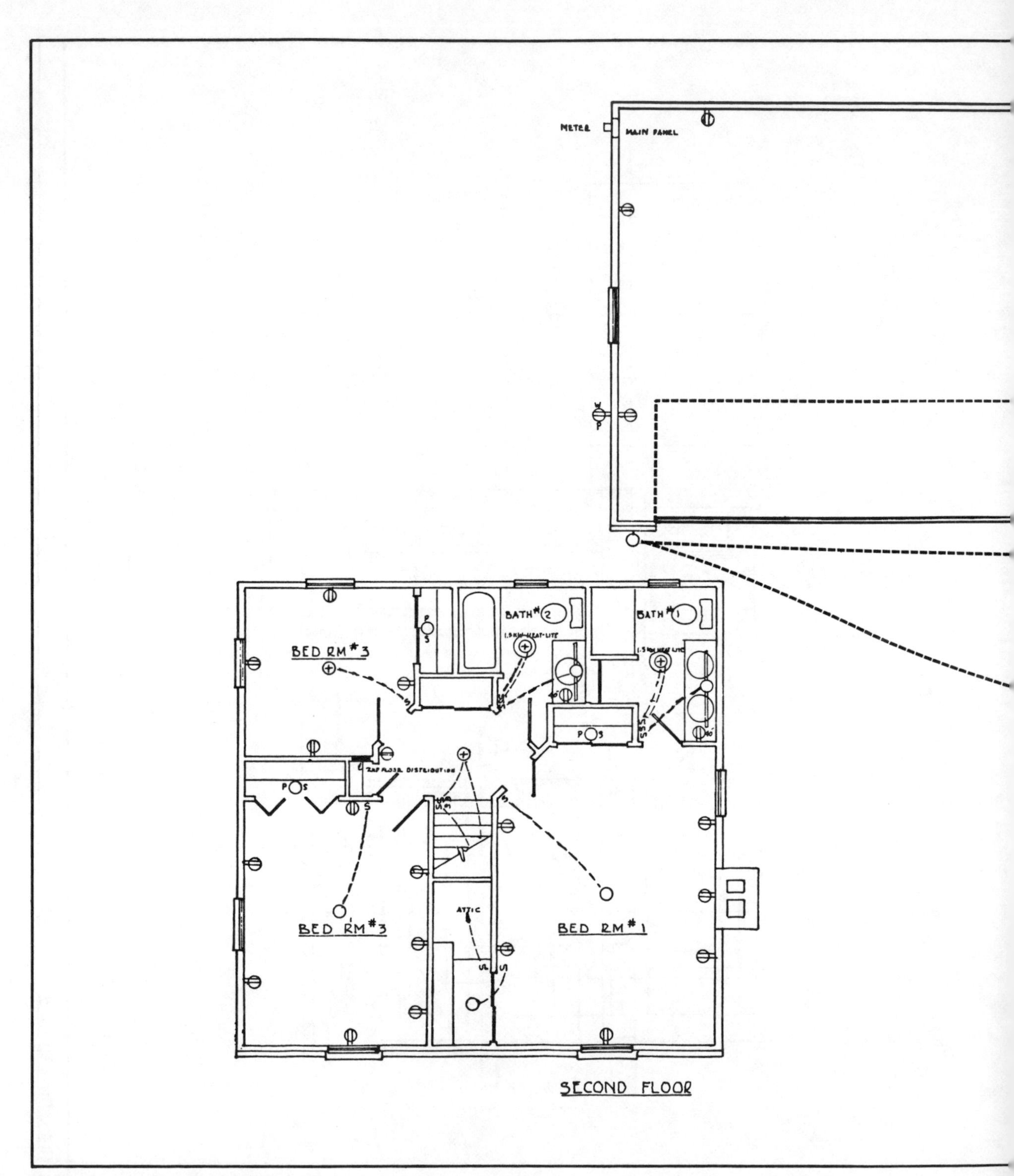

Fig. 19-14. The electrical plans.

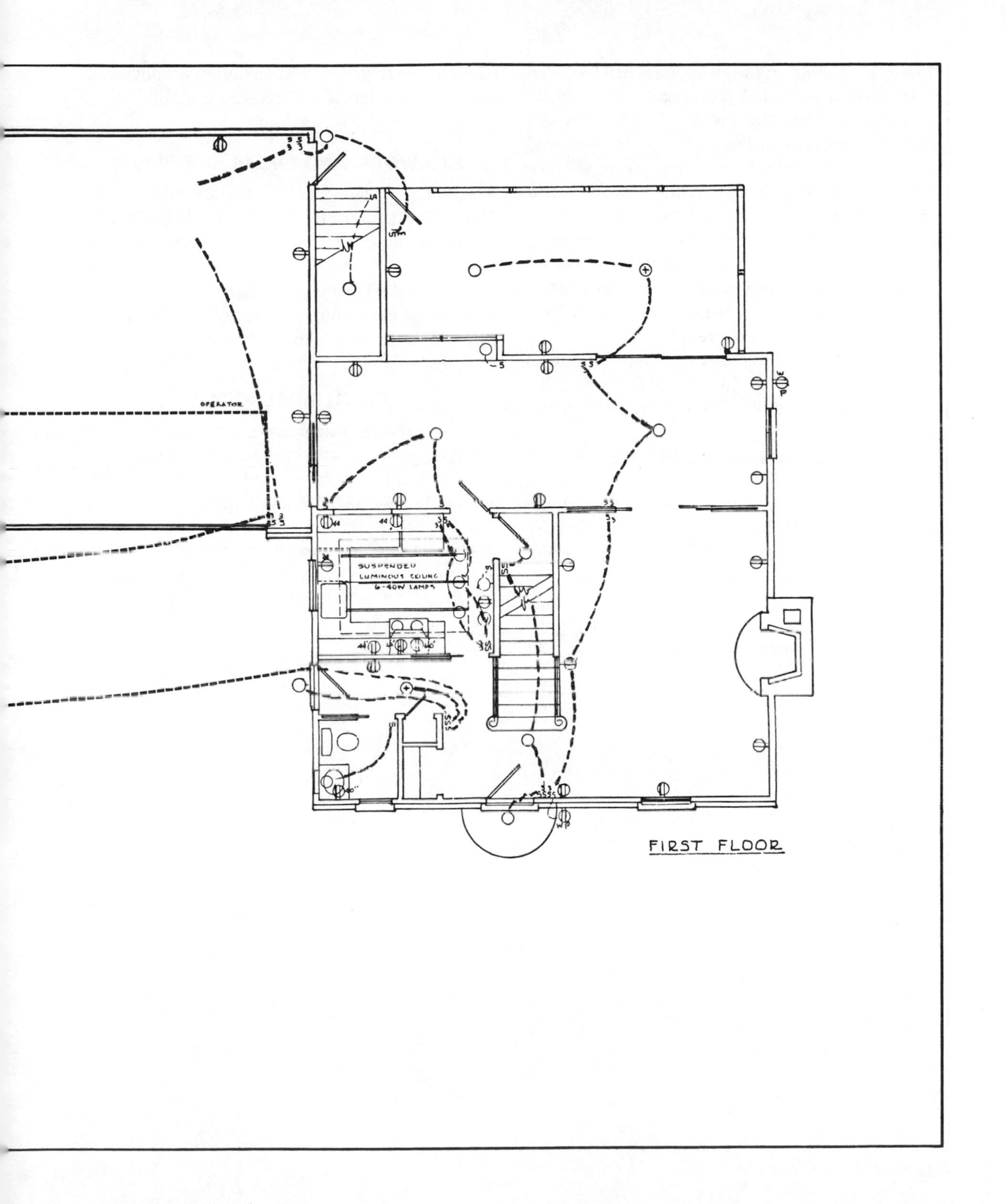
OPERATOR
SUSPENDED
LUMINOUS CEILING
6-40W LAMPS
FIRST FLOOR

moved them about on the elevations until I decided on the proper sizes and positions.

I started with the number 1 bedroom and centered a window in the 13 foot wide front wall and placed the first-floor living room window directly below. I made sure that the front window for the number 2 bedroom was the same distance from the left corner of the house as the window in the number 1 bedroom was from the right corner of the house. This was a matter of symmetrical balance of the exterior. All three bedrooms had to have windows to provide cross ventilation.

Downstairs I centered the kitchen window in the exterior wall (Fig. 19-11). While positioning windows, I was careful to leave room for draperies on each side.

The front entrance of the house presented a minor problem, for it could not be centered on the front of the house. If centered, it would open into the 13 foot wall of the living room. I finally positioned it so that there would be equal amounts of exterior wall on each side of the front living room window.

Basement windows are desirable for light and ventilation. Whenever possible, I kept basement windows *above* ground level to avoid window wells which collect debris and require gratings.

MILLWORK AND FIREPLACE PLANS

The millwork plans for the house are shown in Fig. 19-12. They show different views of the kitchen and the bathrooms. I drew all the plans to the same scale.

Figure 19-13 shows plans for the fireplace—an exterior elevation, a construction section, and a first-floor interior elevation.

ELECTRICAL PLANS

After I finished the millwork drawings, I concentrated on the electrical plans (Fig. 19-14). I drew an electrical plan for the first floor and one for the second. I made sure that all rooms would have adequate lighting and that all appliances would have the necessary convenience outlets. I double-checked to ensure that my plans met all local codes.

Plumbing and heating/cooling plans can be taken care of by contractors. So I didn't have to draw those plans.

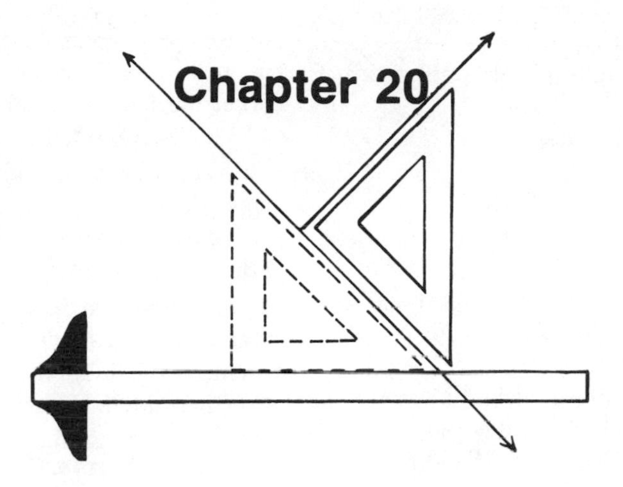

Specifications

It is the purpose of specifications, commonly called *specs*, to instruct a builder about details of materials, construction requirements, and workmanship that cannot be shown graphically or by notes on the drawings. Along with the drawings, specs become a legal document forming part of the contract between an owner and a building contractor.

Specifications help to prevent misunderstandings, so they should conform to the information given on the drawings. If there is a difference between the drawings and the specifications, the specs take precedence over the drawings.

On large projects, specs must go into considerable detail and are often the size of a book. On most residential jobs, they may be shortened by referring to some standard, such as the *Minimum Property Standards for One- and Two-Family Dwellings*, prepared by the U.S. Department of Housing and Urban Development.

Many residential planners use prepared specifications containing information common to the type of building being considered. Such readymade specs have blank spaces where specific information can be written.

In specs, it is also common for the owner to state a price limit for various materials. Because a builder can often buy at a discount, specs often specify whether the limit refers to the actual cost to the contractor or the list price.

When writing specs use simple, direct language. Avoid technical and legal terms. Keep specs as short as possible. Short ones are more likely to be read and followed than long, complicated ones. Many good houses have been built with specs of only one or two pages long.

Specifications should be divided into categories based on the various trades involved. And the categories should be listed in the order in which they will be considered on the job. This order will vary in different areas, but the following is usual:

1. General
2. Site Work—Excavation
3. Concrete
4. Masonry—Fireplaces
5. Stucco

6. Waterproofing
7. Structural Iron
8. Carpentry—Rough—Finish—Millwork
9. Doors—Windows—Glazing
10. Roofing—Sheet Metal
11. Insulation
12. Dry Wall—Lath and Plaster
13. Ceramic Tile and Marble
14. Cabinets (Factory-Made)
15. Specialties—Resilient Flooring, etc.
16. Painting and Decorating
17. Plumbing—Rough—Finish
18. Electrical—Rough—Finish—Fixtures
19. Heating and Cooling—Rough—Finish
20. Landscaping—Driveway—Walks
21. Other

The following are specifications written for the small house of Chapter 18. These specifications are for a building site in a frost-free area.

1. General. Unless stated otherwise on the drawings or in these specifications, the *Minimum Property Standards for One- and Two-Family Dwellings*, 1973 Edition, of the U.S. Department of Housing and Urban Development, shall be the required minimum standards for materials and construction on this job.

2. Insurance. The owner will take out appropriate builder's liability, fire, and storm insurance to protect himself and the contractors. Each contractor and subcontractor shall carry his own liability, Workmen's Compensation, and unemployment insurance as required by law, and each shall be responsible for sales taxes, payroll taxes for social security, etc. on his portion of the work. The general contractor shall present satisfactory evidence to the owner of his compliance. If requested, subcontractors shall do the same. Should the owner purchase materials not included in the contract for incorporation in this job, he will be responsible for all taxes in connection therewith.

3. **Site**. This house is to be built on lot located at 1776 McKay Creek Drive, Largo, Florida, McKay Creek Drive is the first street left off Avocado Drive which turns southerly off Indian Rocks Road just north of McKay Creek bridge.

The lot is clear and ready for building. There are two trees to remain; that in front must be boxed and protected. The lot has been surveyed and corner markers are visible.

4. Laying Out. The mason shall lay out the house at the location, with setbacks and clearances shown on the plot plan. The floor level shall be approved by the owner and batter boards shall be set with an accurate builder's leveling instrument.

5. Excavation. Dig for footings shall extend at least 8 inches into solid undisturbed ground. Back filling the bottom of the trench will not be permitted. No soil shall be removed from the site.

6. Concrete. Concrete for footings shall be at least 2500# and all other concrete above the footings shall be 3000# ready-mix from an approved supplier. All concrete shall be placed as dry as possible (5 maximum slump). A sloppy mix will not be permitted.

When concrete is placed in bond beam forms, care shall be taken to rod the concrete where there is vertical block reinforcing to ensure that block cells are completely filled.

Bond beam forms shall be well vibrated to settle the concrete around the reinforcing and to prevent stone pockets. Should any be found when the forms are removed they shall be patched immediately while the concrete is damp.

7. Reinforcing. All concrete shall be reinforced as called for on the drawings. Bars shall be sharply bent around corners. Bar splices shall be lapped at least 18 inches and wired twice. Bars shall be held in proper position on suitable chairs in such as way that they will not be displaced when concrete is poured.

Where vertical wall reinforcing is called for, footing dowels are required. They shall be hooked under the footing bars and project at least 18 inches above the floor slab.

All slabs on the ground shall be reinforced with 6 × 6 10/10 welded wire mesh with side and end laps 6 inches and wired. The mesh shall be properly flattened before it is placed in position and shall be lifted up as the concrete is being poured to ensure proper cover.

8. Masonry Walls. Foundation walls shall be standard concrete units. Walls above the foundation shall be light weight insulating block. Mortar shall be made of one part approved masonry cement and three parts clear sharp masonry sand. No retempering of partly dried mortar will be permitted.

9. Fill Under Slab. Fill beneath the slab shall be clean local pit sand deposited and spread in layers and properly compacted, preferably by a mechanical tamper. When spread by a tractor, its continual passage over the fill is acceptable if the corners and edges are well hand tamped.

10. Soil Poisoning. After underslab plumbing is completed and the fill is recompacted over trenches, the soil shall be poisoned to meet HUD requirements by a firm specializing in that work. They shall present evidence of compliance.

11. Moisture Barrier. Install a moisture barrier under all slabs, including the carport, of 4 mil polyethylene sheeting. Use the widest sheets possible and anticipate the flow of concrete so that 6 inch laps will not be disturbed. Take care to make tight joints around pipes coming up through the slab and seal with asphalt mastic.

12. Floor Slabs. Pour floor slabs as specified, taking care to keep them true to grade or level. After screeding, tamp with an open mesh tool and finish with a machine float. Finish is to be smooth inside and a light broom finish for the carport and other outside slabs.

13. Concrete Cure. As soon as a slab attains initial set (about 24 hours) brush on a heavy coat of approved concrete curing liquid.

14. Bond Beam. Form, place reinforcing, and pour the bond beam. Note that it extends across the front of the carport to act as a girder. Install 1/2 × 8 inch hooked bolts for plate anchors as soon as concrete can hold them (48 required). Use a 6 inch wide strip of felt paper to prevent concrete from filling block cells except where there is vertical reinforcing.

15. Insulating Fill. Before the bond beam is formed, fill all the cells of the block walls with vermiculite or other approved insulating fill. (This does not apply to places where there is to be vertical reinforcing.) This may be done in stages as the wall is laid if preferred.

16. Stucco. Stucco is to be two-coat follow-up work made of waterproof cement and masonry sand, 1 to 3 mix. Skim coat the inside walls of the laundry room.

17. Rough Carpentry. All lumber in contact with masonry or concrete shall be pressure treated yellow pine. Lumber for the exposed ceiling joists and ridge girder shall be 1500F western fir D4D, selected for good appearance. As far as possible, fastenings shall be concealed. Use Weldwood glue on the girder. Other framing lumber to be 1200F grade Fir or yellow pine. Use Teco or similar galvanized framing anchors where called for. Where the girder passes through masonry gable ends, the wood shall be protected by a wrap of polyethylene sheet or black felt paper.

Exterior exposed wood shall be red wood or cypress, and all exterior fastenings shall be galvanized.

18. Insulating Roof Deck. Shall be Homosote 2 3/8 inches thick or approved equal installed according to the manufacturer's directions. Use 16d galvanized nails, 5 at each bearing. Pairs of planks come wrapped, so care shall be taken to avoid marring the underside finish. Just before the deck is installed, staple kitchen type waxed paper to the top of the roof joists. This is to protect the deck finish when the joist is being painted. Cut ends of plank must be sealed.

19. Furring. The interior sides of all masonry block walls, but not including the laundry, are to be furred with PT 1 × 2 inch strips vertically on 16 inch centers and by PT 1 × 4 inch strips horizontally above all windows (except kitchen and bath) and above the sliding glass door. Extend at least 16 inches each side of opening. Use 1 × 6 PT boards back of studs contacting masonry walls. Fasten with cut nails, about 24 inch centers, or by power driven studs.

20. Interior Finish. Interior finish and moldings shall be soft pine, thoroughly seasoned and free from defects or tool marks. Install in long lengths, but where necessary bevel cut joints will be required. Free ends, if any, shall be copied to the pattern of the mold.

Closet shelves to be 11 1/2 inches wide set on a hook strip with aluminum coat hooks on 12 inch centers. Closets to have chrome-plated poles, and when more than 48 inches long, are to have center supports. Shelves wider than 11 1/2 to be glued up stock.

21. Insulation. Where wood panelling is called for, it shall be 1/4 inch prefinished stock selected by the owner. Moldings to match if needed. Allow $24 for two 4 × 8 inch sheets. To be glue fastened.

Insulate all exterior walls with 1 inch foil-faced glass wool stapled between the furring and carried to the ceiling. Avoid cutting at electric boxes and elsewhere as far as possible to eliminate air passage. Fill the space between the roof joists above the walls with heavy insulation.

22. Dry Wall. Dry wall shall be 1/2 inch thick, installed and finished in accordance with standard practice. Nails to be 13 or 14 gauge annular or threaded type 1 1/8 inch long with 1/4 inch flat heads. Use metal corner bead on outside corners if any. Use tile backer board around the tub and back of the lavatory in the bathroom. All joints to be taped and sanded and left ready for the painter.

23. Roofing and Flashing. Immediately after the roof deck is installed, cover it with 30# dry felt-tin-tagged or otherwise temporarily secured until the roofing can be finished with 2G5# 3-1 asphalt shingles. Nails must be annular or threaded type for use with insulating deck. Place galvanized drip strips along the rakes and eaves, and flash around the roof scuttle with 26 gauge galvanized metal. Lead flashing at the vents will be supplied by the plumber. The owner to select the color and approve the make of shingles. A light color is preferred.

24. Aluminum Windows and Door. All windows shall be Anderson locking type awning windows with DSB glazing, except in bath which shall be obscure. All to have screens. Operators to be right-hand, except where double windows require a left-hand operator at one side.

The double sliding glass door shall be glazed with 7/16 inch safety glass and shall be provided with a suitable locking pin in addition to the pull locks.

25. Doors. All exterior doors to be 1 3/4 inches thick. The two doors to the laundry room to be pine panelled and glazed. The two house doors to be solid core, flush, luan stain grade. The front door to have a 12 × 12 inch one-way type glass light as shown.

All interior doors to be 1 3/8 inches thick, hollow core, luan stain grade. Two bedroom closets to have 4-panel 1 1/8 inches thick, full louvered, bi-fold type doors complete with hardware and pulls.

Cupboard doors in hall and bath to be 3/4 inches luan plywood.

26. Rough Hardware. All rough hardware is to be included in the contract. Furnish the mason with 48 1/2 × 8 inch hook bolts with 2 inch washers and nuts for installation in the bond beam. Also give him 6 galvanized 10 gauge metal straps (2 × 48 inches) for tiedowns to be mortared into the block walls for securing the ridge girder.

27. Finish Hardware. Both house doors to have double security locksets with dead bolts. The front is to have thumb latch and handle. The others with knobs only. The laundry room doors to have cylinder locks only, but all four doors are to be keyed alike. Bedroom doors to have privacy locksets. Others to be closet type. Aluminum finish. Owner shall approve make and design. All exterior doors to have extruded aluminum thresholds with replaceable gaskets. House doors to have adjustable stops with weather strips.

Exterior doors to be hung on 3 LP rustless 4 × 4 inch butts. Inside doors on 2 LP 3 1/2 × 3 1/2 inch butts.

Closets to be provided with rustless rods and aluminum clothes hook on 12 inch centers. Owner to select the cabinet hardware.

28. Ceramic Tile. Ceramic tile to be owner's selection of standard grade 4 1/2 × 4 1/2 inch or 1 inch mosaic, set in adhesive where called for on the drawings. Provide cap where necessary. Furnish and install matching bath fittings, including two soap and grabs where shown. If requested, install a safety bar to be furnished by the owner.

Install marble stools at all windows, properly set in cement mortar.

29. Vinyl Floor Tile. Install owner's selection of medium grade vinyl floor tile and base on bath and kitchen using adhesive or prefixed tiles. Provide a stainless steel or aluminum edge strip where necessary. If house is not to be carpeted, install tile in hall. Otherwise, hall tile shall be omitted. Quote additional cost for hall tile.

30. Cabinets. Cabinets shall be either shop or job built by competent workmen. All visible surfaces are to be covered with plastic laminate, the pattern and colors to be selected by the owner.

If job-built, drawers shall be made of molded seamless polystyrene requiring only a front matching other cabinetwork. Use 3/4 inch plywood or particle board. Plywood shall be waterproof at sink and lavatory. Joints shall be blind nailed, screwed and glued as necessary. Door hinges to be concealed pin type and have touch-lock fasteners. Pulls to be selected by the owner. All interiors to be spray coated washable enamel.

Install pan rack, sliding towel bars, and any other cabinet fittings furnished by the owner. Note that one 27 inch wide base cabinet at the range must be removable for access to the concealed water heater.

31. Bath Cabinets and Tub Enclosure. Furnish and install a Miami-Carey Duette Princess #2015-L with light. This fits the 30 inch space above the lavatory.

Furnish and install a plastic panel accordion folding bathtub enclosure. Owner's selection.

32. Painting and Decorating. The owner shall approve all colors. And if requested, the painter shall prepare sample panels for approval. Ordinarily the owner will make his selections at a paint store designated by the painter.

All workmanship shall be first class. The work shall be properly prepared by sanding, filling nail heads, etc. where required.

Caulk and make tight all open joints on the exterior.

Immediately after erection, give all exterior woodwork a heavy coat of Cabot's or Olympic stain and a second coat after other trades have finished.

All hardware shall be removed before finish is applied. Exterior stucco is to be painted with two coats of medium grade latex.

Living area doors, both outside and inside, are to be stained and wiped to show the grain. They shall be finished with two coats of urethane varnish.

Other inside trim to be painted or stained as the owner prefers.

Interior dry walls are to be painted with two coats of medium grade latex.

The bathroom above the tile is to be covered with waterproof and washable, fabric-backed, vinyl wall covering applied as directed.

If the living area is not to be carpeted, it shall be painted with one coat of Tufftop concrete floor enamel. Paint all other concrete slabs with one coat of the same Tufftop. Concrete may require acid etching, so if the inside floor is to be enameled, it should be done as soon as possible before inside trim is installed and a second coat when other trades are finished.

33. Electrical. The electrician shall obtain and pay for all permits in connection fees and make any required deposit in the owner's name. The owner reimburses him for such deposit and the general contractor is to pay for electricity used during construction.

As soon as the job is ready to start and he is notified, the electrician shall set an approved power pole with meter ring and fused disconnect where directed by the contractor and have the power company make a connection.

All work shall meet the requirements and inspections of the local and NEC codes.

Wiring to be done with 3-wire nonmetallic cable with all outlets properly grounded. Switches and duplex outlets to be first grade. Switches silent type.

Provide suitable connection for the following appliances:

- ☐ Range.
- ☐ Range hood.
- ☐ Dishwasher.
- ☐ Disposal.

- ☐ Water heater.
- ☐ Bathroom heater (1.5 kW).
- ☐ Automatic washer.
- ☐ Dryer.
- ☐ Heat pump with strip.

Install bell wiring with lighted push buttons at two entrance doors. Chimes are to be included in the fixture allowance.

Install suitable TV wiring for two outlets.

Install phone wiring for two outlets.

Distribution panel is to have automatic circuit breakers.

Porcelain lamp sockets with pull switches are to be furnished under this contract and are not included in the fixture allowance of $100.

All wires to be copper and none shall be less than #12 gauge.

The fixture allowance is $150 and includes the bathroom heater-light and door chime but not the job-built kitchen fixture. The electrician shall furnish and install the kitchen fixture (two 2-light 40 W fluorescent fixtures).

All electric lamps are to be furnished in contract.

34. Plumbing. The plumber shall obtain and pay for all permits, connection fees, and deposits required for his portion of the work. Deposits shall be taken out in the owner's name, and the owner will reimburse the plumber upon request. The general contractor shall pay for all water used during construction.

When the job is ready to start and he is notified, the plumber shall provide a hose connection near the meter with a key operated faucet.

As soon as the slab fill is in and compacted, the plumber shall install all rough plumbing. Soil pipe 3 inches and over shall be service weight cast iron with leaded joints. Pipe 2 inches and under may be galvanized iron with drainage fittings. Water lines from a main shutoff outside the building line shall be type K copper with solder joints. All risers are to have 12 inch extension air chambers to inhibit water hammer.

Quote the price difference if code-approved PVA piping is used instead of above.

Obtain all required inspections.

Furnish and install the following fixtures or owner-approved equals:

A-S (American-Standard) 60 inch bath: Contour recess, non slip, shower, diverter.

A-S 20 × 17 inch lavatory Aqualon, china, self-rimming.

A-S toilet: Cadet, water saver, solid plastic seat.

Kitchen sink: double compartment, 32 inch, 18 gauge stainless steel, single lever faucet and spray.

Laundry sink: Durastone, 20 × 24 inches with swing laundry faucet.

Water heater: 40 gallon under-counter type with double elements, 7 1/2 year guarantee.

Garbage disposal: Waste King Model 6000 or an approved equal.

Provide two outside hose connections.

Provide brass plug cleanout on soil line outside the front wall and extend to street sewer with 4 inch cement-asbestos or approved pipe.

35. Heating and Cooling. Heating and cooling is to be accomplished by means of a heat pump located in the attic space above the bathroom. Access for installation and servicing of equipment will be through a hatch of suitable size on the roof. Air intake and discharge must be above the roof. Return air through a grill in the hall ceiling or otherwise. Conditioned air is to be circulated to four grills with adjustable dampers for proportioning the discharge. Thermostat on inside wall of living room.

The electrician will provide a fused disconnect at the power panel in the laundry and run suitable wires to a disconnect in the attic near the machine. 230 volts.

The machine shall be supported to minimize vibration and noise, and unless the machine is provided with proper condensation drainage, a galvanized pan for running drainage outside shall be provided.

The system shall be designed to suit conditions for this area and to maintain an inside temperature of 72 °F.

In addition to any manufacturer's guarantee on equipment, this contractor shall guarantee all parts and workmanship for a period of one year from date of occupancy of the house. He shall balance the system to ensure proper air distribution and provide all necessary service, without charge, during that period.

Alternate number 1:

A suitable outdoor heat pump at the rear wall with discharge into the attic duct system and return air in a masonry type duct beneath the floor slab. This duct to be installed at this contractor's cost.

Alternate number 2:

Provide a suitable upflow heating unit that can be placed in the 24 inch square space marked brooms on the floor plan. Heat to be oil or LP gas. Suitable tanks to be provided. Cooling to be by coils above the heat unit with a condenser outside. Ducts as required.

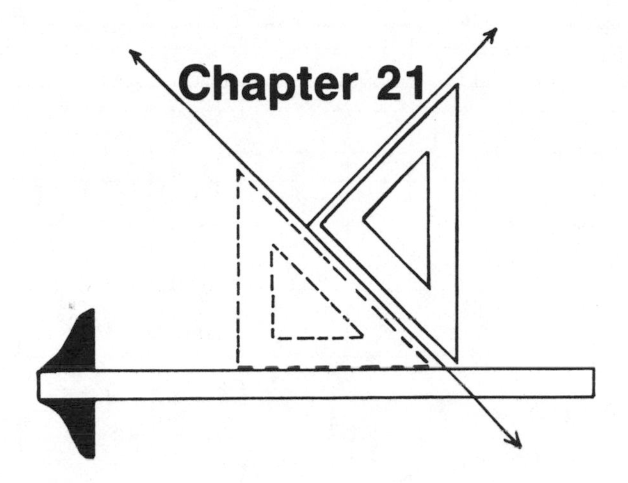

Materials Lists and Estimates

Whether you expect to do your own construction or let a general contract, there is no better way for an inexperienced planner to check his drawings and specifications than to make a list of all the materials that would be required to do the job. Pricing the materials and estimating the labor affords the most accurate way of determining the probable cost of a building. That is how most experienced contractors prepare their bids.

The estimator's bible (it looks a bit like one) is Walker's *Estimator's Reference Book.* It tells how to estimate necessary building materials and labor for all kinds of construction jobs—houses, large buildings, etc. It is kept up to date by a new edition every three or four years. This is an essential reference work for anyone compiling a materials list and estimating costs.

If you have completed plans for a house, you have a good idea of the materials you'll need, so a materials list is no more than a piece-by-piece count of the things needed.

You may take your plans to a building supply firm and have them make a materials list for you. They, of course, expect you to buy materials from them. But they cannot afford the time to make a precise list, so they tend to be overly generous, listing more materials than you actually need.

But if you want to save some money and keep a tight rein on expenses, you can make your own materials list.

General contractors, using the same house plan many times, frequently buy the materials and are very careful about the quantities sent to a job. They send just enough. And it's up to the workmen to complete the house from what has been delivered.

The following pages show a typical materials list. The work categories are listed in the order in which they would be considered on a job. I have made entries showing my estimates of materials and labor for the small house of Chapter 18. I have employed standard symbols: ′ for feet, ″ for inches, LF for linear feet, SF for square feet, CF for cubic feet, CYD for cubic yards, HR for hours, SX for sacks, and GAL for gallons.

		AMOUNT	UNIT COST	MATERIAL	LABOR	SUB-CONT.
	DATA – AREA - 32.0 × 26.67 = 853.5 ◻'					
	12.0 × 14.0 = 168.0					
	SLAB ON FILL 1021.5 ◻'					
	SLAB ON					
	12.0 × 16.67 = 200					
	6.0 × 4.0 = 24					
	12.0 × 3.0 = 36					
	260 ◻'					
	FOUNDATION PERIMETER					
	44 + 44 + 26.67 + 26.67 + 14 = 155.3 LF					
	LESS 6 × .67 - 4.0					
	NET LENGTH 151.3 LF					
1	SITE					
2	LAYOUT - BATTER BOARDS 2×4 × 8/0	4 PCS				
	1×4 × 12/0	4 "				
	2 MEN MASON	2 HR				
	" HELP	2 "				
3	FOOTINGS - HAND DIG 151 LF					
	LABOR	10 HR				
4	REINFORCING SET REINF #5 BARS 21	420 LF				
	WIRE CHAIRS	50				
	TIE WIRE	5#				
	LABOR	6 HR				
INS	INSPECTION					
5	POUR FOOTINGS 3000# CONC	5½ CYD				
	MASON	4 HR				
	" HELP	8 "				
6	FOUNDATION WALL 152 LF.					
	8×8×16 REG BLOX	120				
	" " " HEADERS	110				
	8×4×16 HORIZ. CUTS	10				
	MASON SAND	3 CYD				
	" CEMENT	6 SX				
	MASON	8 HR				
	" HELPER	8 "				
7	FILL UNDER SLAB	40 CYD				
	LABOR	4 HR				
8	PLUMBER - ROUGH PIPING 40					

Fig. 21-1.

	PLUMBING INSPECTION					
	BACK FILL & RECOMPACT LABOR	2 HR				
9	SOIL POISONING	1000 ◻'				
10	MOISTURE BARRIER					
	.004" VISQUEEN 12' × 100'	1200 ◻'				
	2 CUT & PLACE LABOR	2 HR				
11	SLAB REINF. 6×6 10/10 WWM 5'					
	3 - 100' ROLLS	1500 ◻'				
	CUT & PLACE LABOR	8 HR				
	FORM CARPORT SLAB 1×8 × 10/0	5				
	LABOR	2 HR				
	INSPECTION					
12	POUR & FINISH ALL SLABS					
	1240 ◻' 3000# CONC.	16½				
	PLACE & SCREED, MASON	12 HR				
	" HELP	36"				
	MACHINE FINISH - 1240 ◻'	8 "				
13	BLOCK WALLS 8×8×16 REG.	980				
	8×8×8 ½ REG	90				
	8×4×16 HOR. CUTS	25				
	CEMENT BRICK	30				
	PRECAST SILLS 24½"	2				
	32½"	1				
	48½"	3				
	80½"	2				
	MASONRY CEMENT	202 V				
	BLOCK LAYING - MASON	40 HR				
	HELP	40'				
14	BOND BEAM - REINF #5 BARS	780'				
	WIRE CHAIRS 2"	40				
	" " 10"	40				
	6" FELT STRIP	300'				
	PLACE REINF. LABOR	16 HR				
	FORMS 16" WIDE × 8' 300' } RENT					
	" CLAMPS 50 } RENT					
	INSPECTION					
	CONCRETE 3000# MIX	4½ C				
	PLATE BOLTS ½" 6 × 8'	80				
	TIE DOWN STRAP 10GA × 1½	24'				

Fig. 21-2.

	#WD	
14	BOND BEAM - POUR MASON	8 HR
	LABOR	24 "
15	EXTERIOR FRAME & FINISH	
	PLATES PT. 2x8x 12/0	1
	" " 16/0	4
	RIDGE GIRDER #1 2x12x 12/0	1
	" 14/0	1
	" 16/0	2
	" 18/0	2
	POST 4x4 x 10/0	1
	GALV. POST ANCHOR	1
	ROOF JOISTS 4x8x 16/0	18
	" x 18/0	6
	LOOKOUTS (2' LENGTH) 2x8x 14/0	2
	FASCIA & RAKE - RED WD 2x8x 10/0	1
	" " 12/0	1
	" " 14/0	2
	" " 16/0	4
	" " 18/0	2
	" 1x4x 12/0	14
	AC SCUTTLE " 2x8x 8/0	2
	" 5/4x4x 8/0	2
	#2 2x6x 10/0	2
	3/4" W.P. PLY WD	1
	LABOR @40/M CARPENTER	64 HR
	" LABOR	32 "
16	ROOF DECK	
	2 3/8 HOMOSOTE 2'x8'	100 PCS
	16d GALV. NAILS -1500	35#
	SEALANT - DAP TUBES	8
	WAXPAPER ROLL 200LF	2
	LABOR PLACING CARP.	72 HR
	" HELP	36 "
17	ROOFING LAYER 30# FELT	4 ROLLS
	APPLYING LABOR	4 HR
	ASPHALT SHINGLES 265#	16 SQS
	1 1/4 ANNULAR NAILS	35#

Fig. 21-3.

17	ROOFING CON'T	
	GALV. DRIP STRIP	120'
	METAL FLASHING	20'
	MASTIC	2 GAL
	LABOR CARP	80 HR
18	EXTERIOR - FINISHING	
A	FRAME WALL-AT LAUNDRY	
	PT 2x4x 10/0	3
	" 10/0	8
	5/8x4x8 TEXTURE 1-11 SHEATHING	4
	LABOR CARP	4 HR
B	GLASS GABLES	
	PT. 2x10x 10/0	2
	#1 2x8x 12/0	4
	GLASS BEAD MLD	150'
	LABOR CARP	16 HR
	GLAZING	
C	TRIM BTW ROOF JOIST	
	RED WD 1x8x 8	11
	LABOR CARP.	8 HR
D	MAKE & SET DOOR FRAMES	
	PT 2x6x 14/0	3
	4 1/2" JAMBS W/STOPS 2/8	1
	LABOR	6 HR
E	INSTALL WINDOWS & SL. GL. DOOR	
	PT 1x2x 8/0	10
	LABOR CARP	10 HR
F	FIT & HANG EXT. DOORS	
	INC. LOCKSETS ENT. DOOR 2	12 HR
	LDY. " 2	10 "
	#3 GRADE CEDAR SHINGLE-WEDGES	1 BDL
19	STUCCO	130 SYD

Fig. 21-4.

No.	Item	Qty.
20	WINDOWS – ALUM. AWNING - W/SCREENS	
	M-11	1
	M-½ 32	3
	" MULL	1
	M-½ 33	2
	" MULL	1
	M-32	3
	DBL SLIDING GLASS DOOR W/SCR. 6/0 X 6/8	1
21	EXTERIOR DOORS ALL 1¾"	
	FRONT - W/1 WAY GLASS 3/0 X 6/8	1
	SIDE - S.C. FLUSH 2/8 X 6/8	1
	LAUNDRY - PINE PANEL W/GL 2/8	2
	WOOD STOPS ½ X 1½ X 12/0	1
	" X 14/0	3
22	INTERIOR FRAMING	
	FURRING P.T. 1 X 2 X 10/0	45
	" " 12/0	25
	" 1 X 4 X 8/0	6
	" " 10/0	2
	" 1 X 6 X 10/0	3
	" " 12/0	2
	FLOOR PLATES PT 2 X 4 X 8/0	5
	" " 10/0	2
	" 2 X 6 X 10/0	1
	STUDS, ETC #2 2 X 4 X 8/0	36
	" " 10/0	30
	" " 12/0	25
	" " 14/0	3
	" " 16/0	1
	" 2 X 6 X 10/0	8
	" X 12/0	3
	" X 14/0	2
	ATTIC FLOOR ⅝ PLY WD 4 X 8	4
	1 X 4 X 14/0	2
	LABOR - FURRING CARP.	12 HR
	FRAMING "	24 "
	LABOR	8 "

Fig. 21-5.

No.	Item	Qty.
23	ROUGH ELECTRICAL	
INS	ELECTRICAL INSPECTION	
24	HEATING ROUGH	
25	INSULATION	
	FOIL BACK 1" BLANKET X 15"	900'
	4" " X 15"	100'
	LABOR CARP	8 HR
26	DRY WALL	
	½" GYP. BOARD 4' X 8'	14
	4 X 10	14
	¼ TILE BACKER 4' X 8'	4
	TAPE	
	CEMENT	
	LABOR	18 HR
27	INTERIOR MILLWORK	
	DOORS 1⅜ FLUSH LUAN STAIN GR.	
	" 1/8 X 6/8	2
	" 2/4 "	1
	" " 3/0 "	3
	" 1⅛ BI-FOLD-LOUVRED-4 PL 6/0	2
	JAMBS - 4½ W/STOPS 2/8 SETS	3
	" " NO " 6/0 "	2
	DOOR TRIM SETS TO 2/8"	11
	" " 1 SIDE 3/0	1
	" " 1 " 6/0	5
	POCKET SLIDING DOOR FRAME 2/6	2
	LABOR CARP	26 HR
	TRIM - BASE	160'
	" " SHOE	160
	" BTW JOISTS P.P. 1 X 8 X 8/0	8
	" RABBETED P.P 1 X 3 X 12/0	5
	LABOR CARP.	24 HR
	CLOSET SHELVING 1 X 12 X 8/0	5
	" HOOK STRIP 1 X 3 X 8/0	4
	LABOR CARP	6 HR

Fig. 21-6.

No.	Item	Quantity
28	KITCHEN CABINETS	
	BASE UNITS 27"	2
	24" DRAWER	2
	36" SINK FRONT	1
	WALL UNITS 15" HI × 30" WIDE	2
	18" " × 36" "	1
	30" " × 24" "	1
	" × 33 "	1
	" × 48" BLIND COR	1
	COUNTER TOP PER DRAWING	13.5'
	INSTALLATION - CARP.	11 HR
	" HELPER	11 HR
28	HALL-BATH CLOSETS #1 HI BR	
	3/4 PARTICAL BD 4×8	1 PC
	3/4 LUAN PLYWD "	1 "
	1×4 P.P 12/0	3 "
	1/4 LUAN PLYWD	1 "
	1×10 PP 12/0	1 "
	1/4 PLY WD 4×4	1 "
	USE IN LDY ALSO 1/8" PEG BOARD 4×8	3 "
	CARP	16 HR
29	ODDS & ENDS	
	RANGE HOOD INSTALATION	2 HR
	BATH TUB ENCLOSURE INSTALLATION	1 HR

Fig. 21-7.

No.	Item	Quantity
30	CERAMIC TILE	
	BATH 4¼×4¼ CTN.40	14 CTN
	2" CAP	20'
	KITCHEN 4¼×4¼	8 CTN
	BATH FIXTURE SET	1
	MATCHING 2' BAR	1
	ADHESIVE	1 GAL
	MARBLE STOOLS 24½"	2
	32½"	1
	48½"	3
	80½	2
	INSTALLATION - MECH.	20 HR
	" HELP'R	20 HR
31	VINYL FLOOR & BASE	
	BATH 12"×12"	25 PCS
	KITCHEN "	50 "
	HALL "	36 "
	BASE 4'	36'
	SS. EDGE STRIP	15'
	INSTALLATION MECH	8 HR
	ADHESIVE	1 GAL
32	PLUMBING - FINISH	
33	ELECTRICAL FINISH	
	FINAL ELECTRICAL INSPECT.	
34	HEATING & COOLING - FINISH	
35	PAINTING & DECORATING	
	MATERIALS	
	LABOR PAINTER	80 HRS
36	WASH WINDOWS - CLEAN-UP LABOR	16 HRS
37	FINISH GRADING	
	BULLDOZER	4 HR
38	DRIVE WAY 4" CONC.	
	EDGE FORM - USED 2×4S	80'
	EXPANSION JOINT ½"×4"×10'	2 PCS
	REINFORCING 6×6 10/10 MESH	400 SF
	CONCRETE - 2500# MIX	5 CYD
	LABOR FINISHER	8 HR
	LABOR	8 "

Fig. 21-8.

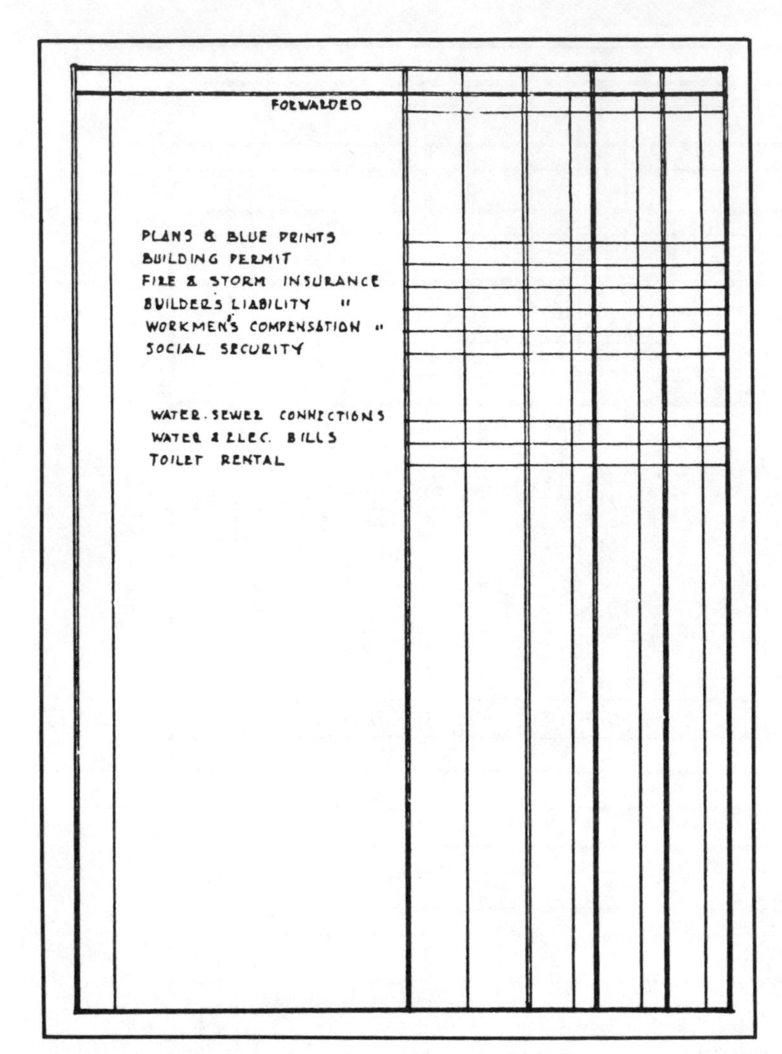

Fig. 21-9.

While you're compiling a materials list and estimating costs, you should also consider one additional item—insurance.

You must have fire and storm insurance to protect yourself and your subcontractors. The cost of such insurance is usually based on the value of the finished house.

You should also have builder's liability insurance to protect yourself from claims for injuries occurring on the job. Commonly in the amount of $100,000 to $300,000.

Chapter 22

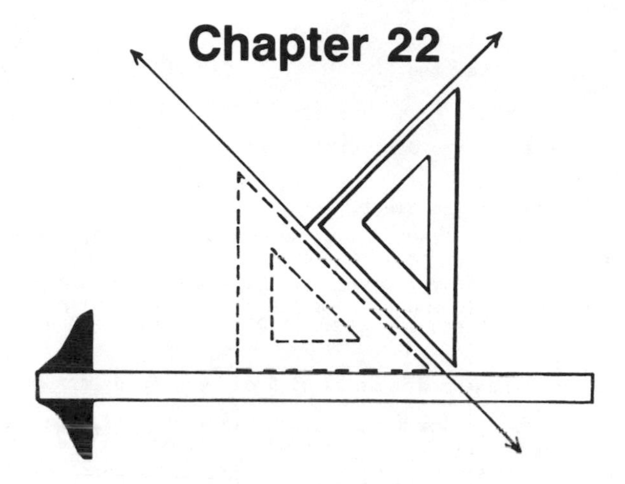

Building Contracts

There are several ways to go about building a house. You can have a general contractor build one for you. Or you can subcontract the work, that is, you can contract with the individual trades (plumbers, masons, etc.) to do the various jobs required. If you subcontract the work, you must oversee the building of your own house.

But whatever arrangements you make, contracts will be involved. The dictionary defines a contract as an agreement between two parties for doing (or not doing) something specific. A contract, of course, is enforceable by law.

A building contract consists of four documents: (1) the agreement, (2) the drawings, (3) the specifications, and (4) the general conditions. A building contract between the owner and a general contractor requires only one agreement document. But if work is to be subcontracted, each subcontractor (plumber, electrician, etc.) must have his own agreement document.

An agreement document includes the names and addresses of the contractor and the owner, the general terms of the agreement, a list of contract documents (drawings, specs, etc.), and other important details. The following is a sample agreement form (for an agreement between an owner and a general contractor).

A general conditions document defines the owner's and contractor's obligations and duties. The following is a typical general conditions document—one used in contracts between owners and general contractors.

GENERAL CONDITIONS

The Owner, the Contractor, and the Designer are those mentioned as such in the Agreement. They are treated throughout the Contract Documents as if each were of the singular number and masculine gender.

Where the Drawings and Specifications have been supplied by the Owner and there is no Designer of record, the Owner will assume the responsibilities of the Designer.

The term Subcontractor as employed herein includes only those having direct contract with the Contractor and includes one who furnishes mate-

AGREEMENT FORM

THIS AGREEMENT, made the day of 19__
by and between

hereinafter called the Contractor, and

hereinafter called the Owner.

WITNESSETH, that Contractor and the Owner for the consideration hereinafter named, agree as follows:

ARTICLE 1. SCOPE OF THE WORK. The Contractor shall furnish all of the materials and perform all of the work shown on the Drawings and described in the Specifications entitled

prepared by (Name and address of Designer, Architect or other)

and shall do everything required by this Agreement, the General Conditions of the Contract, the Specifications and the Drawings.

ARTICLE 2. TIME OF COMPLETION. The work to be performed under this Contract shall be commenced within______ weeks and shall be substantially completed within_________weeks.

ARTICLE 3. THE CONTRACT SUM. The Owner shall pay the Contractor for the performance of this Contract, subject to additions and deductions, in current funds the sum of _____.

ARTICLE 4. PROGRESS PAYMENTS. The Owner shall make payments on account of the Contract, as follows:

__% When floor slab is poured
__% When the building is dried in
__% When ready for finish plumbing and electrical
__% When completed

ARTICLE 5. ACCEPTANCE AND FINAL PAYMENT. Final payment shall be due within seven days after substantial completion of the work, provided the work be then fully completed and the Contract fully performed.

Upon receipt of notice that the work is ready for final inspection, the Owner or his representative shall promptly make such inspection and when he finds the work acceptable under the Contract and the Contract fully performed, he shall make the final payment.

Before the final payment is made, the Contractor shall present evidence that all payrolls, material bills, and other indebtedness connected with this work have been paid, or when payment is to be completed from the final sum due, the Contractor shall make a sworn statement to that effect, or present other evidence satisfactory to the Owner that all payments will be made.

If after the work has been substantially completed and full completion is delayed through no fault of the Contractor, the Owner shall, without terminating the Contract, make payment for the balance due for that portion of the Work fully completed and accepted.

ARTICLE 6. CONTRACT DOCUMENTS. The General Conditions of the Contract, the Specifications and the Drawings, together with this Agreement, form the Contract, and they are as fully a part of the Contract as if hereto attached and herein repeated.

IN WITNESS HEREOF the Parties hereto have executed this Agreement the day and year first above written.

Contractor Owner

Fig. 22-A.

rial worked to a special design according to the Plans and Specifications but does not include those who merely furnish material not specially worked.

The term work includes labor or materials or both.

All the time limits stated in the Contract Documents are of the essence of this Contract.

The laws of the place of building shall govern the construction of this Contract.

All materials and construction shall equal or exceed the requirements of local and area Codes as well as the *Minimum Property Standards for One and Two-Family Dwellings* of the U.S. Department of Housing and Urban Development.

Should any question of intent or of error on the part of the Plans and Specifications be found before the Contract is signed, it shall be settled with the Owner or Designer and clarifying notations shall be made on the appropriate Documents.

Should any question of intent or of error be found during the progress of the work and not be settled by amicable agreement between the parties, then the Designer shall rule and if this is not acceptable by either Party, it shall be settled by Standard Arbitration Procedure and the party against whom the decision is rendered shall pay the costs, if any, of arbitration. It is mutually agreed that the progress of the work will not be held up during arbitration but will proceed as directed by the Owner.

In addition to the two signed copies in the Contract Documents, one to be retained by the Owner, one by the Contractor, the Owner will furnish to the Contractor, free of charge, up to seven additional sets of Plans and Specifications for use as follows:

Two Permit sets, one with the Permit approval to be kept on the job at all times during working hours. One set on the job for use by workmen. One set for Contractor's office use. Three sets for use by the mechanical and other trades as needed.

Other than the two signed Contract Documents, all plans are the property of the Designer and shall not be used for any other work and shall, if requested, be returned to him on completion of the Contract.

Unless otherwise stipulated, the Contractor shall provide and pay for all materials, labor, tools, equipment, water, lights, power, transportation, and other facilities necessary for the execution and completion of the work.

Unless otherwise specified, all materials shall be new, and both materials and workmanship shall be good quality. The Contractor shall, if requested, furnish satisfactory evidence as to the kind and quality of materials.

The Contractor shall at all times give proper superintendence to the work, enforce strict discipline and good order among his employees, and shall not employ on the work any unfit person or anyone not skilled in the work assigned to him.

The Owner shall furnish all surveys unless otherwise specified. The property shall be identified by permanent markers set at all pertinent corners and properly identified so that there can be no question as to the correct site of construction.

The floor level of the building shall be established by the Designer or Owner before the Contract is signed by reference to some permanent object at the site.

The Contractor shall obtain and pay for all Permits, fees, and deposits required by law or otherwise and shall receive any refunds that may be due upon completion of the Contract.

Whenever the law of the place of building requires a sales, consumer, use, or other similar tax, the Contractor shall pay such tax.

The Contractor shall be responsible for the protection of all his work from damage from all sources and the Owner's property from injury or loss arising in connection with this Contract. He shall make good any such damage, loss, or injury except those beyond the Contractor's control or not due to his fault or negligence.

The Contractor shall take all necessary precautions for the safety of his employees and others and shall comply with all lawful safety regulations.

The Contractor shall be responsible for the erection and lighting of barricades if needed and provide warning signs if needed.

The Contractor shall install adequate toilet facilities for the use of his workmen and maintain them properly.

The Owner and his representatives shall, at all times, have access to the work whenever it is in progress or preparation.

The Contractor shall make all required meter deposits in the name of the Owner, but any such that are returnable shall be reimbursed to the Contractor upon completion of the Contract.

The Owner shall carry sufficient Fire and Wind insurance to protect himself, the Contractor, and his Subcontractors for 100% of the insurable value as their several interests may appear.

The Contractor and each of his Subcontractors shall carry Contractor's Liability Insurance and Workmens Compensation as required by law.

Before construction is started, the Contractor shall present Certification satisfactory to the Owner, evidencing full compliance with all the above requirements.

The Contractor shall be liable for and make good any vandalism occurring in the building during construction and may protect himself by insurance or otherwise as he deems best.

The Owner shall maintain such insurance as will protect him from contingent liability to others for damages of every sort, which may arise from operations under this Contract, and any other liability for which the Contractor is required to insure under the provisions of this Contract.

The Contractor shall, as soon as possible after the execution of the Contract, notify the Owner in writing of the names of the Subcontractors proposed for the principal parts of the work, and shall not employ any that the Owner may, within a reasonable time, object to as incompetent or unfit.

The Contractor agrees that he is fully responsible for the acts or omissions of his Subcontractors or any persons employed directly or indirectly by them as he is for the acts or omissions of persons employed directly or indirectly by himself.

Nothing contained herein shall create any contractural relation between any Subcontractor and the Owner.

The Owner shall have the right to make changes, additions, or omissions during the progress of the work but such shall be made only on duplicate written order signed by both parties and stating what is to be done and the entire additional cost or credit to be allowed for such change. Where such extra cost is not immediately obtainable and the parties mutually agree, then this extra cost work may proceed on a cost plus basis as follows: Net cost of materials including all discounts plus 10%. Actual extra payroll plus 10% for supervision and profit and 15% for overhead and insurance, a total of 25% on payroll. No extra charge shall be made for supervision or use of tools and equipment unless such change requires the hiring of outside equipment not normally used by the Contractor and then only if agreed on by the Owner.

The Owner shall have the right to have other Contractors perform work not included in the original Contract and these shall have the full cooperation of the General Contractor.

Neither party to this Contract shall assign the Contract or sublet it as a whole without the written consent of the other, nor shall the Contractor assign any monies due or to become due to him hereunder without the previous written consent of the Owner.

If the Contractor should be adjudged bankrupt or otherwise be unable to prosecute the work, or if he should persistently refuse or fail, except where an extension of time has been granted, to supply enough skilled workmen or proper materials, or if he should fail to make prompt payment to Subcontractors when due or in any other respect so that the Owner has reason to believe that sufficient cause exists to justify such action, the Owner may, upon 7 days written notice, handed personally to the Contractor or mailed by registered letter to his address, terminate the employment of the Contractor and take possession of the premises and of all materials, tools, and equipment thereon and finish the work by whatever method he may deem expedient. In which case the Contractor shall not be entitled to receive any further payment until the work is finished. If the unpaid balance of the Contract price shall exceed the expense of finishing the work, such excess shall be paid to the Contractor. If the expense shall exceed such unpaid balance, the Contractor shall pay the difference to the

Owner. If the Contractor does not agree as to the expense incurred by the Owner, it shall be settled by Standard Arbitration Procedure as heretofore outlined.

If the work should be stopped under any order of legal or public authority for a period of 30 days through no act or fault of the Contractor or anyone employed by him, then the Contractor may, upon 7 days written notice to the Owner and handed to him personally or mailed by registered letter to his address, terminate this Contract and recover from the Owner payment for all work executed and for any proven loss sustained on equipment or materials and reasonable profit and damages.

Should the Owner fail to make payment, through no fault of the Contractor within 7 days after the Contractor's written request for payment becomes due or any sum awarded by arbitration, then the Contractor may upon 7 days written notice to the Owner, stop the work or terminate the Contract as set out in the preceding paragraph.

Payment shall be made on the terms stated in the Agreement. When such payments become due, the Contractor shall make written application to the Owner for each payment and shall, if required, present receipts, vouchers, or other satisfactory evidence showing his payments for labor, materials, and Subcontractors.

The Owner may withhold payment for all or any part of any payment to the extent necessary to protect himself from loss on account of

1. Defective work not remedied
2. Claims filed or reasonable evidence indicating probable filing of claims
3. Failure of the Contractor to make payments to Subcontractors or for labor or materials
4. A reasonable doubt that the Contract can be completed for the balance then due
5. Damage to another Contractor

When any of the above grounds are removed, payment shall be made in the amount withheld.

Should the Owner fail to pay the sum due under the terms of the Agreement or in any award of arbitration upon demand, the Contractor shall receive in addition to the sum due, interest thereon at 7 1/2% for the overdue period.

No payments made to the Contractor nor partial or entire use or occupancy of the work by the Owner shall be acceptance of any work or material not in accordance with the Contract.

The making and acceptance of the final payment shall constitute a waiver of all claims by the Owner, other than those arising from unsettled liens or from faulty work appearing after final payment and also of all claims by the Contractor, except those previously made and still unsettled.

The final payment shall not become due until the Contractor, if requested, shall deliver to the Owner a complete release of all liens arising out of the Contract, or the receipt in full in lieu thereof. The Contractor may, if he prefers, furnish a bond satisfactory to the Owner to indemnify him against any lien or claim.

Each trade shall do its own cleanup work and shall immediately remove all debris, surplus materials, and equipment from the site upon completion of his part of the work.

The Contractor shall be responsible for proper job housekeeping and for the final cleanup which shall include the removal of all debris and surplus materials from the premises, the washing of all glass, cleaning and polishing of floors, paint touchup, and leaving the premises fully ready for occupancy and use. All to be completed before the final payment is made.

The Contractor hereby guarantees all work and materials, except as follows, for a period of one year from the date of completion and request for final payment, except as follows:

Mechanical and other equipment that is covered by the guarantee of the manufacturer. This guarantee does not cover damage due to wear and tear. The Contractor does guarantee the installation done by him or his Subcontractors as well as any damage to such equipment during construction.

It is the obligation of the Contractor when unpacking and installing items under the manufacturer's guarantee to preserve and present to the Owner before final payment, all such guarantees and books of instruction as to their installation and

We, the Subcontractor, hereby propose to furnish all of the material and perform all of the labor to complete a certain portion of a dwelling to be constructed at ______________________________

____

for ______________ the Owner whose address is ______________ all as described and called for on drawings, specifications and general conditions prepared by ______________.

The Subcontractor and the Owner agree that the materials to be furnished and the work to be done by the Subcontractor are as follows:

____

____

____

____

____

____

____

____

____

The Subcontractor agrees to commence his parts of the work without delay and in full cooperation with the Owner and with others when notified by phone or in writing.

The Owner agrees to pay this Subcontractor for the performance of his work the sum of ______________ $__________ subject to additions and deductions for changes that may be agreed on in writing and to make payments on account thereof in accordance with the terms of the General Conditions as follows: ______________________________

Accepted and signed ______________. Submitted and signed ______________.

______________ ______________

Owner Subcontractor

Fig. 22-1. A subcontractor's proposal and agreement.

No.________

Date________197 Job Location ________________

Owner____

Contractor ____

Change Proposed by___________________________________

Describe Change in Detail with drawings if necessary

Make in Duplicate

Cross out Price Basis not applicable

Pricing Basis; Agreed Price $ ____________
Cost plus 10% on Materials, 15% on Labor

____________________ ____________________

Accepted

____________________ ____________________

Contractor Owner

Fig. 22-2. A change order.

ROOFING	PAYMENTS DATE	V#	CONTRACT	PAYMENT	BALANCE
ADVANCE ROOFING CO			284 40		
214 STATE ST	7-12-77	127		200 —	84 40
PAINESVILLE, OHIO	8-6-77	156		84 40	—
584-7776					
BID 4-3-77 $284.40					
SIGNED BY JOHN KING					
SHINGLES SELECTED					
BIRD 235#					
COLOR - MEADOW					
WORK STARTED 7-1-77					

Fig. 22-3. A typical account card.

operation. Failure to present such guarantees to the Owner shall make the Contractor liable to the extent of the terms of the lost guarantee.

SUBCONTRACTORS

An agreement document for subcontractors is shown in Fig. 22-1.

The old adage "You can't make a silk purse out of a sow's ear" is true in the building business. It is very difficult to make an unqualified or unscrupulous contractor do a first-class building job, so whether you let a contract for a complete job or subcontract part or all of the work, you can avoid heartaches and wasted money if you know something of the reputation of the people you employ. Do not hesitate to ask any prospective contractor to refer you to some of his recent jobs.

If you act as your own general contractor, choose your subcontractors for their record of satisfactory work.

It is usually necessary to ask subcontractors to quote or bid on their particular part of the whole. It is well to have them understand exactly what part of the work they are being asked to do. In any case, unless they know you personally, you should inform them as to your ability to pay. A letter from your credit union or bank will do. A bid price is more likely to be low if the subcontractor knows he will be paid when his job is completed and approved.

Get every price quotation in writing. Usually a contractor will have his own agreement form which should be signed and submitted in duplicate. Your signature completes the agreement document, which becomes a legal contract enforceable by law.

After subcontracts are let, any changes should be noted in writing on a *change form* (Fig. 22-2). The change form should be signed in duplicate with the price stated if possible. Failure to have a proper change order can void the original agreement.

Maintain an orderly accounting system. It need be no more than a 3 × 5 file card headed by the subcontractors name, address, and type of work (Fig. 22-3). Enter the bid price at the top and list the payments and the date they are made. Pay only by check and list the check number. In these days of taxes and fees you may be required to present true if you hire direct labor, for you will be required to maintain Social Security records, payroll records, and tax records. You will probably be required to make payments to the state Workmen's Compensation Fund, and your payroll will probably be audited when your house is finished.

It saves a lot of paper work if you avoid direct employment, but you are obligated to see that each of your subcontractors is in compliance with all such laws. If he fails to pay either for the materials he uses on your job or the men who work on it, you could have a lien placed on your property. Employ established and competent contractors.

Appendices

Appendix A
Building Plan Checklist

Col. 1 Permit Plans
Col. 2 VA-HUD Plans
Col. 3 Complete Plans

PLOT PLAN

	1	2	3
1. Small but suitably scaled lot outline with bearings and property line dimensions. North point, scale and name of adjacent streets.	x	x	x
2. Setbacks at front, rear and sides. Location of all new and existing buildings.	x	x	x
3. Locate and specify all walks, driveway, parkings and their material. Easements, steps, terraces, patios, retaining walls, curbs, steep slopes, banks, ditches, power poles, underground power connections, swimming pools, covered or uncovered tanks.	x	x	x
4. Indicate grading on hillside lots, cut, fill and drainage.	x	x	x
5. Legal description of the lot. Lot number, tract or block number,			

	1 2 3
subdivision name. County, state, name of community and page number and volume number of record book if available.	x x x
6. Location and size of septic tank, seepage pit, and leach lines if required and permitted.	x x x
7. Buildings within 5 feet of property line on adjacent lots.	x x
8. Size, material and location of water and sewer lines. Invert elevation at point of connection.	x x
9. Existing and finish grade elevations at corners of building and of lot, centerline of street and curb elevations at extension of side property lines. Finish floor elevation.	x x
10. Pitch and direction of all slopes and drainage.	x x
11. Complete curb, pad, fence, retaining wall and garden wall construction with elevations of top of wall and top of footings.	x
12. Location of existing trees to be retained.	x

SLAB FLOOR FOUNDATION PLAN

1. Complete dimensions verified with floor plan.	x x x
2. Locate and specify porches, patios, garage, planters, piers, thickened slabs, depressed areas, steps.	x x x
3. Thickness and reinforcing of slab. Moisture barrier, soil poisoning, compacted fill, etc.	x x x
4. #5 vertical dowel bars if masonry walls. Foundation bolts if frame. Set bolts in 1 inch from corners and door openings and not to exceed 60 inch centers between.	x x x
5. Show #3 horizontal dowels 24 inch centers to tie porch slab to foundation.	x x x
6. Moisture proof membrane. Soil poison. Compact.	x x
7. Extend garage slab 2 inches in front of garage door for apron.	x x
8. Location, size and depth of underfloor ducts, outlets, tunnels, pits, etc.	x x

	1	2	3
9. Show elevation of stepped footings if any.		x	x

WOOD FLOOR FOUNDATION PLAN

	1	2	3
1. Complete dimensions verified with floor plan.	x	x	x
2. Locate and specify porches, patios, garage, planters, piers, steps, etc.	x	x	x
3. Show direction, size and spacing of all framing members, piers, girders, etc. Double joists under partitions. Solid blocking over girders. Bridging when joist span is over 10 feet.	x	x	x
4. Locate and dimension fireplace footing. To be 6″ wider all around than masonry, and 12″ into undisturbed soil. Show reinforcing #4 or #5 bars 8 inches c/c both ways.	x	x	x
5. Foundation bolts 1 foot in from corners and doors, and not to exceed 60 inches c/c between.	x	x	x
6. Location of vents and foundation access holes. Vent area not less than 1/150th of floor area screen with 1/4 inch galv. hardware cloth, 3 sides.	x	x	x
7. Galv. straps to posts where wind uplift is involved.	x	x	x
8. Extend garage slab 2 feet in front of garage door for apron.		x	x
9. Exterior piers not to exceed 3 times least dimension unless reinforced, interior piers 10 times least dimension unless reinforced.		x	x
10. Indicate elevation of finish floor, stepped footings,etc.		x	x
11. Show location of each foundation bolt, strap, etc.		x	x

FLOOR PLAN

	1	2	3
1. Name rooms.	x	x	x
2. Complete dimensions. Verify with foundation plan.	x	x	x
3. Size, direction and spacing of all framing.Ties every 48 inches if joists are not parallel to rafters.	x	x	x
4. Beam and post sizes. Connection details. Soffits and furred ceiling.	x	x	x

1 2 3

5. Doors, glass doors, garage doors, and windows type, size. Show at opening if possible or a schedule on same sheet as floor plan. Glass area 10% of floor area. Opening for ventilation 5% of floor area, LR, DR,BR. Both 5% for kitchen and baths. x x x

6. Show all plumbing fixtures to scale. Indicate disposal, water heater, laundry equipment, hose faucets, gas and water shutoffs. x x x

7. If no special electrical plan, indicate all switches, lights, plugs, main and branch panels, TV, phones, intercom, etc. x x x

8. If stairs, indicate rise, run and head room and hand rails. x x x

9. Show and name equipment to be included. Range, refrigerator, built-in equipment such as shelves, book case, desk, pullmans, etc.

10. Furnace, fixed heaters, air cond., boiler, burners, oil tank, etc. x x x

11. Vent for range thru roof or wall. Dryer vent. Combustion air for fired heaters. x x x

12. Access panel to tub plumbing. Access to attic min. 2 × 3 feet. x x x

13. If attic furnace, provide light with switch at access. Built-in ladder or steps to access. Provide means for replacement. Solid floor. x x x

14. Show garage framing, ties, bracing, permanent vent openings. x x x

15 One hour's fire resistive wall between garage and house. Solid core self-closing door garage to house where required. x x x

16. Room finish schedule showing floor, walls, and ceiling finishes. x

17. Completely engineered electrical plan for large houses, circuit layout, panel diagram. x

18. Complete separate plumbing diagram showing size and layout of all soil and vent lines. Water lines, gas lines, oil lines. x

EXTERIOR ELEVATIONS

1. Show at least 2 exterior elevations with all openings, wall finish materials, fascia, planter, veneer, exterior trim, flashing, exterior grade, floor grade and ceiling heights, roof pitch and material, garage, foundation and attic vents. x x x

	1	2	3
2. Show and specify hand rails, steps, posts, beams, ties, etc.	x	x	x
3. Show chimney, material, lining, flashing, saddle or cricket, cleanout doors etc.	x	x	x
4. Dimension all overhangs, ceiling heights, indicate type of windows, sliding doors etc.	x	x	x
5. Indicate diagonal bracing at corners unless sheath bracing for frame walls. Show #5 bars in concrete filled block at corners where reqd.	x	x	x
6. Air intakes, windows etc. must be 10 feet from any plumbing vent.	x	x	x
7. Show gutters, downspouts, deflectors, etc.		x	x
8. Indicate present and finish grade, depth of footings if not otherwise shown.		x	x
9. Sections and concealed sides not shown on exteriors elevations.			x
10. Door design and material. Window glass grade, plate, crystal, obscure, heat resistant, etc.			x
11. Indicate material pattern, texture, finish appearance of wood, brick, rock, etc.			x
12. Show section symbol and detail door and window heads, jambs, sills, etc.			x

CONSTRUCTION SECTIONS

	1	2	3
1. Draw to at least 1/2 inch scale typical wall sections from footing to roof and name size and material, pitch of roof, untypical constructions, fireplace, planter, steps, etc.	x	x	x
2. Fully dimensioned section through stairways showing headroom and details, railing, etc.		x	x
3. Details of septic tank and leaching system if required and permitted.		x	x
4. Show all structural details, piers, posts, girders, hangers, beam pockets, steel details if any, concrete reinforcing if any, all referenced to plans and elevations.			x
5. Show all interior, exterior, door and window details, casings, apron, cornice moldings interior and exterior trims.			x

CABINET DETAILS

1. No cabinet details required. x

2. Show details for kitchen, bath and other built-in cabinets at scale of 1/2 inch = 1 foot. x x

3. Show all appliances, built in equipment, hoods, accessories. Verify dimensions as to room space. x x

4. Show bathroom tile work if included. x x

5. Show cabinet sections and details, unusal cabinets, special hardware. x

SPECIFICATIONS

1. All concrete shall be at least 2500# mix for footings and 3000# mix for all other work or the equal if job mixed concrete is necessary. x

2. Step foundation where slop exceeds 1 1/2 to 1. Footing shall bear on solid undisturbed soil. Form if necessary. x

3. All framing lumber shall be grade marked by an approved grading agency. All horizontal lumber, joists, rafters, beams, girders, etc. shall be at least 1200f grade unless a stronger grade is required structurally. x

4. All lumber in contact with masonry shall be pressure treated. x

5. Wood siding, exterior stucco, brick veneer shall be placed over approved moisture proof building paper. x

6. Fireplace and chimney shall comply with governing code for the area, and shall not support any load other than its own. x

7. Flash every exterior opening with sheet metal or approved waterproof building paper.

8. No gas piping in or under concrete. x

9. Completely fill in specification form provided by the lending agency. x

Appendix B

Abbreviations

ac—On houseplans, particularly when they appear on a box with a diagonal slash through it, these letters stand for air-conditioning unit. If these letters appear on a wiring diagram or plan, they most likely stand for alternating current. If the wiring plan is for an area that includes an air conditioner, it is a toss up and you better ask for more information.

asph—This means asphalt. Used as topping for driveways at times, but more often makes its appearance as a saturating substance as in asphalt paper or felt.

bldg—Stands for building and only that.

BR—Most commonly used on houseplans to designate bedrooms. Can also stand for brick, as in brick veneer, fireplace, or hearth.

BRM—Space set aside to house brooms, mops, and other such in a broom closet.

BS—On spec. sheets, B.S. means bevel siding.

casmt—Stands for those nifty windows that open outward or inward, but not up and down as most others.

cem—Cement out of which you can make concrete.

CI—Cast iron as in railings or bannisters.

cl—These letters stand for closets.

clg—This means ceiling. It usually appears when a specific finishing material is indicated, as in a sound proof ceiling, or if there is a particular feature to the ceiling, as in a suspended ceiling.

clr—It means clear, as in clear glass or plastic.

CO—A cased opening.

conc—Here is the concrete.

cond—This can have three separate meanings: mean conductor, conduit, or condenser. When in doubt ask.

cop—Means copper and usually is used to indicate copper pipes.

corn—Hold the butter, this refers to a cornice.

crys—This refers to crystal (likely a chandelier).

CT—Crock tile is what it means, but the designation of crock tile itself is something of a conversation stopper. Crock means pottery vessel in one definition; in another it stands for broken pieces of same. So a crock tile floor could be a ceramic tile floor (possibly you have relatively low-fired tiles), or it could be a floor made out of broken pieces of tile in a mosaic sort of way. If you want to be sure, ask.

D—Designates a wonderful appliance—the dryer.
d—Originally it stood for price-per-hundred of nails. It has nothing to do with the price of nails today, but the hardware stores still have 2d nails, 4d nails, and so on. It now refers to size.
dc—On a wiring diagram, it will mean direct current. On an elevation, it most likely refers to drip cap.
DG—Drawn glass. Glass that has been heated and stretched.
diag—Diagonal either in measurement or the way something is applied to something else, as in wood trim applied diagonally.
diam—This is for diameter and refers to something circular.
dim—Dimensions.
div—Divided as in a divided bath.
dn—Down, as in stairs, usually accompanied by an arrow pointing in the right direction and by the word up, plus an arrow going the opposite way.
DR—Dining room.
dr—A door.
dr c—Drop cord. Usually seen on electrical diagrams.
drg—This abbreviation you mostly come across in those cryptical communications from your contractor or builder, the kind that goes like this: inld. 2 drgs for kit. cabnts. installation. Meaning two drawings, etc.
DRR—Usually means that little left over space between the closet and the bathroom that is grandly called the dressing room.
DS—Refers to downspout.
DW—The dishwasher.

ea—Each.
el—Refers to elevation.
ent—Entrance.
ext—Exterior, as in ext. trim.

fin—Finished.
fin ceil—Finished ceiling.
fl—The floor under your feet.
flash—The flashing on your roof.
FMR—The family room.
FP—Santa's entrance—the fireplace.
FT—Foot or feet.
ftg—The footing.

gar—The place where you park your car, theoretically, but which is usually filled up with all kinds of things.
GI—Galvanized iron.
gl—Glass, as in sliding glass doors.
gr—Grade.
gup bd—Gypsum board (also called plasterboard).

HP—High point. Usually refers to the roof.
ht—It means height.

in—Inch or inches.

kit—The kitchen.

L—Center line.
lav—Lavatory.
lin—Linen closet.
lndy—The laundry.
LP—This means low point.
LR—The living room.
lt—Light.

MBR—The master bedroom. Part of the master suite (as they say in the ads).
mldg—Refers to molding.
mull—Refers to mullion; the name given to the slender strips of wood or metal that divide the panes of glass in windows.

O—The letter o stands for built-in oven.
obs—Obscure.
OC—On center. Measured from center of one stud to the center of the next stud. Another of those strange but wonderful rites. Can also stand for outside casing, as in a door or window.

OD—Outside dimensions; as in the outside dimensions of a house. In contrast to the inside dimensions which are, of course, smaller.

OS—Outside.

pl—May stand for plaster or plate (part of your building components inside walls).

pl ht—It stands for plate height.

PWR—This is a quaint euphemism for half bath, meaning powder room.

R—If there's a circle anywhere near it means radius. On a floor plan in a kitchen, it stands for range.

ref'g—Usually found next to the R in the Kit. In case you're still not with us, the refrigerator.

rm—Rooms.

RO—This indicates a rough opening big enough to set in a whole door or window assembly.

RW—Redwood.

S—Sink.

scr—Means screen, as in scr. porch.

sdg—Siding.

specs—The specifications; plans and materials to build your house.

stor—Storage.

T & G—Tongue and groove (as in boards).

TC—Terra-cotta. Usually refers to a special kind of tile, low fired and red in color.

th—The threshold.

typ—Stands for typical (as in typical opening).

ven—Veneer. Like in brick veneer: BR Ven.

VT & G—Vertically grooved and tongued.

W—On a plan usually accompanied by D, it means washer. It can also stand for wide.

wc—Sometimes means wood casing.

WC—Water Closet.

wd—Wood.

WG—Wire glass (glass reinforced with wire).

WI—This stands for wrought iron. Like in a wrought iron railing.

wp—Waterproof.

WR—Wash room.

X—Marks the spot, usually on the plate, where the next stud goes.

yd—Yard. The measurement though it is sometimes used to designate the yard in which grass and flowers grow.

Glossary

absorption field—A system of trenches containing coarse aggregate which supports open-jointed or perforated pipe through which septic tank effluent may seep, or leach, into the surrounding soil. Same as *leaching field*.

across the grain—Perpendicular or at right angle to the grain of the wood.

adhesive—Any of a number of substances used to adhere objects. They come in various types and under many names. Some of the common ones are: glue, (white and otherwise), mastic, contact cement, panel adhesive, ceramic tile adhesive, and resilient tile adhesive.

AIA—The American Institute of Architects (1735 New York Avenue, N.W. Washington D.C. 20006).

air-dried lumber—Lumber that has been left to dry for a length of time.

airway—A space between roof insulation and roof boards; necessary for movement of air.

alligatoring—A cracking pattern in paint that resembles the expensive reptile hide (considered undesirable by most people). *See* checking.

anchor bolts—Steel bolts embedded in fresh concrete and protruding through holes in the sill. When the concrete is cured, the framing material is fastened to these anchor bolts and held down by nuts.

antihammer—A piece of water pipe installed in the line, about 1 foot long, one size larger than the line. The antihammer is capped at one end. The purpose of this bit of plumbing is to prevent hammering noises in the line.

apron—A piece of finish board. One type is a board located under a window that covers the joint between wall and sill. The other is as trim below a stool. *See* stool cap.

architect—Commonly used to describe a person who designs buildings and draws plans. It is also a legal term referring to a person who has been professionally licensed by one or more states to practice architecture.

asbestos board—A fire-resistant material made of asbestos fiber and portland cement.

asbestos joint runner—An asbestos jig with a metal clamp that is used to keep molten lead in place while caulking soil pipes in a horizontal position.

asbestos mastic—Sounds like an adhesive but it isn't. It is used to waterproof surfaces such as outside basement walls and such.

ash-pit cleanout—An opening with a metal door somewhere near the base of the chimney and used for removing ashes.

ash-pit dump—A metal trap door on the fireplace floor that allows ashes to fall into a container in the basement or through a pipe to the ash-pit cleanout outside the building.

asphalt—Technically, it is a residue of petroleum evaporation. It is used in roof shingles, in backing for exterior siding, and under finished floors.

asphalt roofing paper—Heavy paper used for insulating and waterproofing.

asphalt shingles—Composition roof shingles made of asphalt-impregnated felt covered with mineral granules.

backer boards—The boards that partially overhang the plate and run parallel with the ceiling joists. The boards provide the nailers for ceiling boards at opposite sides of the room.

backfill—The replacement of excavated earth back into a trench or around a building foundation.

balustrade—A porch, roof, or stairway railing that is made of a top rail baluster and often a bottom rail.

balusters—Vertical parts, often turned, or 1 1/4-×-1 1/4-inch connecting top and bottom rails of banister.

band—A fancy piece of millwork that is usually used as molding.

bargeboard—A finish board on the face side of the gable framing.

baseboard—Finish board nailed onto the wall at floor level to hide the joint between the two. Also called simply *base*.

basement—An area that is below grade level. Usually below the entire first floor of a building and accessible from inside the house.

base molding—The fancy band that goes on some baseboards as decoration or potential dust catcher (it depends on your viewpoint).

base shoe—Quarter-round or other molding nailed to the baseboard at rug or floor level.

batter board—Horizontal boards nailed to posts and placed just outside the corners of a building so strings can be stretched from corner to corner to indicate the outlines of a foundation.

batten—Narrow strips of finish or hardboard used to cover panel joints on exterior walls to get a finished appearance.

batten boards—A rough pair of boards nailed to stakes at right angles to each other. Set at the corners of excavations to show the first floor level and also to serve as supports for different guidelines used before framing in a building.

bay—The distance between two principal column lines.

beam—A structural member supporting a load along its length.

bearing plate—A steel plate that provides support for a structural member.

bearing posts—Partial supports of the girders of a building.

bearing seats—Recesses in the foundation walls for partial support of girders.

bearing walls—A wall or partition that supports another load in addition to its own weight. That includes ceiling joists, floors, and roof.

bed molding—Applied between any vertical and horizontal surface. For example, between eaves and an exterior wall. *See also* Cornice trim.

bell tile—The sectional tile with one bell-shaped end is used to lay ditches that carry waste water to a lower level or to the sewer.

bell wire—Small-gauge copper wire you use in wiring your doorbells or chimes.

benchmark—A surveyors' term meaning a known point of elevation above sea level. From this, the meaning has been expanded to include any known or given measure on which one can base other measures. A certain level of performance or excellence which has to be met or duplicated.

bevel cut—A board or timber cut or shaped to a desired angle.

beveled siding—A special type of siding that is made by sawing boards on the diagonal and so producing two wedge-shaped boards.

bird's mouth—Carpenterese for the notch sawn into rafters to allow full seating on the plate of the wall.

blind nailing—Nailing through wood in such a way that the nail head won't show. You can do that with tongue and groove boards, for instance.

blueprint—Originally the copy of a drawing. Usually a tracing of the working plans which are drawn in pen or pencil. Now used synonymously with plans or working drawings.

board and batten—Vertical siding made of boards with batten covering the joints.

board foot—The waterloo of many a do-it-yourselfer. Actually a measurement that denotes a given size of a piece of lumber that is a board 1 inch thick, 12 inches long and 12 inches wide is said to be a board foot. A 1 × 12 a foot long contains one board foot. A 2 × 12 6 inches long also contains a board foot.

boot—A sheet-metal receptacle that makes a connection between some ductwork and a register. An example is between the air-supply duct and the register in your air-conditioning system.

Boston ridge—Overlaying and blind-nailing shingles on the ridge of a roof.

boxed cornice—A decorative boxy overhang on a building.

box gutter—A box-shaped gutter.

box sill—A built-up sill on the foundation that has the sole plate resting on the floor joists and not the sill.

brace—A board or a stud set at an angle to add support and strength.

brad—A small, thin nail with a deep head. It is used when a finishing nail or a casing nail would be too large.

branch circuit—Any of the electrical circuits that are connected to the service panel.

branch drain—A piece of pipe that is the drain line for only one fixture.

branch water line—A water line that serves only one fixture.

bridging—A technique for stiffening floor joists, ceiling joists, and partitions in which small wooden braces are put between the larger wooden pieces on the diagonal. Sometimes wood of the same size is used and is placed at right angles. It must fit tightly between joists or partitions.

brick veneer—A facing of brick fastened to the sheathing of a frame or other wall construction.

buck—A frame for a door.

building code—That body of rules and regulation covering the various phases of building and construction work you better know about before you make your final plans.

building drain—The piece of drain line from the foot of the stack to the sanitary sewer.

building lines—Refers to the outside and inside edges of the foundation. Usually put in solid on building plans.

building paper—Heavy asphalt-saturated paper that is put over sheathing or over subflooring.

built-up roof—A roof composed of from three to five layers of asphalt felt laminated together with coal tar, pitch, or asphalt. The top is finished with a layer of crushed slag or stone embedded in hot tar.

bushing—A cylindrical lining that you screw tightly against the end of a conduit to protect insulated wires.

butt—To put boards together in such a way that all points of the edges, or ends, touch.

butt hinge—Usually a door hinge that has a loose pin that holds two metal leaves together. One leaf is fastened to the jamb and the other to the edge of the door.

butt joint—When two pieces of wood butt together, end to end or at right angles. The edges of each piece are square.

cantilever—A beam, slab, or portion of a building that projects out or overhangs beyond a vertical support.

cap—The upper member of a column, pilaster, door cornice, molding, etc.

capillary attraction—The inside of a copper fitting that attracts melted solder and so makes a sealed joint between pipe and fitting.

casement windows—Windows hinged on one side that open either in or out (usually out).

casing—Trim for a door, window, or other framed opening attached to both the jamb and the wall.

casing nail—Similar to a finishing nail, but slightly smaller in gauge and with a slightly tapered head.

cat—Reinforcing or blocking between joists, studs, or rafters to assure a solid backing for nails or screws. Usually 2 × 4s nailed horizontally between studs to act as nailers for wood-board paneling.

caulking—Using soft, pliable material to seal seams, joints, and cracks for waterproofing and weatherproofing.

caulking compound—Used to make seams, cracks or joints watertight.

caulking gun—A metal frame for holding and applying tubes of caulking compound.

ceiling joists—These members are fastened to the top plates and run across the narrow span of a house. They are called the main ceiling joists. Short joists running at right angles from the last ceiling joist to a top plate are referred to as stub ceiling joists.

ceiling nailer—Usually a 2-inch strip of wood nailed to the top plate with an overhang for ceiling material to be nailed to eventually.

ceiling tile—Square or rectangular blocks. Usually interlocking with a tongue and groove edge and some sound- deadening characteristics.

cement—A powder of calcined rock or stone used in mixing concrete.

central vacuum system—The system has the tank (and noise) located in the basement along with the electric power unit. There are outlets in each room for a plug-in hose that uses the normal array of attachments. If your house doesn't have a basement, the power unit and tank can be located in a garage or utility room.

chalk line—This is an important gadget in the building trade to make sure things are plumb and true. Actually all it is is a piece of cord covered with chalk that is napped against a wall or over lumber to form a line. Incidentally that's where the expression of walking the chalk line originates.

chase—A vertical space in a building wall for ducts, pipes, or wiring.

checking—Cracks in a paint job that look like small squares. *See* alligatoring.

checkrails—The joint of the bottom of the top window sash and the top of the bottom window sash beveled to make an airtight joint.

cheek cut—A side cut at the end on both sides of rafters.

chimney—A vertical tube with a flue that carries the smoke and gases away from an open flame or fire.

clapboards—As on New England houses, a siding made out of beveled boards that overlap horizontally.

cleanout—A removable piece of pipe in a line, usually right by a drain, that allows cleaning out a stopped-up drain pipe.

cleat—A wedge of wood that's nailed on to serve as a check (as at the end of a drawer) or support.

cold chisel—A wedge-shaped steel tool with a cutting edge used for cutting or shaping metal without heating the metal to soften it.

collar beam—Boards, usually 1 or 2 inches thick, that connect opposite roof rafters. They are used to keep the rafters from spreading apart and to keep the roof from sagging under extra heavy weights such as snow.

column—A vertical support piece. Also called a pillar or post.

common rafter—A rafter that goes at a right angle from the ridge to the plate.

concrete—A mix of cement, sand, water and rocks (aggregate) that hardens to rock-like consistency.

concrete blocks—Blocks made out of concrete and used for building purposes.

concrete nail—A nail made to fasten wood to concrete (all others won't do).

conduit—A pipe or tube, usually metal, through which electrical wire is carried.

contemporary—Any modern house which doesn't follow any special traditional or special style, but sets its own character through the juxtaposition of materials and structural features.

coped joint—A joint in which the end of one piece is cut to fit the surface of another. For instance, a piece of molding cut to fit against the face of a second piece of such molding.

corner boards—Trim for the external corners of a house.

corner post—Three full-length studs and three spacer blocks (12 to 16 inches long) situated at the corners of wall or partition frames.

cornice—The top of a wall directly under the eaves. Also used more loosely for a board or molding that runs along the top of the wall at ceiling level.

counterflashing—A flashing used on chimneys at the roof line to cover shingle flashing and prevent the entrance of moisture.

coupling—A pipe fitting which joins two pipes.

cove—A wood molding that is used in interior corners or a plastic or ceramic piece. Also concave piece that fits between a countertop and the backsplash. Also used to top a ceramic tile wall that doesn't reach all the way to the ceiling.

crawl space—The area under the floor joists and the ground in a house that has no basement and isn't built on a slab.

cricket—A small sloped rain-diverting structure placed at the junction of surfaces that meet at an angle. Also called saddle.

crimped pipe end—A pipe with small, regular creases pressed into one end.

cripple jack rafter—A rafter that extends from a hip to a valley.

cripple studs—Short studs under or over openings.

cross cutting—Cutting lumber across the grain. The opposite of ripping boards.

cross members—Framing material fastened at right angles to other framing material (like 2- × -4 pieces nailed between studs).

crown—The highest point or the top of a building.

crushed stone—Bits of rock that are too big to go through a 1/4-inch screen.

curing concrete—A way to treat fresh concrete by sprinkling with water at regular intervals.

dad—A groove cut across a board.

deck paint—A wear-resistant enamel designed for use on floors.

deep-seal trap—A neat little device in the main drain pipe designed to prevent sewage from backing up and flooding the basement.

diagonal thickness—Measuring lumber diagonally instead of the usual vertical or horizontal way.

differential settlement—Uneven settlement of a building foundation due to nonuniform earth support.

dimension lumber—Lumber that is 2 to 5 inches thick and from 2 to 12 inches wide, in 2-inch increments.

direct nailing—What you do when you pound in a nail with your hammer and the nail head shows. Also called face nailing.

disconnect switch—A special switch wired into a circuit so that the circuit and the appliance or device connected to it can be disconnected when needed.

dormer—A window set in the framework of a gable-type intersecting roof.

double framing—Used to add strength such as around a stairway.

double-hung windows—Two-window sash where each window can slide up and down in the window frame, passing each other in their separate grooves.

downspout—The tube or pipe that drains the water from the gutter to the ground.

dressed lumber—No coverings added. Actually, the rough lumber is planed down, exposed if you will. That's a 2 × 4 never measures more than 1 1/2 × 3 1/2 as a dressed 2 × 4.

drip—A projecting construction member or exterior finish course for draining off water.

drip cap—A molding at the exterior top of a door or window frame that keeps water from flowing down the face of the frame.

dry wall—Plasterboard used as an indoor wall covering. Comes in sheets. Has replaced the old lathe-and-plaster routine.

dry well—A hole in the ground usually filled with gravel for draining water from gutters.

duct—A shaft, usually metal, used to transmit heating or cooling.

duplex outlet—An electrical wall outlet with two plug receptacles.

easement—The right held by a person or company or government for the use of land belonging to someone else for a single, specific purpose—like setting telephone poles.

eaves—That part of the roof that overhangs the wall.

edge pull—A door pull set flush on the edge of a sliding door.

elbow—Not the thing that gets sore in tennis. A pipe or drain fitting with a 90 degree angle.

elevation—A straight-on view of an object. Usually exterior or interior walls of a house drawn to scale and represented in that manner.

ell—An extension or wing of a building constructed at right angles to the main structure.

end cap—A metal cap that goes at one end of a gutter.

end grain—The grain at the ends of boards.

entourage—An indication of plants on a floor plan, elevation, or perspective drawing.

escutcheon—The piece of metal, usually fancied up a bit, that you see around faucets and pipes, or on both sides of a door lock set, or other such.

excavation—Hole in the ground for your foundation, walls, or footings.

excavation lies—Usually dotted lines shown on building plans that indicated the outside and inside edges of footings or foundation walls.

expansion joint—An asphalt impregnated strip of fiber put in grooves made in concrete to prevent the concrete surface from cracking.

exterior finish—Finishing materials such as siding, clapboard shingles, and so forth.

face brick—Brick of good quality used on the face of a building.

face of rafter—The top side of same indicated by a bevel cut at the end.

face shells—The outer part of a concrete block.

fascia—Finish board covering the ends of roof rafters.

fascia bracket holder—It is a hanger attached to the fascia and holds gutters in place.

felt—Tarpaper used under floors, siding, and roofing materials.

ferrule—A sleeve, usually metal, placed around a tool handle to keep it from splitting.

fieldstone—Natural stone used in construction of foundation and retaining walls. Also used for paths and stepping stones in yards.

field tile—Short lengths of drain with 1/4-inch gaps between them.

filler box—A box form of wood the size of the proposed window or door used in building finished concrete foundation walls.

finish—In painting, the final coat applied of paint, varnish or wax. In carpentry, putting up the exterior trim, interior trim, countersinking nail holes, or final sanding.

finish hardware—The hardware such as knobs, locks and hinges.

finishing nail—A small-headed nail used in wood when countersinking is planned.

firebrick—Brick made from fireclay that is used to line fireplaces.

fireclay—Refractory clay used for bricks and for the mortar that holds bricks together. Refractory refers resistant to heat.

fireplace unit—A prefabricated metal box that lines the inside or your fireplace opening.

fire-retardant wood—Wood treated by an impregnation process so as to reduce its combustibility.

fire-stop—Fire-resistant insulation used to prevent the spread of fire.

five-quarter—A board cut to 1 1/4 inches thick and dressed to 1 1/8 inches.

flagstone—Flat stones used for floors, pathways and sidewalks.

flashing—Strips of nonrusting metals put into both sides of a joint to prevent water from seeping in.

float—A small, flat tool used by concrete finishers.

floating—Smoothing the surface of freshly poured concrete to a rough finish.

flue—The space in a chimney through which smoke, gas, or flames ascend.

flush joint—To be exact; a joint flush with a

masonry surface.

fly rafter—The end rafter of a gabled roof that overhangs the main wall and is supported by lookouts and roof sheathing.

footing—The base on which the foundation of a building rests. Usually twice the width of the thickness of the foundation wall. Can also support pier and beams and other types of pillars or posts.

footing pads—Single footings that provide bases for bearing pillars.

forms—Wooden structures used in pouring concrete to hold it in place until it has hardened.

foundation—Construction below or partly below grade, including the footings, that supports the first floor and structure above.

four-way switch—Used with two three-way switches so that a light or an appliance can be turned on or off from three different locations.

framing—Balloon framing in which wall studs extend from foundation sill to roof line and to which the floor joists are attached. Not used much in these days. Platform framing in which each floor is built separately and serves as a foundation for the walls. That's what we use most today in wood-frame buildings.

framing square—A measuring and leveling tool.

frieze board—A finish board at the top of the wall directly underneath the eaves.

frost line—The depth to which the ground freezes in the winter. The footing must go below the frost line to prevent heaving and movement during freezing and thawing.

furring—A thin wooden framework used to thicken a wall or ceiling or, most commonly, to provide a nailing surface for paneling or tiles.

furring strips—Wood or metal strips fastened to wall, ceiling or other surface to serve as nailers or to provide an air space.

gable—The vertical triangle end of a wall from the eaves to the roof ridge.

gable roof—A roof line at the end of a double sloped roof. Forms a triangle from the peak of the roof to the bottom of each rafter.

gang box—Two or more electric outlet boxes mounted side by side.

gas vent—A chimney designed for the use of gas-burning appliances only.

girder—Horizontal beam carrying one or more intermediate beams. Heavy member, either single, built-up, or iron I beam, used to support heavy loads such as joists or walls over an opening.

glazing—The placing of glass in windows and doors.

grade—The slope of the ground in regard to a specific reference point, usually the top of the foundation or the ground-floor elevation.

gradient—The slope of a surface, road, or pipe.

grain—Direction of fibers running in wood.

grout—A kind of mortar, with or without sand, that is used to fill in between wall tiles. Floor tiles are sometimes set directly into a grout bed and then the spaces in between the tiles are filled in with more grout.

gusset—A plywood or metal plate used to strengthen the joints of a truss.

gutter—A metal trough that carries rain water from the roof to the downspouts.

gypsum board—Also called Sheetrock or plasterboard.

gypsum board sheathing—Panels with a core of gypsum between two covers of water-repellent paper that are used as sheathing.

hanger—A metal strip used to support pipe or the ends of joists.

hardboard—A man-made material made from wood and similar in most aspects to wood. Comes in sheets or panels, usually 4 × 8 feet and in thicknesses of 1/4 of an inch.

header—A wooden beam in a floor or roof that is between two long beams and is supporting the ends of one or more intermediate beams. A beam placed as a lintel over a door or window opening. A beam placed at right angles to joists to form openings—such as skylights, stairwells.

header joist—A floor joist connecting the ends of regular floor joists and forming part of the edge of the floor framing (the opposite of a stringer joist).

hearth—The floor of the fireplace that extends into the room.

hip—The external angle formed by the meeting of two sloping sides of a roof.

hip roof—A roof that is sloped up from four sides, one slope to a wall of the house, that has a short ridge.

hollow-core door—A door that has an air space and some filler between the two outside surfaces.

house sewer—The watertight waste pipe extending from street sewer to a house foundation.

hub—The end of a soil pipe.

I-beam—Steel beam used to support joists running long distances or also used as an extra long header over windows or doors.

insulation—Sound: material made mostly of fiberglass that reduces the transmission of sound through walls, floors, and ceilings. Thermal: material made from fiberglass, mineral wool urethane, or styrene or other similar material that prevents heat loss when placed between exterior and interior walls, in attics, between roof rafters, and under floors.

intersecting roof—One roof passes through another roof seemingly as a gable roof intersecting a hip roof.

interior finish—Materials used to finish interior walls: plasterboard, paneling, wallpaper and wallcloth.

jack rafter—A rafter that spans the distance from a wall plate to a hip, or from a valley to a ridge.

jamb—The two sides and top members of a door or window frame.

joint compound—A plaster-like material, that has glue in it, used to fill in the spaces between plasterboard panels and also the nail head on the panels.

joint gauge—A board with uniform markings to help in laying bricks or concrete blocks with equal-size joints.

joist—One of a number of parallel beams that are supported by girders or bearing walls.

joist hangers—Metal fasteners to secure joist ends directly to a girder or another joist.

keyway—A V-shaped groove cut into the center of the basement wall footing to provide a way of bonding the wall to the footing.

kiln-dried lumber—Lumber that has been dried and cured to produce a better grade lumber than when air dried.

lally column—A concrete-filled pipe column.

landing—A platform that divides a flight of stairs into two sections. This can be a straight stairway or one that makes a right-angle turn.

lap joint—A joint in which two pieces of wood overlap so that they form a single surface on both sides.

lath—Wood, metal, or plasterboard used as a base for plaster.

lath stripping—Thin, narrow strips that hold edges of building paper temporarily in place on the roof decking.

lattice—Framework with crossed wooden or metal strips and air spaces in between.

layout—The arrangement of rooms, the plan of a house, a story or wing of a house.

ledger board—Board on a wall that is to receive one end of the lookouts.

ledger strip—A strip of lumber fastened to the lower part of a beam or girder to which notched joists are attached.

level—A tool to test if a surface is even.

light weight concrete—Concrete having a density of not over 115 pounds per cubic foot.

linear foot—A one-foot measurement along a straight line.

line length—The distance between the center of a ridge and the end of a rafter.

line voltage—The number of volts supplied by the service wires that come from the utility company's power supply. For instance 115 volts to have current for the lights in the house.

lintel—A horizontal support over a door or window opening.

load-bearing wall—A wall designed to support a weight placed on it from above.

lookouts—Short pieces of 2 × 4s set between the lower end of rafters and the ledger board on the side of the wall, set in horizontally.

louver—An opening, usually with a screen, that has slats allowing air passage in and out. Used for ventilation in attics, basements, crawl spaces, and pantry closets.

low-density concrete—Lightweight concrete with a density of 50 pounds per cubic foot or less and a compressive strength of less than 500 pounds per square inch.

lumber (matched)—Lumber that is dressed and shaped on one edge in a grooved pattern, and a tongued pattern on the other edge.

lumber (shiplap)—Lumber that is dress edged to make a close rabbet or lapped joint.

mantel—The shelf above the fireplace. Often includes the trim (wooden or tile), around the fireplace opening.

mastic—A thick flexible sealant or adhesive.

matched lumber—Tongue-and-groove boards.

millwork—All house fixtures made of finished wood in millwork plants. Includes doors, windows and their frames, blinds, porchwork, mantels, panelwork, stairways, moldings, interior trim, and cabinets.

miter box—The gadget used to make miter cuts without tears.

miter joint—A joint made of two pieces of wood each cut to a 45-degree angle that forms a regular right-angle corner.

modern—In architects' jargon, a building erected with current skills and materials. Usually synonymous with contemporary, but can stand for a modern reproduction of, for example, a salt box.

modernistic—Architectese meaning imitation contemporary.

modular concrete blocks—Concrete blocks of one standard size.

molding—Decorative strips of wood used in trim and detail work inside and outside.

monolithic—Concrete construction poured in one piece without joints.

mortise—A hole cut into a piece of wood into which another piece of wood is placed for an exact fit.

mullion—A post frame or double jamb that vertically divides two windows or large panes of fixed glass.

muntin—The horizontal strips dividing windowpanes.

natural finish—A transparent finish which does not seriously alter the original grain and color of the natural wood.

newel—A post to which the end of a stairs railing or bulustrade is fastened.

nominal size—The commercial size of lumber or other material. Nominal size is not the actual size. For example, a 2 × 4 nominal is actually 1 1/2 × 3 1/2 inches.

nonbearing wall—A wall supporting no load other than its own weight.

nosing—Projecting molding. Especially the projecting part of the tread over the riser on stairs.

ogee—A molding with a profile in the form of the letter S.

on center—Used in the spacing of studs and joists. Means measuring from center of stud or joist to the center of next of the same. Standard is 16 inches, but 24 inches is sometimes used.

orthographic view—A single-plane view of an object such as a floor plan or an elevation.

outrigger—An extension of a rafter beyond the wall line.

panel—A thin piece of wood fitted into grooves as in door panels. Thin wood used to finish walls.

paper sheathing—A building material, generally paper or felt, used in wall and roof construction as a protection against the passage of air and moisture.

parge coat—A thin coat of cement plaster applied to a masonry wall for waterproofing or for improving appearance.

particle board—A sheet made by gluing wood chips or particles together under pressure. This comes in various thicknesses and sizes. Once only used for underlay. With the escalating prices of wood, it has become a paneling material as well as a regular building component for shelves and such.

partition—A wall that divides spaces within any story of a building.

partition junction stud—A triple stud that forms an inside corner on both sides of a partition frame.

pegboard—Hardboard with holes. Good when you need ventilation or want to hang up things.

penny—A measurement of nails that formerly indicated price per hundred. Now used as part of a designation for size of nail.

pervious soil—Loose, porous, coarse-grained soil that allows the penetration of water.

pier—A column of masonry, usually heavy, that is used to support a structure.

pier and beam—A type of foundation consisting of a number of piers to which beams are attached. On the beams is built the floor for the first floor. The beams raise the structure for 2 to 4 feet above ground level and the space is left open and only skirted at the walls with siding or trellises.

pilaster—An upright column that's rectangular and is part of a wall.

pitch—A drawing or diagram, usually to scale, that shows the arrangements of the rooms of a house. Also, a drawing of a building seen from above without the roof.

plaster—A lime, sand and gypsum mixture troweled onto lath in several layers to be used as a finished wall.

plasterboard—A type of board used as a substitute for plaster.

plate—A floor plate or sole plate is the bottom horizontal member of a stud wall that is attached to the slab or the subfloor. A top plate is the top horizontal member that supports the roof rafters or the second floor joists.

plenum—A main air duct that either carries cool air from the blower coil of the air-conditioning unit to the cold-air ducts or carries warm air from the branch warm-air duct to the blower coil.

plumb—Exactly vertical or perpendicular.

plywood—Wood made by laminating thin sheets together with the grain running in opposite directions. Comes in various sizes, thicknesses and grades.

polyethylene—Plastic sheeting usd as a vapor barrier. Also used to cover building materials to protect them from the weather during construction of a building.

precast concrete—Plain or reinforced concrete elements cast in other than their final position in a structure.

preservative—A copper- or pentachlorophenol-base liquid that when painted on wood, will keep wood from rotting, at least for a while.

primer—First coat of paint job.

purlin—An intermediate supporting member in a roof at right angles to rafter or truss framing.

quarter-round—A three-sided molding with a curved face used around the bottom of baseboards. A molding one-quarter of a dowel.

quoins—Large squared stones set in the corners of a masonry building for appearance sake.

rabbet—A groove at the end of a board going across the grain cut to receive a second piece of wood.

rafter—A beam, usually about 2 inches thick, that helps support the roof.

rafter tail—A section of a rafter between the plate and the lower end of the rafter.

rail—Horizontal frame piece of a window or paneled door. Also the upper and lower pieces of a balustrade or railing.

rake—Sloped edge of a gabled roof.

random rubble—Stonework having irregular shaped units and no indication of systematic coursing.

ready-mix concrete—Concrete mixed by the experts in special trucks and brought to the site.

reinforcement—Steel rods or mesh placed in concrete to give it added strength.

rendering—A pictorial representation or perspective of a building or some part of one.

resin nail—A nail that is coated to give it greater holding power.

ridge—The horitonal line at the junction of two sloping roof surfaces.

ridge board—A board between the uppermost ends of common rafters spiked together from opposite sides.

ripping—Sawing lumber with the grain.

rise—On stairs, the vertical height of the flight or also the height of one step. The number of inches a roof rises for every foot of run.

riser—The vertical boards between the treads of a stairway.

roof decking—Boards or sheets of plywood that covers the rafters and serve as a nailing surface for shingles or other roof cover.

roofing felt—Asphalt paper fastened to the roof sheathing before applying the shingles.

roll roofing—Roofing material composed of felt saturated with asphalt. It may be coated with mineral granules.

roof sheathing—The boards or sheet material to which the shingles or other roof covering is attached.

rough—Bare framing.

rough fascia—A board nailed to the bottom ends of rafters.

rough hardware—The concealed hardware in a building, such as nails, bolts, hangers, etc.

rough-in measurements—Measurements showing the size and locations of plumbing fixtures and other built-in fixtures.

rough opening—Opening in a frame wall for a door or window, in a floor for stairs or chimney, or in a roof for a skylight.

rubbing brick—A brick especially made to give concrete surfaces a finish.

run—The horizontal dimension of a stair.

saddle—The threshold of a door. A metal, wood, or marble tread running the full width of the opening about 5/8 of an inch high.

sash—The framework that holds the glass in a window frame. A single unit of the frame containing one or more lights of glass.

saturated felt—Felt which is impregnated with tar or asphalt.

scaffold—A temporary structure with a top platform that is used by the workers during building.

scale—A reference standard in measurement. Proportions relating a representation to actual size. Also, a calibrated line that shows such a relationship.

scale drawing—A rendering of a building or parts thereof in precisely reduced proportions. For instance, a scale of 1/4 inch could represent 1 foot.

scratch coat—The first coat of plaster which is scratched to provide a bond for a second coat.

screed—A guide used in smoothing freshly poured concrete.

scribe—Marking with a pointed instrument as a guide for sawing.

sealer—A liquid that seals the wood as a base for paint, varnish, or other finish.

sealing stain—A handy sealer/stain that makes a one step operation out of two.

shake—A thick, wood shingle you can either use on exterior walls or the roof.

sheathing—The rough covering or boarding over the frame of a house.

shed roof—A flat roof with a slight, single slope.

shims—A strip of wood or metal, or a wedge of same, that will fill in a small space. Shingles are often used for that job around windows, doors, and the finished job.

shingles—Roofing material made of tapered wood, metal, slate, or asbestos. They come in bundles or in strips and are applied in an overlapping fashion.

siding—The exterior finish of a wall. Such things as clapboard, shingle, board and batten, aluminum, and vinyl.

silicone sealer—A liquid that will waterproof masonry above the ground level.

sill—The bottom of a window or door.

sill plate—That part of the side wall that sits flat on the foundation.

slab—A concrete section ofen used as a foundation for a house.

sleeper—A board 2 inches thick that fastened to a concrete floor as the base for the wood floor above it.

slump—The degree of wetness of concrete related to workability.

solid bridging—A solid member placed between adjacent floor joists to prevent the joists from twisting.

soffit—Usually the horizontal surface between the end of the rafters and the outside wall of the house. The underside of a cornice or boxed eaves. The meaning has been extended to include that space between your top cabinets and the ceiling in your kitchen which is usually boxed in.

soil stack—Usually abbreviated to stack. A vertical pipe that extends through the roof and vents sewer gases to the outside.

soil tests—Very important sampling of the soil at different depths to see what kind of footings are necessary for a building.

sole plate—The horizontal timber in a frame to which the outside wall studs and partition studs are attached and on which they rest.

solid-core door—A door that has softwood blocks that fill up all the spaces between the two surfaces of the door.

span—The horizontal distance between supports for joists, beams, girders, or trusses.

specifications—Also known as specs. A document usually prepared by the architect, but can be done by the person who designed the building, the owner, or the contractor. Details about the proposed building that can not be included in the working drawings. Examples are hardware for cabinets, style of trim, light fixtures of a specific kind, and so forth.

splash guard—Material placed under the downspout to keep the topsoil from washing away.

starter strip—A metal strip that is used to fasten the bottom edge of roof sheathing. A narrow strip of rolled roofing that is fastened to the lower end of roof sheathing before putting down roof shingles.

stool—The ledge under a window frame.

stop—A molding on a window sash or sliding door frame that stops the window or door from going any further.

storm sash—Insulating windows made of wood or aluminum (wood is preferred) fitted over the outside of the windows of a house. Sometimes applied inside.

straightedge—A board with true edges to draw straight lines.

stringers—The side pieces that form the frame for stairs.

stucco—A siding material, made with cement as a base, that is applied to the outside walls over a metal lath.

studding—Upright framing in walls.

studs—Upright timber, usually 2×4s used in framing. Sometimes metal studs are used.

subfloor—Rough floor made out of planks or plywood nailed to the floor joist. The base for the finished flooring.

suspended ceiling—A ceiling that hangs from the joists by brackets or wires.

tail beam—A short joist supported by a wall on one side and a header on the other, usually around fireplaces.

template—A pattern usually made out of thin wood or metal. It can be cut out of cardboard or heavy paper. Used as a guide for cutting.

termite shield—A shield, usually a noncorrodible metal, placed in or on a foundation or around pipes to prevent the passage of termites.

terrazzo—Finish flooring made of marble chips or small stones embedded in cement.

thickened edge slab—A type of concrete floor slab that is constructed integrally (in one piece) with the foundation.

threshold—The bottom part or sill of a doorway.

tie rods—Special steel rods that hold the concrete forms together.

toenail—A technique for fastening lumber by nailing it at an angle. The nail goes in at an angle.

tongue-and-groove—A very special joint in which one board has a groove along a long edge and the other has the edge of one long side so trimmed that it fits into the groove.

top plate—The horizontal member of a house frame at the top of the studs.

trap—A U-shaped pipe below plumbing fixtures

that creates a water seal and prevents sewer odors from entering habitable rooms.

trim—Finishing material. Also called millwork or woodwork.

trimmer—A beam or joist to which a header is nailed in framing a chimney, stairway, or other opening.

truss—A set of rafters with a collar beam and other connecting lumber in place. They are prebuilt and ready to install to connect opposite walls.

underlayment—Plywood, particleboard, or hardboard installed over the subfloor as a base for floor tile or sheet flooring.

unit rise—The rise, in inches, that the common rafter extends in a vertical direction for every foot of unit run.

valley—The internal angle formed by the junction of two sloping surfaces of a roof.

valley rafter—A diagonal rafter at the junction of intersecting roof slopes.

vapor barrier—A watertight material used to prevent the passage of moisture or water vapor into walls or elsewhere.

veneer—Thin sheets of wood, applied over cheaper, coarser wood for a finished panel or in furniture.

vent pipe—A pipe used to release sewer gas above the roof line and to prevent back pressure and siphonage.

vent stack—A vertical soil pipe connected to the drainage system to allow ventilation and to prevent a vacuum from drawing the water seal from plumbing traps.

walers—Timbers that help hold concrete forms true and straight.

wall, bearing—A wall which supports any vertical load in addition to its own weight.

wall, nonbearing—A wall which supports no vertical load other than its own weight.

wall plate—The covering that hides the inside of your switch or electrical outlet.

water table—A horizontal member or molding extending from the surface of an exterior wall so as to shed rainwater from the wall.

weather strip—A narrow piece of wood, plastic, or some other material placed between a door or window and its frame to act as insulation.

weep hole—Small holes in masonry cavity walls that release any accumulation of inside water.

window unit—Window jambs, outside casings, trim, and window mounted into sash as a unit ready to put into your rough window opening.

wiring diagram—A drawing or plan that explains a wiring system such as the one for your electric stove or your air-conditioning system.

working drawings—Those all-important drawings that show exactly how a building shall be put together.

wrecking bar—A tool that one uses if the workmen do not follow working drawings.

wythe—A single-width masonry wall.

Bibliography

Allen, Edw. and Goldberg, Gale Beth *Teach Yourself to Build*. Cambridge: MIT Press, 1979.

Armstrong, Leslie *The Little House*. New York: Macmillan Publishing Co., Inc., 1979.

Baker, John M. *How to Build a House without an Architect*. New York: J.B. Lippincott Company, 1977.

D'Amelio, Joseph *Perspective Drawing Handbook*. New York: Tudor Publishing Co. 1964.

Dazell, J. Ralph *Plan Reading for Home Builders*. New York: McGraw Hill Book Co., 1972.

DiDonno, Lupe and Sperling, Phyllis *How to Design and Build Your Own House*. New York: Alfred A. Knopf, 1978.

Doblin, Hay *Perspective—A New System for Designers*. New York: Whitney Publications, Inc., 1956.

Eccli, Eugene *Low Cost Energy Efficient Shelter*. Emmanus, Pennsylvania: Rodale Press, 1976.

Glegg, Gordon L. *Making and Interpreting Mechanical Drawings*. Cambridge: Cambridge University Press, 1971.

Goodban, Wm. and Haysett, Jack J. *Architectural Drawing and Planning*. New York: McGraw Hill Book Co., 1971.

Hornung, Wm. J. *Architectural Drafting*. Englewood Cliffs, New Jersey: Prentice-Hall, 1971.

Myller, Rolf *From Idea into House*. New York: Atheneum Publishing Co., 1974.

Olgyay, Victor *Design with Climate*. Princeton: Princeton University Press, 1963.

Roberts, Rex *Your Engineered Home*. New York: M. Evans & Co., Inc., 1964.

Stillman, R.J. *Do It Yourself Contracting to Build Your Own Home*. Radnor, Pennsylvania: Chilton Book Co., 1974.

Tatum, Rita *The Alternative Home*. Los Angeles: Reed Books, 1978.

Wade, Alex and Ewenstein *30 Energy Efficient Houses You Can Build*. Emmanus, Pennsylvania: Rodale Press, 1977.

Waschek, Carmen *Your Guide to Good Shelter*. Reston, Virginia: Reston Publishing Co., Inc., 1978.

Wilson, Roy L. *Build Your Own Energy Saver Home or Upgrade Your Existing Home*. Austin, Texas: Energy Saver Home Co., 1978.

Bibliography

Index

Index